Magaly Koch and Thomas M. Missimer (Eds.)

Water Resources Assessment and Management in Drylands

MDPI

This book is a reprint of the Special Issue that appeared in the online, open access journal, *Water* (ISSN 2073-4441) from 2015–2016 (available at: http://www.mdpi.com/journal/water/special_issues/water-resources-drylands).

Guest Editors
Magaly Koch
Center for Remote Sensing, Boston University
USA

Thomas M. Missimer
U. A. Whitaker College of Engineering, Florida Gulf Coast University
USA

Editorial Office
MDPI AG
St. Alban-Anlage 66
Basel, Switzerland

Publisher
Shu-Kun Lin

Managing Editor
Cherry Gong

1. Edition 2016

MDPI • Basel • Beijing • Wuhan • Barcelona

ISBN 978-3-03842-247-1 (Hbk)
ISBN 978-3-03842-248-8 (PDF)

Table of Contents

Chapter 1: Methods to Assess and Manage Water in Drylands

Chapter 3: Management of Agricultural Water Use in Drylands

List of Contributors

Rania Abdelkhaleq The Ministry of Water and Irrigation, Amman 11181, Jordan.

Yukuo Abe University of Tsukuba, 1-1-1 Tennodai, Tsukuba, Ibaraki 305-8577, Japan.

Miguel Aguila Institute of Agricultural Engineering, Tropics and Subtropics Group, Universität Hohenheim, Garbenstraße 9, Stuttgart 70593, Germany.

Jawad T. Al-Bakri Department of Land, Water and Environment, Faculty of Agriculture, The University of Jordan, Amman 11942, Jordan.

Jose Albiac Department of Agricultural Economics, Centro de Investigación y Tecnología Agroalimentaria–Diputación General de Aragón (CITA-DGA), Avenida Montañana 930, Zaragoza 50059, Spain.

Samir Al-Mashharawi Water Desalination and Reuse Center, King Abdullah University of Science and Technology, Thuwal 23955-6900, Saudi Arabia.

Lorenzo Avella Department of Economics and Social Sciences, Polytechnical University of Valencia, Valencia 46022, Spain.

Ali Behrangi Jet Propulsion Laboratory, California Institute of Technology, Pasadena, CA 91109, USA.

Elena Calvo Department of Economic Analysis, University of Zaragoza, Zaragoza 50018, Spain.

Chao Dai Power China Kunming Engineering Corporation Limited, Kunming 650000, China.

Ariel Dinar School of Public Policy, University of California, Riverside, CA 92521, USA.

Encarna Esteban Department of Economic Analysis, University of Zaragoza, Zaragoza 50018, Spain.

James S. Famiglietti Jet Propulsion Laboratory, California Institute of Technology, Pasadena, CA 91109, USAV; Department of Civil and Environmental Engineering; Department of Earth System Science, University of California, Irvine, CA 92697, USA.

Xinlong Feng College of Mathematics and System Sciences, Xinjiang University, Urumqi 830046, Xinjiang, China.

Marta Garcia-Molla Department of Economics and Social Sciences, Polytechnical University of Valencia, Valencia 46022, Spain.

Ali Ghanim The Ministry of Water and Irrigation, Amman 11181, Jordan.

Dongwei Gui, Cele National Station of Observation and Research for Desert-Grassland Ecosystems, Qira 848300, Xinjiang, China; State Key Laboratory of Desert and Oasis Ecology, Xinjiang Institute of Ecology and Geography, Chinese Academy of Sciences, Urumqi 830011, Xinjiang, China.

Sherif M. Hanafy Department of Earth and Engineering Science, King Abdullah University of Science and Technology, Thuwal 23955-6900, Saudi Arabia.

Rongze Hou Faculty of Engineering and Information Technology, Griffith University, 404 Musgrave Road, Coopers Plains QLD 4018, Australia.

Hongchang Hu Department of Hydraulic Engineering, State Key Laboratory of Hydroscience and Engineering, Tsinghua University, Beijing 100084, China.

Jinting Huang Key Laboratory of Subsurface Hydrology and Ecological Effect in Arid Region of Ministry of Education, Chang'An University, No. 126, Yata Road, Xi'an 710054, China; Department of Hydrogeology and Environmental Geology, Xi'an Institute of Geology and Mineral Resources, No. 438, Youyidong Road, Xi'an 710054, China.

Yuan Huang Xinjiang Institute of Ecology and Geography, Chinese Academy of Sciences, Urumqi 830011, Xinjiang, China.

Koji Inosako Faculty of Agriculture, Tottori University, 4-101 Koyama-Minami, Tottori 680-8553, Japan.

Khan Z. Jadoon Water Desalination and Reuse Center, King Abdullah University of Science and Technology, Thuwal, 23955-6900, Saudi Arabia; Department of Civil Engineering, COMSATS Institute of Information Technology, Abbottabad 22060, Pakistan; Water Desalination and Reuse Center, King Abdullah University of Science and Technology, Thuwal 23955-6900, Saudi Arabia.

Mohamed Taher Kahil International Institute for Applied Systems Analysis (IIASA), Water Program, Laxenburg 2361, Austria.

Magaly Koch Center for Remote Sensing, Boston University, 725 Commonwealth Ave., Boston, MA 02215-1402, USA.

Toshinori Kojima Department of Materials and Life Science, Seikei University, 3-3-1, Kichijoji-kitamachi, Musashino, Tokyo 180-8633, Japan.

Jiaqiang Lei Cele National Station of Observation and Research for Desert-Grassland Ecosystems, Qira 848300, Xinjiang, China; State Key Laboratory of Desert and Oasis Ecology, Xinjiang Institute of Ecology and Geography, Chinese Academy of Sciences, Urumqi 830011, Xinjiang, China.

Congjuan Li Xinjiang Institute of Ecology and Geography, Chinese Academy of Sciences, Urumqi 830011, Xinjiang, China.

Oliver M. Lopez Water Desalination and Reuse Center, King Abdullah University of Science and Technology, Thuwal, 23955-6900, Saudi Arabia.

P. Martin Mai Earth Science and Engineering, King Abdullah University of Science and Technology, Thuwal 23955-6900, Saudi Arabia.

Donglei Mao Cele National Station of Observation and Research for Desert-Grassland Ecosystems, Qira 848300, Xinjiang, China; Chinese Academy of Sciences, Urumqi 830011, Xinjiang, China.

Thomas M. Missimer U. A. Whitaker College of Engineering, Florida Gulf Coast University, 10501 FGCU Boulevard, Fort Myers, FL 33965-6565, USA.

Iqra Mughal Water Desalination and Reuse Center, King Abdullah University of Science and Technology, Thuwal 23955-6900, Saudi Arabia.

Tadaomi Saito Faculty of Agriculture, Tottori University, 4-101 Koyama-Minami, Tottori 680-8553, Japan.

Gerard T. Schuster Department of Earth and Engineering Science, King Abdullah University of Science and Technology, Thuwal 23955-6900, Saudi Arabia.

Sari Shawash The Ministry of Water and Irrigation, Amman 11181, Jordan.

Hideki Suganuma School of Veterinary and Life Sciences, Murdoch University, 90 South Street, Murdoch 6150 WA, Australia; Department of Materials and Life Science, Seikei University, 3-3-1, Kichijoji-kitamachi, Musashino, Tokyo 180-8633, Japan.

Brian F. Thomas Jet Propulsion Laboratory, California Institute of Technology, Pasadena, CA 91109, USA.

Fuqiang Tian Department of Hydraulic Engineering, State Key Laboratory of Hydroscience and Engineering, Tsinghua University, Beijing 100084, China.

Stefan Uhlenbrook Faculty of Civil Engineering and Geosciences, Water Resources Section, Delft University of Technology, PO Box 5048, Delft 2600 GA, The Netherlands; UNESCO-IHE Institute for Water Education, PO Box 3015, Delft 2601 DA, The Netherlands.

Li Wan School of Water Resources and Environment, China University of Geosciences (Beijing), Beijing 10083, China.

Yongdong Wang Xinjiang Institute of Ecology and Geography, Chinese Academy of Sciences, Urumqi 830011, Xinjiang, China.

Guanghui Wei College of Hydraulic and Civil Engineering, Xinjiang Agricultural University, Urumqi 830052, China.

Jochen Wenninger UNESCO-IHE Institute for Water Education, PO Box 3015, Delft 2601 DA, The Netherlands; Faculty of Civil Engineering and Geosciences, Water Resources Section, Delft University of Technology, PO Box 5048, Delft 2600 GA, The Netherlands.

Xinwen Xu Xinjiang Institute of Ecology and Geography, Chinese Academy of Sciences, Urumqi 830011, Xinjiang, China.

Jie Xue Cele National Station of Observation and Research for Desert-Grassland Ecosystems, Qira 848300, Xinjiang, China; University of Chinese Academy of Sciences, Beijing 100049, China.

Pengju Yang Department of Hydraulic Engineering, State Key Laboratory of Hydroscience and Engineering, Tsinghua University, Beijing 100084, China.

Pengnian Yang College of Hydraulic and Civil Engineering, Xinjiang Agricultural University, Urumqi 830052, China.

Zhi Yang Institute of Huai River Water Resources Protection, Bengbu 233001, China; School of Water Resources and Environment, China University of Geosciences (Beijing), Beijing 10083, China; UNESCO-IHE Institute for Water Education, PO Box 3015, Delft 2601 DA, The Netherlands.

Hiroshi Yasuda Arid Land Research Center, Tottori University, 1390 Hamasaka, Tottori 680-0001, Japan.

Fanjiang Zeng State Key Laboratory of Desert and Oasis Ecology, Xinjiang Institute of Ecology and Geography, Chinese Academy of Sciences, Urumqi 830011, Xinjiang, China; Cele National Station of Observation and Research for Desert-Grassland Ecosystems, Qira 848300, Xinjiang, China.

Jianguo Zhang College of Natural Resources and Environment, Northwest A&F University, Yangling 712100, China.

Ying Zhao Chinese Academy of Sciences, Urumqi 830011, Xinjiang, China; College of Natural Resources and Environment; Key Laboratory of Plant Nutrition and the Agri-Environment in Northwest China, Ministry of Agriculture, Northwest A&F University, Yangling 712100, China.

Ruisen Zhong State Key Laboratory of Desert and Oasis Ecology, Xinjiang Institute of Ecology and Geography, Chinese Academy of Sciences, Urumqi 830000, China.

Jie Zhou Cele National Station of Observation and Research for Desert-Grassland Ecosystems, Qira 848300, Xinjiang, China; University of Chinese Academy of Sciences, Beijing 100049, China.

Yangxiao Zhou UNESCO-IHE Institute for Water Education, Department of Water Science and Engineering, P.O. Box 3015, Delft 2601 DA, The Netherlands.

Zia-Khan Institute of Agricultural Engineering, Tropics and Subtropics Group, Universität Hohenheim, Garbenstraße 9, Stuttgart 70593, Germany.

About the Guest Editors

Magaly Koch is a geologist specialized in the application of Remote Sensing and Geographic Information Systems in the study of groundwater resources and environmental change of arid and semiarid regions. She graduated from the University of Cologne, Germany, in 1986 with a M.Sc. in Geology. Her PhD research, on the use of remote sensing in ground water studies, was undertaken at Boston University, USA, and completed in 1993. Subsequently she was awarded a Marie Curie Fellowship by the European Union to undertake post-doctoral research at the Earth Science Institute, CSIC, Barcelona, Spain (1996–1998). More recently she was awarded a Fulbright Fellowship to undertake teaching and research activities at Tohoku University in Japan (2014–2015). Her current post is that of Research Associate Professor at the Remote Sensing Center of Boston University, Boston (MA), USA. In addition, she holds a post as Lecturer at the Civil & Environmental Engineering Department in Tufts University. She has published over 50 peer-reviewed journal papers and about 100 conference papers/technical reports mainly in the subject of RS/GIS applications in arid/semiarid lands. She co-authored (with P.M. Mather) the textbook: Computer Processing of Remotely-Sensed Images: An Introduction, Fourth Edition, Chichester, UK: Wiley-Blackwell (published in 2011).

Thomas M. Missimer is a Visiting Professor at the U. A. Whitaker College of Engineering, Florida Gulf Coast University and President of Missimer Hydrological Services, Inc. (consulting firm). He has practiced as a hydrogeologist for 43 years. He has a B.A. in geology from Franklin & Marshall College, a M.S. in geology (coastal) from Florida State University, and a PhD in marine geology and geophysics from the University of Miami. He began his career with the U. S. Geological Survey and then worked as a consultant for 34 years, founding three firms. He is formerly a visiting professor at King Abdullah University of Science and Technology in Saudi Arabia. He is the author, co-author or editor of 10 books and author of over 400 technical publications of which nearly 100 are peer-reviewed journal papers. His research areas of interest are arid lands hydrology and water management, groundwater hydrology, water policy, desalination, and sedimentology. He is the winner of a number of awards for technical paper presentations and book publications.

Preface to "Water Resources Assessment and Management in Drylands"

Drylands are fragile environments and, therefore, highly susceptible to environmental changes. They cover nearly 50% of the world's land surface and are increasingly being reclaimed by a growing population for food production and urbanization. This makes water resources management in drylands an extremely important issue. The unplanned water resources development may result in aquifer depletion, soil and/or water salinization, loss of water through evapotranspiration due to inadequate irrigation systems, and land degradation such as soil erosion, soil crusting, and sand encroachment to name a few.

Drylands can also serve as excellent indicators of the onset, nature and severity of climate shifts as their ecosystems respond almost immediately to temperature variations and water stresses. Therefore, assessing and managing these ecosystems is vital because they may serve as early warning systems. The combined effect of climate and land use changes can have long lasting and disastrous consequences on these ecosystems that must be adverted as they become increasingly important for food production.

Fortunately technological advances are providing the means to observe and measure almost in real time the status of dryland ecosystems and their resources. Dryland ecosystem models are increasingly built with information derived from a variety of disparate sources and scales, from field surveys and field experiments to satellite Earth observations. Studies on dryland ecosystem function, services and water demands are progressively conducted by drawing insights from multiple disciplines. This is largely made possible due to the use of geospatial information technology and tools which enable surveying and analyzing large amounts of data.

The diversity of problems affecting drylands water resources demands up-to-date scientific information to guide their rational uses. Such information needs to be made accessible to decision makers and the general public to promote transparency and engagement in the decision-making process. Adequate policies can help in mitigating adverse environmental impacts by regulating and managing these precious and scarce natural resources.

This Special Issue on "Water Resources Assessment and Management in Dryland" is intended to provide a collection of articles addressing various aspects of dryland hydrology. Articles about recent scientific discoveries in hydrology / hydrogeology, new emerging technologies and their use in water resources assessment, development, and management are compiled here in three chapters. The first chapter addresses methods and techniques to assess water resources, the second chapter deals with water policy and management issues in drylands, and the third chapter contains examples of agricultural water management in drylands. It is hoped that the selection of papers will give the reader an overview of current research in this field.

Magaly Koch and Thomas M. Missimer
Guest Editors

Editorial to "Water Resources Assessment and Management in Drylands"

Magaly Koch and Thomas M. Missimer

Abstract: Drylands regions of the world face difficult issues in maintaining water resources to meet current demands which will intensify in the future with population increases, infrastructure development, increased agricultural water demands, and climate change impacts on the hydrologic system. New water resources evaluation and management methods will be needed to assure that water resources in drylands are optimally managed in a sustainable manner. Development of water management and conservation methods is a multi-disciplinary endeavor. Scientists and engineers must collaborate and cooperate with water managers, planners, and politicians to successfully adopt new strategies to manage water not only for humans, but to maintain all aspects of the environment. This particularly applies to drylands regions where resources are already limited and conflicts over water are occurring. Every aspect of the hydrologic cycle needs to be assessed to be able to quantify the available water resources, to monitor natural and anthropogenic changes, and to develop flexible policies and management strategies that can change as conditions dictate. Optimal, sustainable water management is achieved by cooperation and not conflict, thereby necessitating the need for high quality scientific research and input into the process.

Reprinted from *Water*. Cite as: Koch, M.; Missimer, T.M. Water Resources Assessment and Management in Drylands. *Water* **2016**, *8*, 8.

1. Introduction

Population and water demand are rapidly growing in the drylands regions of the world. More than 20% of the world's population, at least 1.2 billion, currently live in areas with physical scarcity of water. Arid and semi-arid regions occur in about 30% of the total land area of the Earth and with intensification of desertification caused by global warming and poor land management practices, this percentage is increasing [1]. The future health and well-being of the populations occupying these areas are dependent upon being able to assess the status of the available water resources in real time and to develop and implement policies and management strategies to maintain and grow water supplies.

The papers contained within this Special Issue of *Water*, entitled "Water Resources Assessment and Management in Drylands", describe new methods of water resources assessment and evaluation of drylands hydrology, use of new tools in the implementation of water policies and management strategies, and strategies used to manage agricultural water use which accounts for the largest global water use at over 70% [1]. The papers cover different areas of geography and climatic conditions that highlight a variety of techniques tailored to the unique conditions occurring within regions with differing water use rates and management problems.

2. Contributors

New methods and the application of old ones are required to assess and manage water resources in drylands regions. Thomas *et al.* [2] provide a regional assessment of the impact of rainfall intensity on the recharge rate over a large area of the southwestern United States in their paper entitled "Precipitation Intensity Effects on Groundwater Recharge in the Southwestern United States". They use a combination of the water table fluctuation and master recession curve methods to assess aquifer recharge applied to water level data collected from a large number of observation wells. They apply a double mass curve graphical method to assess the consistency of hydrologic data and an intensity-duration-frequency analysis to develop recharge/precipitation ratios. The techniques applied yield intensity-duration-frequency curves for various rainfall events ranging from three to 48 h in duration. This type of analysis is critical in developing assessments of recharge rates during climate change that may bring changes in duration of rainfall events and drought periods. They emphasize the importance of "characterizing groundwater recharge behaviors over short time periods which are affected by variability in precipitation statistics" to understanding overall recharge for the development of improved groundwater management strategies.

The paper by Jadoon *et al.* [3] entitled "Anthropogenic-induced changes in the mechanism of drylands ephemeral stream recharge, western Saudi Arabia" documents how over-pumping of a shallow aquifer system in drylands can impact recharge. The authors demonstrate that a wadi system in western Saudi Arabia can no longer recharge by conventional infiltration and percolation from a channel during storm events. The natural occurrence of thin clay layers in the unsaturated zone and a pumping-induced water table position more than 10 m below surface no longer allows infiltrated water to reach the water table. Recharge can now occur only around the wadi perimeter where permeable sediments intersect with fractured rocks that receive rainfall. The only water management remedy to make the aquifer system sustainable is to curtail pumping until the water table reaches a level that can be maintained by rainfall recharge and a lesser degree of pumping.

In their paper "Predicting soil infiltration and horizon thickness for a large-scale water balance model in an arid environment", Saito *et al.* [4] use a model calibrated from double-ring infiltrator data, horizon thickness measurements, and vegetation surveys to assess infiltration over a 30 km × 50 km area of Western Australia. They developed a set of type curves relating cumulative infiltration with time and related the curves to vegetative biomass. They concluded that a strong correlation occurs that relates cumulative infiltration and horizon thickness with biomass and canopy coverage. The predictive equations developed can be used with vegetation distribution maps derived from a combination of field surveys and satellite images to make landscape-scale infiltration estimates.

Re-forestation is a method commonly used to help slow or mitigate desertification. Huang *et al.* [5], in their paper entitled "Simulation of water use dynamics by *Salix* bush in a semi-arid shallow groundwater area of the Chinese Erdos Plateau", evaluate the impacts of using this plant to develop shelterbelts and to return farmland to forest. They used the Hydrus-1D model to evaluate the contributions of groundwater and the plant to evapotranspiration (ET). This model allowed investigation of the heat flux on soil water flux and evaluation of the impact of *Salix* on evapotranspiration. They concluded that the water use of *Salix* is dependent on rainfall infiltration and, in the driest period, more groundwater is used. Also, groundwater contributions to ET were 26.9% and 40.6% with and without heat, which causes groundwater contribution to be over-estimated when thermally-driven water vapor flow is not taken into account.

A major water management strategy in drylands regions is the use of managed aquifer recharge to balance water supply imbalances [6]. In two companion papers, Lopez *et al.* [7] in their paper "Method of relating grain size distribution to hydraulic conductivity in dune sands to assist in assessing managed aquifer recharge projects: Wadi Khulays dune field, western Saudi Arabia" and Mughal *et al.* [8] in their paper "Experimental measurement of diffusive extinction depth and soil moisture gradients in a dune sand aquifer in western Saudi Arabia: Assessment of evaporation loss for design on an MAR system" deal with two important issues in assessing the feasibility and design of Managed Aquifer Recharge (MAR) systems in a hyper-arid region. The location of MAR projects in stabilized dune fields requires screening of large landscape areas to assess potentially feasible sites. Lopez *et al.* [7] developed a mathematical method using grain size distribution data to estimate hydraulic conductivity. Their research documents the changes of hydraulic conductivity and effective porosity with grain size distribution across dunes and allows the development of saturated water storage to be made without the collection of large amounts of data. This research could be coupled to remote sensing methods to allow landscape screening and assessment of potential MAR sites. A critical design aspect for MAR project development in a dune sand aquifer is the prevention of diffusive evaporative loss of water from storage during system operation. Mughal *et al.* [8] determined that the diffusive evaporative extinction depth is about 1 m in dune sands within the size range evaluated by Lopez *et al.* [7]. Therefore, if a MAR system were to be developed, a 1 m cover of sand above the water table position would be sufficient to prevent water loss. They determined that it took 56 days of soil diffusion to reach the extinction depth.

It is necessary to make improvements in water policy and management in arid and semi-arid regions, especially in consideration of natural climatic variability and anthropogenic warming. Kahil *et al.* [9], in their paper "Improving the performance of water policies: Evidence from drought in Spain", use a hydro-economic model that links hydrological, economic and environmental elements to efficiency, sustainability, and equity requirements. They conclude that water pricing and water markets in a river basin located in Spain are an economic means that work well where private markets control water, but not as well where the government controls the resources. The water pricing policy favored by the European Water Framework Directive (WFD) tends to be detrimental to farmers by increasing their losses during drought conditions, whereas stakeholder cooperation lessens the impacts to individuals by spreading losses more equitably.

Yang *et al.* [10], in their paper "Simulation of groundwater-surface water interactions under different land use scenarios in the Bulang Catchment, northwest China", evaluate the current use of water within the vegetative landscape and four proposed changes in the composition of the vegetation. The different land use scenarios were evaluated using a simple calculated water balance, a steady-state groundwater model, and a transient groundwater model. They found that within the current landscape condition, 91% of the precipitation is consumed by crops, leaving 9% to become groundwater recharge which sustains stream discharge in the observed year of record. Four scenarios were evaluated, including (1) the status quo vegetation types; (2) the previous natural state of vegetation which was desert grasses; (3) a change of crop types to dry resistant types; and (4) an optimal landscape consisting of dry resistant crops and desert grasses. The optimal scenario was found to increase groundwater recharge and increase river discharge.

Maintenance of oases is a critical issue in arid regions because these areas commonly have the only source of freshwater, a high diversity of vegetative types, and support rural human populations. Xue *et al.* [11], in their paper "Quantification of environmental flow requirements to support ecosystem services of oasis areas: A case study in Tarim Basin, northwest China", evaluate the necessary environmental flow requirements (EFRs) to maintain riverine ecosystem health, assurance of oasis-desert ecotone and riparian forest stability, and restoration of ecotone groundwater resources. They quantified the environmental flow requirements, divided the flow into those portions needed for maintenance of various ecosystem functions, and assessed the response of environmental flow requirements to natural runoff. The EFRs for maintenance of the oases ecosystem function and groundwater flow restoration were deterred as a percentage of natural river discharge.

Vegetative belts are commonly used to stabilize wind-transported sand in proximity to major highways or other infrastructure. This vegetation is maintained by using the lowest quantity and quality of water possible which can cause salinization and loss of the vegetative buffer. Huang *et al.* [12], in their paper "Spatiotemporal distribution of soil moisture and salinity in the Taklimakan Desert highway shelterbelt", assess the impact of the irrigation on plant growth to prevent salinization and maintenance of the vegetation buffer for the future. They found that the soil texture played an important role in controlling the plant growth and concluded that the currently designed drip irrigation system would allow the contained growth of the plants within the shelterbelt.

Since agriculture is the largest percentage user of water in the world, new irrigation management techniques are quite important in conserving and managing water. Al-Bakri *et al.* [13], in their paper "Geospatial techniques for improved water management in Jordon", used groundwater pumping data, estimated crop consumption data, and Landsat data to develop an irrigation water auditing technique. They applied the method in three water basins in Jordan to assess if the pumping schemes being used are sustainable. Overall, pumping was found to be in excess of sustainable yields by between 144% and 360%. Also, it was found that the monitoring technique could be used to reveal violations of water use practices and could be incorporated into changes in water law in Jordan.

In arid regions crop ET losses consume a considerable amount of irrigation water. Tian *et al.* [14], in their paper "Partitioning of cotton field evapotranspiration under mulched drip irrigation based on a dual crop coefficient model", used the dual crop model SIMDualKC to estimate actual crop ET and basal crop coefficients over a cotton field located in northwestern China. They found that the model used is capable of providing accurate estimates for the cotton crop ET and could be used to establish efficient irrigation schedules. Also, they found that plastic mulch had a positive effect on reducing irrigation water use.

Development of an efficient irrigation schedule can have significant benefits in terms of reducing crop irrigation water use requirements. Yang *et al.* [15], in their paper "Winter irrigation effects in cotton fields in arid inland irrigated areas in the north of the Tarim Basin, China", investigated the use of winter irrigation to assess water and salt management practices. Their research resulted in the development of a specific range of irrigation applications per unit area and a recommendation to delay application to early December. Irrigation application by using several stages, along with the change in time of application delivery, improved salt control.

3. Conclusions

From the information found in this collection of papers, it is apparent that new evaluation methods will be needed to manage water sustainably in drylands regions.

The 14 papers contained herein incorporate a diverse group of technologies and methods including field measurements, modeling, and remote sensing to evaluate water resources to resolve regional and local issues in drylands regions. Policies and management strategies that reduce water use, control salinization, maintain ecosystems, and maintain sustainable water use for public water supply, crop irrigation, and industrial use are developed from the types of research information herein presented. The diverse technologies and global geography of applications will be useful to all entities involved in drylands water management.

Acknowledgments: The authors of this editorial, who served as Guest Editors of this special issue of *Water*, thank the journal editors for their time and resources, the many authors of the papers for their contributions, and the numerous referees for their hard work that improved the various versions of the manuscripts leading to high quality published papers.

Conflicts of Interest: The authors declare no conflict of interest.

References

1. Maliva, R.G.; Missimer, T.M. *Arid Lands Water Evaluation and Management*; Springer-Verlag: Berlin, Germany, 2012.
2. Thomas, B.F.; Behrangi, A.; Famiglietti, J.S. Precipitation intensity effects on groundwater recharge in the southwestern United States. *Water* **2016**, *8*, 90.
3. Jadoon, K.J.; Al-Mashharawi, S.; Hanafy, S.M.; Schuster, G.T.; Missimer, T.M. Anthropogenic-induced changes in the mechanism of drylands ephemeral stream recharge, western Saudi Arabia. *Water* **2016**, *8*, 136.
4. Saito, T.; Yasuda, H.; Sugannuma, H.; Inosako, K.; Abe, Y.; Kojima, T. Predicting soil infiltration and horizon thickness for a large-scale water balance model in an arid environment. *Water* **2016**, *8*, 96.
5. Huang, J.; Zhou, Y.; Hou, R.; Wenninger, J. Simulation of water use dynamics by *Salix* bush in a semi-arid shallow groundwater area of the Chinese Erdos Plateau. *Water* **2015**, *7*, 6999–7021.
6. Megdal, S.B.; Dillon, P. (Eds.) *Policy and Economics of Managed Aquifer Recharges and Water Banking*; MDPI: Basal, Switzerland, 2015.
7. Lopez, O.M.; Jadoon, K.Z.; Missimer, T.M. Method of relating grain size distribution to hydraulic conductivity in dune sands to assist in assessing managed aquifer recharge projects: Wadi Khulays dune field, western Saudi Arabia. *Water* **2015**, *7*, 6411–6426.
8. Mughal, I.; Jadoon, K.Z.; Mai, P.M.; Al-Mashharawi, S.; Missimer, T.M. Experimental measurement of diffusive extinction depth and soil moisture gradients in a dune sand aquifer in western Saudi Arabia: Assessment of evaporation loss for design on an MAR system. *Water* **2015**, *7*, 6967–6982.
9. Kahil, M.T.; Albiac, J.; Dinar, A.; Calvo, E.; Esteban, E.; Avella, L.; Garcia-Molla, M. Improving the performance of water policies: Evidence from drought in Spain. *Water* **2016**, *8*, 34.

10. Yang, Z.; Zhou, Y.; Wenninger, J.; Uhlenbrook, S.; Wan, L. Simulation of groundwater-surface water interactions under different land use scenarios in the Bulang Catchment, northwest China. *Water* **2015**, 5959–5985.
11. Xue, J.; Gui, D.; Zhao, Y.; Lai, J.; Feng, X.; Zeng, F.; Zhou, J.; Mao, D. Quantification of environmental flow requirements to support ecosystem services of oasis areas: A case study in Tarim Basin: Northwest China. *Water* **2015**, *7*, 5657–5675.
12. Huang, Y.; Wang, Y.; Zhao, Y.; Xu, X.; Zhang, J.; Li, C. Spatiotemporal distribution of soil moisture and salinity in the Taklimakan Desert highway shelterbelt. *Water* **2015**, *7*, 4343–4361.
13. Al-Bakri, J.T.; Shawash, S.; Ghanim, A.; Abdelkhaleq, R. Geospatial techniques for improved water management in Jordan. *Water* **2016**, *8*, 132.
14. Tian, F.; Yang, P.; Hu, H.; Dai, C. Partitioning of cotton filed evapotranspiration under mulched drip irrigation based on a dual crop coefficient model. *Water* **2016**, *8*, 72.
15. Yang, P.; Zia-Khan, S.; Wei, G.; Zhong, R.; Aguila, M. Winter irrigation effects in cotton fields in arid inland irrigated areas in the north of the Tarim Basin, China. *Water* **2016**, *8*, 47.

Chapter 1:
Methods to Assess and Manage Water in Drylands

Precipitation Intensity Effects on Groundwater Recharge in the Southwestern United States

Brian F. Thomas, Ali Behrangi and James S. Famiglietti

Abstract: Episodic recharge as a result of infrequent, high intensity precipitation events comprises the bulk of groundwater recharge in arid environments. Climate change and shifts in precipitation intensity will affect groundwater continuity, thus altering groundwater recharge. This study aims to identify changes in the ratio of groundwater recharge and precipitation, the R:P ratio, in the arid southwestern United States to characterize observed changes in groundwater recharge attributed to variations in precipitation intensity. Our precipitation metric, precipitation intensity magnification, was used to investigate the relationship between the R:P ratio and precipitation intensity. Our analysis identified significant changes in the R:P ratio concurrent with decreases in precipitation intensity. The results illustrate the importance of precipitation intensity in relation to groundwater recharge in arid regions and provide further insights for groundwater management in nonrenewable groundwater systems and in a changing climate.

Reprinted from *Water*. Cite as: Thomas, B.F.; Behrangi, A.; Famiglietti, J.S. Precipitation Intensity Effects on Groundwater Recharge in the Southwestern United States. *Water* **2016**, *8*, 90.

1. Introduction

Sustainable groundwater management requires an effective framework that includes, in part [1], an understanding of groundwater system fluxes. Challenges to sustainable groundwater management are directly linked to the estimation of outflow (*i.e.*, pumping, baseflow) and inflow (*i.e.*, recharge) [2] and are thus linked to the flux of water that infiltrates to the saturated zone. As described by Theis [3], the rate of groundwater recharge is dependent upon the rate of the addition of water to the system and the rate at which available water can infiltrate to a depth thus escaping evaporation. Throughout the world, groundwater depletion has spurred actions which include recharge augmentation such as importing surface water [4] and rainwater capture [5]. The volume of recharge to an aquifer system alone does not correspond to the portion of available groundwater storage for withdrawals. Such an antiquated idea, conceptualized as safe yield [6,7], was replaced by sustainable groundwater management theories which account for ecosystem services provided by groundwater storage including baseflow to streams [8] and the maintenance of groundwater storage.

Global observations and groundwater simulations have evaluated changes to recharge based on IPCC projections [9,10] or have evaluated groundwater recharge to assess resilience [11]. Model-based approaches [9,12–17] range from simplistic soil budget models to fully coupled global climate models and groundwater flow models to study groundwater recharge. Such approaches typically seek to determine long-term groundwater recharge as a function of changes in precipitation intensity and timing.

Groundwater recharge is an important variable for effective groundwater management [1,6,10], especially in arid regions with little recharge [11,18]. Groundwater resources in arid environments have been identified as under stress resulting from the imbalance between groundwater recharge and the combination of natural rates of discharge and groundwater withdrawals. Groundwater storage in arid regions is considered non-renewable [18,19] at timescales of groundwater management planning horizons and are often impacted by anthropogenic activities resulting in dramatic changes in groundwater resources [20]. An example illustrating the influence of anthropogenic impacts to natural aquifer discharge rates is the Nubian aquifer in northeast Africa: prior to extensive development, studies documented the natural rate of groundwater depletion of 0.79 km^3/year [21] to 2.7 km^3/year [22] while more recent depletion rates of 6.1 km^3/year have been observed from 2003 to 2013 [11]. Future imbalances of groundwater recharge are expected to occur as a result of variability attributed to climate change, especially in arid regions [9,23].

In arid and semi-arid environments, groundwater recharge tends to occur during extreme precipitation events [24] as compared to other environments which are characterized as having a combination of constant-rate and episodic behaviors [25]. Groundwater recharge varies spatially as a result of physical attributes including soil type, depth to groundwater, vadose zone porosity, hydrogeology and precipitation patterns. Because of the episodic nature of recharge in arid environments, small changes in precipitation intensity have been suggested to influence groundwater recharge [26,27]. Recent studies have sought to quantify episodic recharge [28] or identify regional groundwater recharge [10,11,23]. Such evaluations, however, did not identify influences in observed groundwater recharge as a function of precipitation, the primary goal of this study.

The arid southwestern United States has experienced prolonged drought conditions [29,30]. Castle *et al.* [31] evaluated groundwater storage changes during the drought and identified significant groundwater depletion in the Lower Colorado River Basin. In that study, the authors attributed significant changes of groundwater depletion rates to discrepancies between surface water (*i.e.*, reservoirs) and groundwater management resulting from the drought. Alley and Konikow [32] argue that a portion of the groundwater storage declines may result from natural groundwater discharges. In reality, the likely scenario of groundwater storage loss in the region includes a combination anthropogenic influences, natural groundwater discharges and climate. In this study, we seek to evaluate changes in the ratio of recharge to precipitation (R:P ratio) which may account for rates of groundwater declines observed in the region. For example, changes in recharge to the aquifers may result in short-term fluctuations in both natural and anthropogenic fluxes from the aquifer [33]. Our evaluation holds important implications for groundwater management in arid environments as groundwater resources are used to meet more than 70% of water demands [34] while climate projections predict reduced precipitation in arid regions [35]. In our study, we evaluate changes in the R:P ratio using a double mass curve approach [26,36] in combination with satellite-observed precipitation data to attribute changes in precipitation statistics to observed changes in groundwater recharge at locations across the arid southwestern United States (Figure 1).

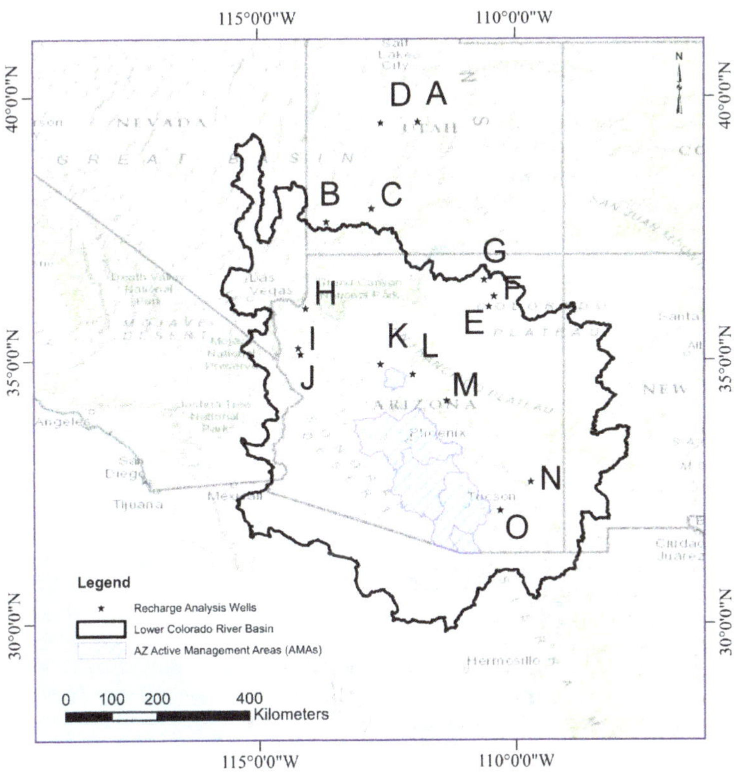

Figure 1. Site map illustrating locations of groundwater observation wells.

2. Data

2.1. TRMM

Data from the latest version of the Tropical Rainfall Measuring Mission (TRMM) merged precipitation product (3B42-V7; [37]) was used to account for precipitation rates within the 0.25-degree grid surrounding each groundwater observation with a focus on the period 2000–2015. The TRMM product is bias adjusted using monthly ground gauge data from the Global Precipitation Climatology Centre (GPCC) product [38]. The 3-h, 0.25-degree product has been found valuable for hydrologic modeling (e.g., [39]) and successfully applied in the southwestern United States to assess monsoon precipitation behavior [40,41].

2.2. Groundwater Hydrographs

Observations of water table depth were collected from the U.S. Geological Survey (USGS) [42] and the Arizona Department of Water Resources (ADWR) [43]. Well locations as illustrated in Figure 1 were selected based on length and completeness of record, time between observations (15-min to 6-h intervals) and for geographical representation throughout the study region. Wells within Active Management Areas in Arizona were not included in our study as groundwater replenishment activities are often used to augment groundwater resources.

2.3. USGS Streamflow Data

Streamflow data from USGS gages [42] were collected for watersheds either in near proximity to or enveloping groundwater observation wells. Gage stations were selected for sites with fewer than 5 consecutive days of no flow readings and a maximum of 5% missing data. Hydrograph recessions were isolated by identifying the beginning of a streamflow recession when a 3-day moving average begins to decrease and ends when a 3-day moving average begins to increase [44,45].

3. Methods

3.1. Estimation of Recharge

The estimation of groundwater recharge using *in situ* data can be conducted using multiple approaches, including the water table fluctuation (WTF) method [46] and the master recession curve (MRC) approach [47]. One would anticipate that in arid environments, groundwater elevations would typically be in a near steady-state decline over time due to the natural recession behavior of aquifer systems without external influences (*i.e.*, pumping) [3]. Periodic recharge would occur during this natural recession decline as a result of intense precipitation events. The MRC approach of [47] is thus advantageous in our evaluation of recharge as the analysis accounts for natural and anthropogenic changes in groundwater hydrograph recessions over time.

The MRC approach provided as an executable program [47] was used to isolate groundwater recharge for select wells in the arid southwestern US (Figure 1) while accounting for rates of decline in observed groundwater elevation data. The MRC approach is an extension of the WTF method which estimates groundwater recharge as a function of water table height with respect to water table elevation which would occur in the absence of recharge (ΔH_t). Groundwater recharge for each time step t (R_t) is calculated by

$$R_t = S_y \times \Delta H_t \quad (1)$$

where S_y is the specific yield. As described by [47], a limitation of the MRC approach is the estimate of the specific yield; thus, recharge was only estimated at wells for which a robust estimate of specific yield could be identified using a relation between streamflow and groundwater elevations [48]. As described by [49], one may estimate specific yield (S_y) as

$$S_y = \frac{q}{\Delta h} \quad (2)$$

where q represents the average streamflow during a hydrograph recession event as a depth per unit of catchment area and Δh represents the average decline in groundwater elevations during the streamflow hydrograph recession period.

Nimmo *et al.* [28] extend the MRC approach to account for episodic recharge events. In that study, selected episodic recharge events from wells were used to estimate recharge. In our analysis, we wish to characterize a change in the relationship between recharge (R) and precipitation (P). Thus, we employ the MRC approach to produce a continuous time series of recharge estimates. The bin-averaged approach was used to isolate recharge in individual groundwater observation wells with recharge rates estimated at the same time interval as groundwater readings (Δt = 15 min to 6 h). Because of the small Δt between observations, a large bin number was used (n = 1000) to ensure accurate representation of average groundwater elevation decline rates expressed in the MRC. Output from the MRC analysis includes an estimate of recharge, cumulative recharge and a representation of the groundwater hydrograph recession behavior used to construct the MRC for each individual well.

3.2. Double Mass Curve

The double mass curve (DMC) is a simple graphical method to evaluate the consistency of hydrological data [36,50]. The DMC approach plots the cumulative data of one variable against the cumulative data of a second variable. A break in the slope of a linear function fit to the data is thought to represent a change in the relation between the variables. This approach provides a robust method to determine a change in the behavior of precipitation and recharge in a simple graphical method. The DMC approach has been used to detect changes in the relation between groundwater recharge and precipitation [26] by plotting the sum of recharge against the sum of precipitation. Such an approach is used here since the change in the fraction of precipitation resulting in observed recharge at a well, the R:P ratio, indicates a change in groundwater continuity [51].

To assess a change in the R:P ratio, we use a single-phase regression (SPR) analysis. The parametric approach takes advantage of the optimal properties of linear regression whereby the paired time series (RP) are divided into 2 groups (RP_{tc},RP_{tn}) given by

$$\sum R_t = \begin{cases} \beta_{O1} + \beta_1 \sum P_{tc} + \varepsilon_{tc} \; 1 \leqslant R_t \leqslant t_c \\ \beta_{O2} + \beta_2 \sum P_{tn} + \varepsilon_{tn} \; 1 + t_c \leqslant R_t \leqslant t_n \end{cases} \tag{3}$$

where t_c and t_n represent the time at which the paired time series is split into 2 groups and β_1 and β_2 represent the slope between the sum of R and the sum of P. The residuals given in Equation (3), ε_{tc} and ε_{tn}, are correlated, an expected result given that the summed time series R_t and P_t are serially correlated, thus failing to meet requirements of ordinary least squares regression. For this analysis, we employed a bootstrap approach to estimate the coefficients β_1 and β_2 in addition to the coefficient standard error estimates ($S_{\beta 1}$ and $S_{\beta 2}$). Our test statistic to identify a significant change of slope in the R:P ratio, T_c, is defined by

$$T_c = \frac{\hat{\beta_1} - \hat{\beta_2}}{\sqrt{S^2_{\beta_1} - S^2_{\beta_2}}} \tag{4}$$

Our statistic T_c follows as student's t distribution. Any reference to significant changes in the slope of R:P in our analysis implies a p-value < 0.001 unless otherwise noted.

3.3. *Intensity-Duration-Frequency*

Intensity-duration-frequency (IDF) analysis provides a statistical graphical representation of rainfall intensity for a given exceedance probability [52,53]. Analyses of statistical precipitation events are often used in the design of engineering structures involving hydrologic flows [54] while statistical representations of IDF results have become commonplace in hydrology textbooks [55,56]. Creating IDF curves begins by isolating annual maximum precipitation depths for the duration of interest, and using a statistical distribution to identify precipitation intensities for specific return periods. The 3-h TRMM data for each groundwater well location was used to identify duration periods of 3, 6, 12, 24, 36 and 48 h by summing precipitation depths over the time window of interest. As described by Stedinger *et al.* [50], the extreme value (EV) distributions are often used to fit observed hydrologic random variable behavior. The Gumbel distribution is one example of an EV distribution often used for rainfall [57] or floods [58,59]. Given a set of X-hour rainfall ($P_1, \ldots, P_n$), let the random variable P be the annual maximum of P_i, where X represents the time window of interest. If P are independent and identically distributed random variables with no upper bound, we can assume that P has a Gumbel distribution. Return intervals for annual maximum values of P can then be estimated using the Gumbel distribution given by the quantile function

$$\chi_p = \varepsilon - \alpha \times \ln(\ln(-p)) \tag{5}$$

for any probability p along with the first moment (*i.e.*, mean) and second moment (*i.e.*, variance)

$$\mu_x = \varepsilon + 0.57721 \times \alpha \tag{6}$$

$$\sigma_x^2 = \pi^2 \alpha^2 / 6 \tag{7}$$

where ε and α in Equations (6) and (7) can be estimated using various approaches including L-moments [60] while probability p can be estimated using the Gringorten plotting position ($p = (i - 0.44)/(n + 0.12)$) where i represents rank and n is the number of observations. IDF analysis employing the Gumbel distribution provides a statistical graphical representation of rainfall intensity for a given exceedance probability [50,51]. IDF curves have been used to evaluate groundwater recharge [61–64] which may be impacted by changes in precipitation patterns as a result of climate change [35]. Of these studies, only Dourte *et al.* [61] evaluated IDF changes which may be related to deviations in groundwater recharge.

4. Results

4.1. Temporal Changes in Groundwater Elevations

Groundwater hydrograph recession behaviors are captured in the predicted rate of decline of groundwater elevations, dh/dt, as estimated in the MRC recharge estimation approach. The data are used to create the master recession curve for each well, synonymous with master recession curves used to characterize streamflow hydrograph behavior [65,66]. Of interest is the general observance of groundwater elevations over time which are used to quantify groundwater recharge. An analysis was conducted using the SPR approach to identify a change in the behaviors of dh/dt (Figure 2) by plotting the sum of dh/dt against time. It is important to note that regions with little change in the cumulative sum of dh/dt represent times during which recharge occurs, thereby negating decreases in dh/dt over time. For wells located within the Lower Colorado River Basin (Figure 1), significant changes in groundwater behaviors where identified in the period ranging from August 2008 to July 2010, well within the timeframe identified by Castle *et al.* [31]

who detected a significant change in groundwater storage behaviors using data from the Gravity Recovery and Climate Experiment (GRACE) satellites to isolate a groundwater storage time series. In that study, changes in groundwater storage were attributed to an intensification of the drought in 2012 and unsustainable groundwater withdrawals. Our results illustrated increased rates of groundwater storage depletion over time, thus corroborating the observed reductions in storage over the region [31,67]. The results illustrated in Figure 2 are important as the general increased rate of decline (*i.e.*, active groundwater depletion) is accounted for in the estimate of groundwater recharge.

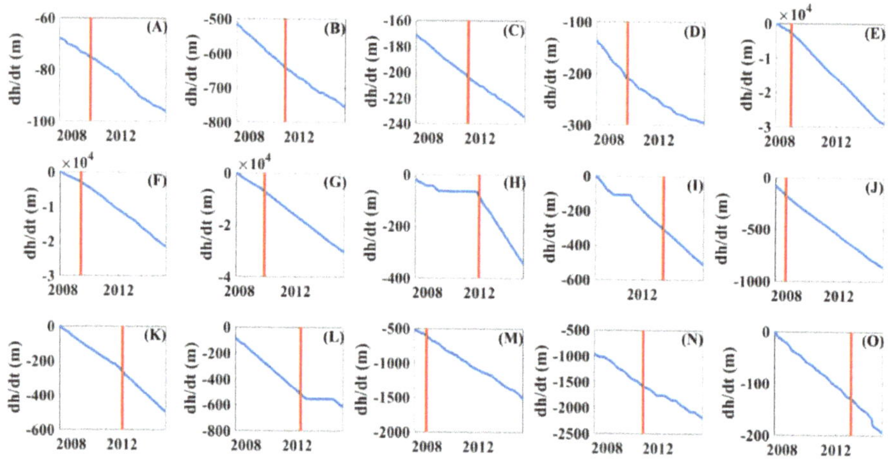

Figure 2. The cumulative rate of change in groundwater elevations (dh/dt) (in **blue**) and the timing of the maximum T_c statistic indicating a significant ($p < 0.001$) change in the behavior of dh/dt (in **orange**). Panels correspond to wells identified in Table 1 and Figure 1.

Table 1. Well information.

Well ID	Figure Label	GW Level Δt	R:P Date
393143111523301	A	6-h	October 2007
373735113394101	B	6-h	September 2007
375244112463201	C	6-h	July 2007
393020112362201	D	6-h	August 2008
360055110304001	E *	15-min	April 2009
361225110240701	F *	15-min	April 2009
363143110355001	G *	15-min	April 2009
355855114043501	H *	6-h	November 2008

Table 1. *Cont.*

Well ID	Figure Label	GW Level Δt	R:P Date
351401114132401	I *	6-h	August 2010
350633114103701	J *	6-h	August 2009
345524112374401	K *	6-h	August 2010
344359111591101	L *	6-h	January 2009
341402111215101	M *	6-h	August 2009
324113109415401	N *	6-h	September 2007
320901110175301	O *	6-h	September 2008

Note: The * identifies wells located within the Lower Colorado River Basin.

4.2. Recharge

Groundwater recharge as estimated from groundwater head changes, groundwater hydrograph recessions and estimated specific yield are illustrated in Figure 3 as monthly estimated recharge as a percent of precipitation. Differences in recharge between sites can be largely explained by variability in hydrogeological conditions including various depths to the potentiometric surface. The results in Figure 3 clearly illustrate the influence of summer monsoonal precipitation events which result in greater rates of recharge as a function of precipitation. The influence of the lag between observations of precipitation and the episodic recharge event in the potentiometric surface is also evident. For example, intense precipitation events were observed at Site A in October 2008 with negligible recharge in the groundwater hydrograph. However, in November 2008, despite little observed precipitation, a large recharge event was identified resulting from the time necessary to infiltrate through the vadoze zone and result in an observed change in groundwater storage in the well, thus causing a large apparent ratio between R and P.

4.3. R:P Ratio

The relation between precipitation and recharge is represented through the double mass curve (DMC) as illustrated in Figure 4. The episodic nature of recharge, depicted by a characteristic step pattern, is clearly evident for all sites. Sites with notable increased linearity in the DMC represent sites with larger recharge estimates (*i.e.*, Figure 4e) [26]. Further, the timing of recharge estimated for the wells coincides with monsoonal precipitation events as depicted in Figure 3.

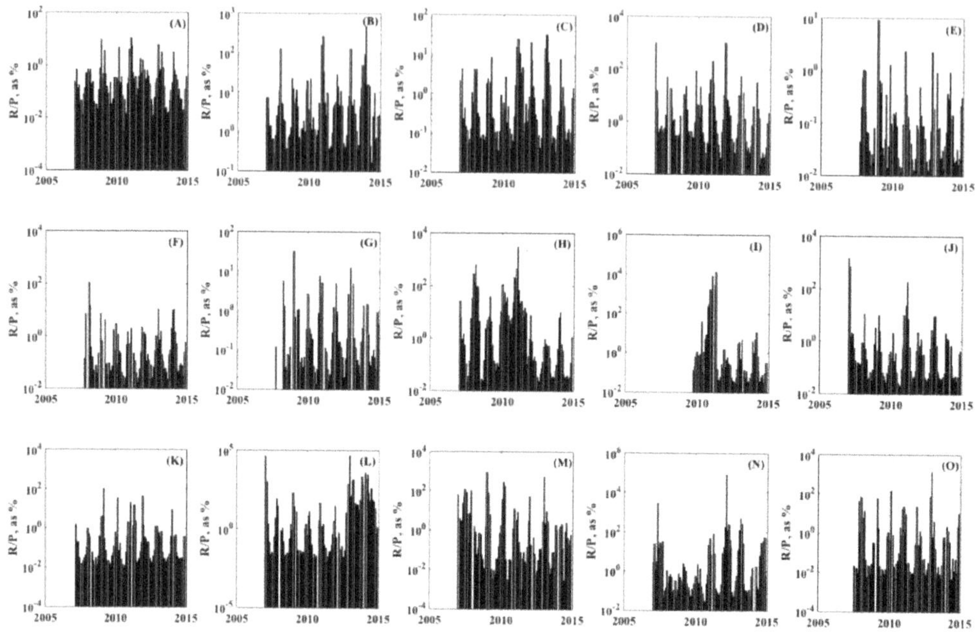

Figure 3. Monthly ratios of recharge (R) to precipitation (P). It is important to recognize the influence of infiltration lag, which produces high ratios of R/P on a monthly time scale.

The SPR approach was used to detect changes in the R:P ratio. The results illustrated in Figure 4 and summarized in Table 1 demonstrate that significant changes in slope were identified in all wells, with changes corresponding to dates ranging from July 2007 to August 2010. This significant change in slope suggests an imbalance in the physical recharge behaviors in the system. To evaluate the potential imbalance, we ponder the climatic factors which could alter the ratio of R:P, including antecedent soil moisture conditions, changes in soil moisture storage, changes in evapotranspiration (ET) rates or changes in precipitation intensity [27]. As the region is arid, it is unlikely that antecedent soil moisture conditions would vary greatly between episodic recharge events. ET rates could potentially change as a result of increased temperatures in the region [68]; however, as potential evapotranspiration (PET) greatly exceeds precipitation on an annual basis, it is unlikely that variability in ET would create a shift in the R:P ratio. To assess our hypothesis, we evaluated changes in the temporal mean and variance of the difference in monthly P and ET for regions around each well as simulated in the National Land Data Assimilation System (NLDAS) [69,70] model including NOAH [71], Variable Infiltration Capacity (VIC) [72] and MOSAIC [73]. The difference, P-ET, represents the flux of water between land and atmosphere. Thus, any significant change in P-ET would suggest that the changes in groundwater recharge could be due to portioning of water to streamflow, soil moisture or infiltration. Our results documented no significant change in P-ET for each 1/8-degree grid encompassing each well. Also, soil moisture likely decreases to levels below the wilting point as a result of high PET between rare episodic precipitation events. Thus, we anticipate that small changes in precipitation intensity, which can have a considerable impact on recharge [27] may result in a change in the R:P ratio. It is important to recognize that changes in groundwater storage, as evaluated in the MRC (Section 3.1) is not a factor in shifts in the R:P ratio as changes in dh/dt have been accounted for prior to the estimation of recharge from the groundwater hydrograph.

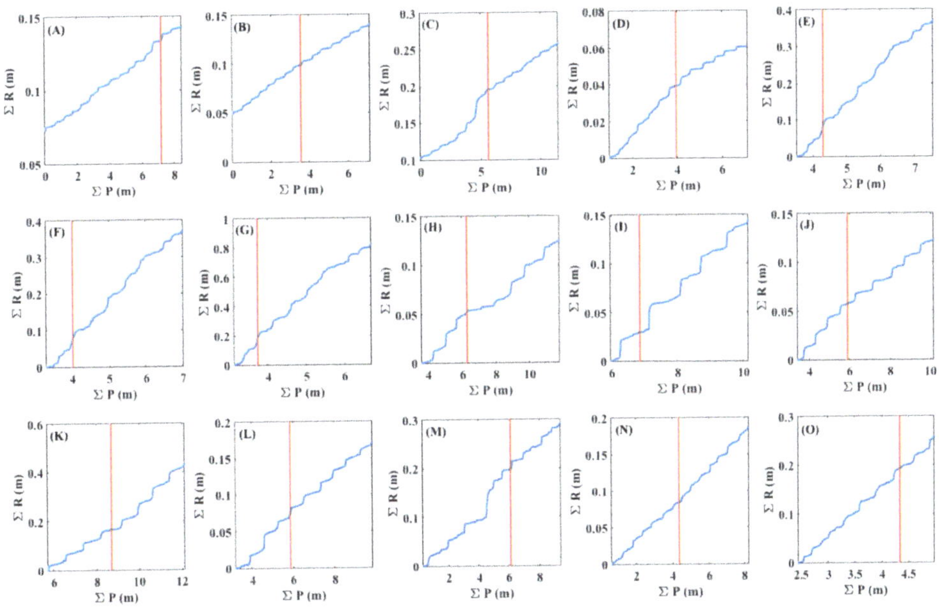

Figure 4. Double mass curves illustrating sum of precipitation (P) *versus* sum of recharge (R). The significant change in the R/P ratio is illustrated as a vertical orange line. The step pattern in the DMC is typical of episodic recharge.

4.4. IDF Curves

To evaluate the influence of precipitation intensity to alter recharge, IDF curves for return periods of 2, 5 and 10 years for 3-, 6-, 12-, 24-, 36- and 48-h precipitation duration events were estimated as illustrated in Figure 5. The IDF curves in Figure 5 illustrate precipitation behaviors either prior to the change in the R:P ratio (Section 4.2) or after the change. We can identify a change in precipitation intensity, especially for short (*i.e.*, 2- and 5-year) return periods. To assess these apparent shifts in precipitation intensities, we evaluate the precipitation intensity magnification, which represents the factor by which the posterior precipitation intensity at a return interval would need to be multiplied by to obtain the preceding precipitation intensity. For example, a magnification factor of 0.5 for a 4-year return period implies that the posterior precipitation intensity has been reduced by 50%. The precipitation intensity magnification is defined as

$$M_p = \frac{I_{p_{tn}}}{I_{p_{tc}}} \tag{8}$$

where I_p is the intensity for precipitation while factors t_n and t_c represent either posterior or preceding time, respectively. Results summarized in Table 2 suggest slight changes in precipitation intensity with magnification factors generally less than 1 for return periods of 2, 5 and 10 years. Such changes in precipitation intensity have been shown to dramatically alter recharge [25,27,74]. Magnification factors for several sites, Site E for example, exhibit reduced magnitudes for short-duration events and increased magnitude for longer-duration events. Such changes illustrate the importance of short-term and high intensity precipitation events in the region which typically result from summer monsoon events [75].

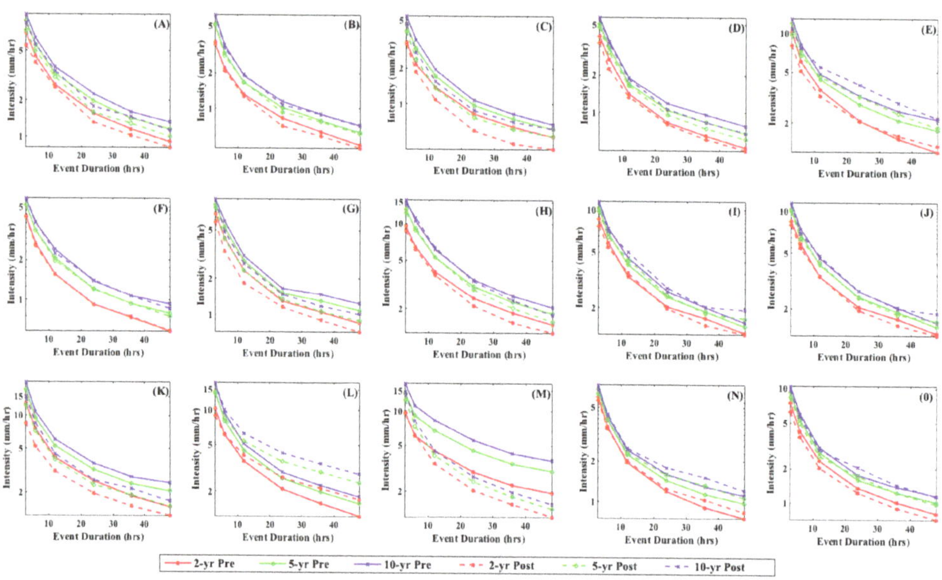

Figure 5. Intensity-Duration-Frequency curves preceding the change in the R/P ratio (solid line) compared to posterior IDF curves for 3-, 6-, 12-, 24-, 36- and 48-h event durations. In general, the observed intensity for posterior precipitation events were lower than preceding precipitation events.

Table 2. Precipitation intensity magnification as estimated using Equation (8).

Duration	Site A			Site B			Site C			Site D			Site E		
	2-year	5-year	10-year	2-year	5-year	10-year	2-year	5-year	10-year	2-year	5-year	10-year	2-year	5-year	10-year
3-h	0.80	0.83	0.85	1.05	1.01	1.00	0.97	0.89	0.86	0.89	0.94	0.97	0.78	0.82	0.84
6-h	0.89	0.89	0.88	0.95	0.95	0.95	0.87	0.81	0.79	0.83	0.92	0.96	0.84	0.91	0.93
12-h	0.95	0.94	0.93	0.96	1.01	1.03	0.78	0.79	0.80	0.94	0.96	0.97	0.89	1.05	1.14
24-h	0.85	0.81	0.79	0.86	0.91	0.94	0.72	0.79	0.82	0.96	0.92	0.90	0.99	1.15	1.22
36-h	0.89	0.90	0.91	0.92	0.97	0.99	0.72	0.81	0.86	0.94	0.89	0.87	1.06	1.13	1.16
48-h	0.90	0.87	0.86	0.95	0.98	0.99	0.80	0.87	0.92	0.95	0.90	0.88	1.10	1.05	1.02
	Site F			**Site G**			**Site H**			**Site I**			**Site J**		
	2-year	5-year	10-year	2-year	5-year	10-year	2-year	5-year	10-year	2-year	5-year	10-year	2-year	5-year	10-year
3-h	0.99	1.00	1.01	0.86	0.94	0.98	0.92	0.93	0.93	0.89	0.95	0.98	0.94	0.97	0.98
6-h	0.97	1.00	1.01	0.77	0.80	0.82	0.96	1.03	1.06	0.92	0.94	0.95	0.93	0.93	0.92
12-h	0.99	0.96	0.94	0.79	0.83	0.86	0.94	1.00	1.03	1.06	1.09	1.10	1.01	1.03	1.04
24-h	0.99	1.01	1.01	0.89	0.89	0.89	0.87	0.94	0.97	0.97	1.03	1.06	0.94	0.98	1.00
36-h	1.01	1.00	0.99	0.86	0.82	0.80	0.83	0.89	0.92	0.89	0.97	1.01	0.90	0.96	0.99
48-h	1.01	0.95	0.92	0.85	0.82	0.81	0.86	0.86	0.86	0.98	1.14	1.22	0.97	1.09	1.16
	Site K			**Site L**			**Site M**			**Site N**			**Site O**		
	2-year	5-year	10-year	2-year	5-year	10-year	2-year	5-year	10-year	2-year	5-year	10-year	2-year	5-year	10-year
3-h	0.66	0.71	0.74	0.87	0.97	1.01	0.99	0.88	0.83	0.94	0.93	0.93	0.84	0.91	0.94
6-h	0.69	0.76	0.79	1.01	1.05	1.07	1.01	0.81	0.74	0.97	1.01	1.03	0.90	0.93	0.95
12-h	0.75	0.74	0.74	1.12	1.20	1.23	0.77	0.60	0.54	1.03	1.03	1.03	0.86	0.92	0.95
24-h	0.77	0.73	0.71	1.25	1.39	1.46	0.69	0.54	0.48	1.04	1.09	1.12	0.91	1.08	1.16
36-h	0.80	0.79	0.78	1.34	1.47	1.53	0.68	0.52	0.46	1.14	1.16	1.17	0.89	0.99	1.04
48-h	0.83	0.73	0.68	1.39	1.51	1.56	0.61	0.47	0.42	1.10	1.09	1.09	0.89	0.95	0.98

5. Discussion

The groundwater recharge literature has few studies that seek to characterize episodic recharge in arid regions. [76] studied episodic recharge in the Murray Darling basin, later identifying the nature of episodic recharge across the region [26]. Limited studies [26,28] have assessed *in situ* groundwater data to characterize changes in recharge. This study evaluated groundwater elevations within the arid southwestern United States to understand potential changes in recharge which may have led to decreases in groundwater storage which Castle *et al.* [31] and Konikow [67] identified within the Lower Colorado River basin and Arizona, respectively. Our results clearly document changes in the relationship between recharge and precipitation identified using the SPR analysis in the double mass curve (Figure 3). Further, the timing of the significant change in the R:P ratio occurred concurrent to noticeable changes in precipitation intensity identified by the precipitation magnification intensity index (Equation (8)). We thus attribute changes in the R:P ratio (Figure 4) to changes in precipitation intensity (Figure 5) across the region. Such results have been predicted for constant recharge sources [77] whereby changes to any of the water budget variables (here precipitation) can lead to changes in the infiltration of water to the saturated zone.

The episodic nature of groundwater recharge [24,25] was evident in Figure 4 highlighted by the step pattern of recharge and precipitation in the double mass curve. Through a simple approach to isolate a change in the relation between recharge and precipitation, the SPR approach, we identified a significant shift in recharge behavior. Given the aridity of the region, we hypothesized that changes in precipitation intensity have altered recharge in the region. Based on the time of the observed change in the R:P ratio (Figure 2, Table 1), we isolated precipitation data to represent precipitation intensity before and after the event. An IDF analysis employing a metric to account for precipitation intensity, a precipitation intensity magnification, identified a slight decrease in intensity posterior to the change identified in the R:P ratio (Table 2, Figure 5). Our findings support previous studies which identified precipitation intensity as an important factor in groundwater recharge [26,27,74,78] whereby small changes in precipitation intensity result in magnified changes in groundwater recharge. Thus, as argued by Alley and Konikow [32], a decrease in groundwater storage can, in part, be explained by a change in the recharge behaviors in wells across the southwestern United States. It remains unknown, however, if the apparent change in the precipitation intensity is a function of climate-scale variability [79] or climate change [80].

Groundwater resources in arid regions of the globe are considered nonrenewable [18] and under stress [11]; therefore, a full understanding of the factors by which stress is identified, primarily use (natural and anthropogenic) and availability (a function of both recharge and groundwater storage), is necessary to effectively and sustainably manage the resource. This study documents the fact that changes in precipitation intensity can result in a change in recharge behaviors, a factor of great importance for groundwater management [81–83]. The change in recharge behaviors was identified during a prolonged drought, thus illustrating the importance of climate variability for recharge resulting from La Nina/El Nino [84] or ENSO/PDO [85]. Reference [80] summarized global groundwater recharge studies over historical time frames, highlighting the need to understand recharge for sustainable groundwater management. Our study provides further evidence of the importance of characterizing groundwater recharge behaviors over short time periods which are affected by variability in precipitation statistics.

Acknowledgments: The authors wish to thank state water resource agencies in the study area for data access. The research was carried out at the Jet Propulsion Laboratory, California Institute of Technology, under a contract with the National Aeronautics and Space Administration. Support from the GRACE Science Team and the Jet Propulsion Laboratory Research and Technology Development programs is gratefully acknowledged.

Author Contributions: Brian F. Thomas conceived and designed the experiments, performed the analysis and analyzed the data, and Ali Behrangi provided guidance on use of TRMM data. All authors (Brian F. Thomas, Ali Behrangi and James S. Famiglietti) contributed to writing of the paper.

Conflicts of Interest: The authors declare no conflict of interest.

References

1. Pandey, V.P.; Shrestha, S.; Chapagain, S.K.; Kazama, F. A framework for measuring groundwater sustainability. *Environ. Sci. Policy* **2011**, *14*, 396–407.
2. Sophocleous, M. Groundwater recharge and sustainability in the High Plains aquifer in Kansas, USA. *Hydrogeol. J.* **2005**, *13*, 351–365.
3. Theis, C.V. The source of water derived from wells. *Civil Eng.* **1940**, *10*, 277–280.
4. Thomas, B.F.; Famiglietti, J.S. Sustainable groundwater management in the arid Southwestern US: Coachella Valley, California. *Water Resour. Manag.* **2015**, *29*, 4411–4426.
5. Sharda, V.N. Watershed Management and Water Harvesting as Strategic Tools for Groundwater Augmentation. Available online: http://publications.iwmi.org/pdf/H039314.pdf (accessed on 3 March 2016).
6. Bredehoeft, J.D. The water budget myth revisited: Why hydrogeologists model. *Groundwater* **2002**, *40*, 340–345.
7. Alley, W.M.; Leake, S.A. The journey from safe yield to sustainability. *Groundwater* **2004**, *42*, 12–16.
8. Zhou, Y. A critical review of groundwater budget myth, safe yield and sustainability. *J. Hydrol.* **2009**, *370*, 207–213.
9. Döll, P.; Flörke, M. *Global-Scale Estimation of Diffuse Groundwater Recharge: Model Tuning to Local Data for Semi-Arid and Arid Regions and Assessment of Climate Change Impact*; Institute of Physical Geography, Frankfurt University: Frankfurt am Main, Germany, 2005.

10. Wada, Y.; van Beek, L.P.; van Kempen, C.M.; Reckman, J.W.; Vasak, S.; Bierkens, M.F. Global depletion of groundwater resources. *Geophys. Res. Lett.* **2010**, *37*.
11. Richey, A.S.; Thomas, B.F.; Lo, M.H.; Reager, J.T.; Famiglietti, J.S.; Voss, K.; Swenson, S.; Rodell, M. Quantifying renewable groundwater stress with GRACE. *Water Resour. Res.* **2015**, *51*, 5217–5238.
12. Loáiciga, H.A.; Maidment, D.R.; Valdes, J.B. Climate-change impacts in a regional karst aquifer, Texas, USA. *J. Hydrol.* **2000**, *227*, 173–194.
13. York, J.P.; Person, M.; Gutowski, W.J.; Winter, T.C. Putting aquifers into atmospheric simulation models: An example from the Mill Creek Watershed, northeastern Kansas. *Adv. Water Resour.* **2002**, *25*, 221–238.
14. Allen, D.M.; Mackie, D.C.; Wei, M. Groundwater and climate change: A sensitivity analysis for the Grand Forks aquifer, southern British Columbia, Canada. *Hydrogeol. J.* **2004**, *12*, 270–290.
15. Hughes, D.A. Incorporating groundwater recharge and discharge functions into an existing monthly rainfall–runoff model/Incorporation de fonctions de recharge et de vidange superficielle de nappes au sein d'un modèle pluie-débit mensuel existant. *Hydrol. Sci. J.* **2004**, *49*.
16. Green, T.R.; Bates, B.C.; Charles, S.P.; Fleming, P.M. Physically based simulation of potential effects of carbon dioxide–altered climates on groundwater recharge. *Vadose Zone J.* **2007**, *6*, 597–609.
17. Crosbie, R.S.; McCallum, J.L.; Walker, G.R.; Chiew, F.H. Modelling climate-change impacts on groundwater recharge in the Murray-Darling Basin, Australia. *Hydrogeol. J.* **2010**, *18*, 1639–1656.
18. Margat, J.; Van der Gun, J. *Groundwater around the World: A Geographic Synopsis*; CRC Press: Boca Raton, FL, USA, 2013.
19. Kløve, B.; Ala-Aho, P.; Bertrand, G.; Gurdak, J.J.; Kupfersberger, H.; Kværner, J.; Muotka, T.; Mykrä, H.; Preda, E.; Rossi, P.; *et al.* Climate change impacts on groundwater and dependent ecosystems. *J. Hydrol.* **2014**, *518*, 250–266.
20. Famiglietti, J.S. The global groundwater crisis. *Nat. Clim. Chang.* **2014**, *4*, 945–948.
21. Konikow, L.F. Contribution of global groundwater depletion since 1900 to sea-level rise. *Geophys. Res. Lett.* **2011**, *38*.
22. Voss, C.I.; Soliman, S.M. The transboundary non-renewable Nubian Aquifer System of Chad, Egypt, Libya and Sudan: Classical groundwater questions and parsimonious hydrogeologic analysis and modeling. *Hydrogeol. J.* **2014**, *22*, 441–468.
23. Hashemi, H.; Uvo, C.B.; Berndtsson, R. Coupled modeling approach to assess climate change impacts on groundwater recharge and adaptation in arid areas. *Hydrol. Earth Syst. Sci.* **2015**, *19*, 4165–4181.
24. Gee, G.W.; Hillel, D. Groundwater recharge in arid regions: Review and critique of estimation methods. *Hydrol. Process.* **1988**, *2*, 255–266.
25. Lewis, F.M.; Walker, G.R. Assessing the potential for significant and episodic recharge in southwestern Australia using rainfall data. *Hydrogeol. J.* **2002**, *10*, 229–237.

26. Crosbie, R.S.; McCallum, J.L.; Walker, G.R.; Chiew, F.H. Episodic recharge and climate change in the Murray-Darling Basin, Australia. *Hydrogeol. J.* **2012**, *20*, 245–261.
27. Barron, O.V.; Crosbie, R.S.; Dawes, W.R.; Charles, S.P.; Pickett, T.; Donn, M.J. Climatic controls on diffuse groundwater recharge across Australia. *Hydrol. Earth Syst. Sci.* **2012**, *16*, 4557–4570.
28. Nimmo, J.R.; Horowitz, C.; Mitchell, L. Discrete-Storm Water-Table Fluctuation Method to Estimate Episodic Recharge. *Groundwater* **2015**, *53*, 282–292.
29. Piechota, T.; Timilsena, J.; Tootle, G.; Hidalgo, H. The western US drought: How bad is it? *Eos Trans. Am. Geophys. Union* **2004**, *85*, 301–304.
30. Cayan, D.R.; Das, T.; Pierce, D.W.; Barnett, T.P.; Tyree, M.; Gershunov, A. Future dryness in the southwest US and the hydrology of the early 21st century drought. *Proc. Natl. Acad. Sci. USA* **2010**, *107*, 21271–21276.
31. Castle, S.L.; Thomas, B.F.; Reager, J.T.; Rodell, M.; Swenson, S.C.; Famiglietti, J.S. Groundwater depletion during drought threatens future water security of the Colorado River Basin. *Geophys. Res. Lett.* **2014**, *41*, 5904–5911.
32. Alley, W.M.; Konikow, L.F. Bringing GRACE down to earth. *Groundwater* **2015**, *53*, 826–829.
33. Alley, W.M.; La Baugh, J.W.; Reilly, T.E. Groundwater as an element in the hydrological cycle. *Encycl. Hydrol. Sci.* **2005**.
34. Siebert, S.; Burke, J.; Faures, J.M.; Frenken, K.; Hoogeveen, J.; Döll, P.; Portmann, F.T. Groundwater use for irrigation—A global inventory. *Hydrol. Earth Syst. Sci.* **2010**, *14*, 1863–1880.
35. Trenberth, K.E.; Dai, A.; Rasmussen, R.M.; Parsons, D.B. The changing character of precipitation. *Bull. Am. Meteorol. Soc.* **2003**, *84*, 1205–1217.
36. Searcy, J.K.; Hardison, C.H. *Double-Mass Curves*; U.S. Government Printing Office: Washington, DC, USA, 1960.
37. Huffman, G.J.; Bolvin, D.T.; Nelkin, E.J.; Wolff, D.B.; Adler, R.F.; Gu, G.; Stocker, E.F. The TRMM multisatellite precipitation analysis (TMPA): Quasi-global, multiyear, combined-sensor precipitation estimates at fine scales. *J. Hydrometeorol.* **2007**, *8*, 38–55.
38. Schneider, U.; Becker, A.; Finger, P.; Meyer-Christoffer, A.; Rudolf, B.; Ziese, M. *GPCC Full Data Reanalysis Version 7.0 At 2.5deg: Monthly Land-Surface Precipitation from Rain-Gages Built on GTS-based and Historic Data*; Global Precipitation Climatology Centre: Boulder, CO, USA, 2015.
39. Behrangi, A.; Khakbaz, B.; Jaw, T.C.; AghaKouchak, A.; Hsu, K.; Sorooshian, S. Hydrologic evaluation of satellite precipitation products over a mid-size basin. *J. Hydrol.* **2011**, *397*, 225–237.
40. Wall, C.L.; Zipser, E.J.; Liu, C. A regional climatology of monsoonal precipitation in the southwestern United States using TRMM. *J. Hydrometeorol.* **2012**, *13*, 310–323.
41. Xu, W.; Zipser, E.J. Properties of deep convection in tropical continental, monsoon, and oceanic rainfall regimes. *Geophys. Res. Lett.* **2012**, *39*.
42. U.S. Geological Survey. USGS Water Data for the Nation. Available online: http://waterdata.usgs.gov/nwis (accessed on 7 March 2016).

43. Arizona Department of Water Resources. Groundwater Site Inventory. Available online: http://gisweb.azwater.gov/waterresourcesdata/gwsi.aspx (accessed on 7 March 2016).
44. Vogel, R.M.; Kroll, C.N. Regional geohydrologic-geomorphic relationships for the estimation of low-flow statistics. *Water Resour. Res.* **1992**, *28*, 2451–2458.
45. Thomas, B.F.; Vogel, R.M.; Kroll, C.N.; Famiglietti, J.S. Estimation of the base flow recession constant under human interference. *Water Resour. Res.* **2013**, *49*, 7366–7379.
46. Healy, R.W.; Cook, P.G. Using groundwater levels to estimate recharge. *Hydrogeol. J.* **2002**, *10*, 91–109.
47. Heppner, C.S.; Nimmo, J.R. A computer program for predicting recharge with a master recession curve. US Geological Survey: Menlo Park, CA, USA, 2005.
48. Olmsted, F.H.; Hely, A.G. *Relation between Ground Water and Surface Water in Brandywine Creek Basin, Pennsylvania (No. 417-A)*; U.S. Government Printing Office: Washington, DC, USA, 1962.
49. Risser, D.W.; Gburek, W.J.; Folmar, G.J. Comparison of recharge estimates at a small watershed in east-central Pennsylvania, USA. *Hydrogeol. J.* **2009**, *17*, 287–298.
50. Merriam, C.F. A comprehensive study of the rainfall on the Susquehanna Valley. *Eos Trans. Am. Geophys. Union* **1937**, *18*, 471–476.
51. Lee, C.H.; Chen, W.P.; Lee, R.H. Estimation of groundwater recharge using water balance coupled with base-flow-record estimation and stable-base-flow analysis. *Environ. Geol.* **2006**, *51*, 73–82.
52. Stedinger, J.R.; Vogel, R.M.; Foufoula-Georgiou, E. *Frequency Analysis of Extreme Events*; U.S. Government Printing Office: Washington, DC, USA, 1993.
53. Smith, J.A. *Precipitation, Handbook of Hydrology*; McGraw-Hill, Inc.: New York, NY, USA, 1993; Volume 3.
54. McCuen, R.H. *Hydrologic Design and Analysis*; Prince Hall: New Jersey, NJ, USA, 1998; Volume 814.
55. Chow, V.T.; Maidment, D.R.; Mays, L.W. *Applied Hydrology*; McGraw-Hill: New York, NY, USA, 1988.
56. Dingman, S.L. *Physical Hydrology*; Waveland Press: Long Grove, IL, USA, 2015.
57. Koutsoyiannis, D. Statistics of extremes and estimation of extreme rainfall: I. Theoretical investigation/Statistiques de valeurs extrêmes et estimation de précipitations extrêmes: I. Recherche théorique. *Hydrol. Sci. J.* **2004**, *49*.
58. Gumbel, E.J. *Statistics of Extremes*; Courier Corporation: New York, NY, USA, 2012.
59. Cuanne, C. Unbiased plotting positions—A review. *J. Hydrol.* **1987**, *37*, 205–222.
60. Hosking, J.R.M. Algorithm AS 215: Maximum-likelihood estimation of the parameters of the generalized extreme-value distribution. *Appl. Stat.* **1985**, *34*, 301–310.
61. Dourte, D.; Shukla, S.; Singh, P.; Haman, D. Rainfall intensity-duration-frequency relationships for Andhra Pradesh, India: Changing rainfall patterns and implications for runoff and groundwater recharge. *J. Hydrol. Eng.* **2012**, *18*, 324–330.
62. Wang, H.; Gao, J.E.; Zhang, M.J.; Li, X.H.; Zhang, S.L.; Jia, L.Z. Effects of rainfall intensity on groundwater recharge based on simulated rainfall experiments and a groundwater flow model. *Catena* **2015**, *127*, 80–91.

63. Camera, C.; Masetti, M.; Apuani, T. Rainfall, infiltration, and groundwater flow in a terraced slope of Valtellina (Northern Italy): Field data and modelling. *Environ. Earth Sci.* **2012**, *65*, 1191–1202.
64. Machusick, M.; Welker, A.; Traver, R. Groundwater mounding at a storm-water infiltration BMP. *J. Irrig. Drain. Eng.* **2011**, *137*, 154–160.
65. Rutledge, A.T. *Computer Programs for Describing the Recession of Ground-Water Discharge and for Estimating Mean Ground-Water Recharge and Discharge from Streamflow Records: Update*; US Department of the Interior: Washington, DC, USA; US Geological Survey: Reston, VA, USA, 1998.
66. Posavec, K.; Bačani, A.; Nakić, Z. A visual basic spreadsheet macro for recession curve analysis. *Groundwater* **2006**, *44*, 764–767.
67. Konikow, L.F. *Groundwater Depletion in the United States (1900–2008)*; US Department of the Interior: Washington, DC, USA; US Geological Survey: Reston, VA, USA, 2013; p. 63.
68. Dobrowski, S.Z.; Abatzoglou, J.; Swanson, A.K.; Greenberg, J.A.; Mynsberge, A.R.; Holden, Z.A.; Schwartz, M.K. The climate velocity of the contiguous United States during the 20th century. *Glob. Chang. Biol.* **2013**, *19*, 241–251.
69. Mitchell, K.E.; Lohmann, D.; Houser, P.R.; Wood, E.F.; Schaake, J.C.; Robock, A.; Cosgrove, B.A.; Sheffield, J.; Duan, Q.; Luo, L.; *et al.* The multi-institution North American Land Data Assimilation System (NLDAS): Utilizing multiple GCIP products and partners in a continental distributed hydrological modeling system. *J. Geophys. Res. Atmos.* **2004**, *109*.
70. Xia, Y.; Mitchell, K.; Ek, M.; Cosgrove, B.; Sheffield, J.; Luo, L.; Alonge, C.; Wei, H.; Meng, J.; Livneh, B.; *et al.* Continental-scale water and energy flux analysis and validation for North American Land Data Assimilation System project phase 2 (NLDAS-2): 2. Validation of model-simulated streamflow. *J. Geophys. Res. Atmos.* **2012**, *117*.
71. Koren, V.; Schaake, J.; Mitchell, K.; Duan, Q.Y.; Chen, F.; Baker, J.M. A parameterization of snowpack and frozen ground intended for NCEP weather and climate models. *J. Geophys. Res. Atmos.* **1999**, *104*, 19569–19585.
72. Liang, X.; Lettenmaier, D.P.; Wood, E.F.; Burges, S.J. A simple hydrologically based model of land surface water and energy fluxes for general circulation models. *J. Geophys. Res. All Ser.* **1994**, *99*, 14415–14428.
73. Koster, R.D.; Suarez, M.J. Modeling the land surface boundary in climate models as a composite of independent vegetation stands. *J. Geophys. Res.* **1992**, *97*.
74. French, R.H.; Jacobson, R.L.; Lyles, B.F. Threshold precipitation events and potential ground-water recharge. *J. Hydr. Eng.* **1996**, *122*, 573–578.
75. Mock, C.J. Climatic controls and spatial variations of precipitation in the western United States. *J. Clim.* **1996**, *9*, 1111–1125.
76. Crosbie, R.S.; Dawes, W.R.; Charles, S.P.; Mpelasoka, F.S.; Aryal, S.; Barron, O.; Summerell, G.K. Differences in future recharge estimates due to GCMs, downscaling methods and hydrological models. *Geophys. Res. Lett.* **2011**, *38*.

77. McCallum, J.L.; Crosbie, R.S.; Walker, G.R.; Dawes, W.R. Impacts of climate change on groundwater in Australia: A sensitivity analysis of recharge. *Hydrogeol. J.* **2010**, *18*, 1625–1638.
78. Kendy, E.; Zhang, Y.; Liu, C.; Wang, J.; Steenhuis, T. Groundwater recharge from irrigated cropland in the North China Plain: Case study of Luancheng County, Hebei Province, 1949–2000. *Hydrol. Process.* **2004**, *18*, 2289–2302.
79. Griffin, D.; Woodhouse, C.A.; Meko, D.M.; Stahle, D.W.; Faulstich, H.L.; Carrillo, C.; Leavitt, S.W. North American monsoon precipitation reconstructed from tree-ring latewood. *Geophys. Res. Lett.* **2013**, *40*, 954–958.
80. Cook, B.I.; Seager, R. The response of the North American Monsoon to increased greenhouse gas forcing. *J. Geophys. Res. Atmos.* **2013**, *118*, 1690–1699.
81. Gleick, P.H. Water in crisis: Paths to sustainable water use. *Ecol. Appl.* **1998**, *8*, 571–579.
82. Scanlon, B.R.; Keese, K.E.; Flint, A.L.; Flint, L.E.; Gaye, C.B.; Edmunds, W.M.; Simmers, I. Global synthesis of groundwater recharge in semiarid and arid regions. *Hydrol. Process.* **2006**, *20*, 3335–3370.
83. Pietersen, K. Multiple criteria decision analysis (MCDA): A tool to support sustainable management of groundwater resources in South Africa. *Water SA* **2007**, *32*, 119–128.
84. McCabe, G.J.; Dettinger, M.D. Decadal variations in the strength of ENSO teleconnections with precipitation in the western United States. *Int. J. Climatol.* **1999**, *19*, 1399–1410.
85. Cayan, D.R.; Redmond, K.T.; Riddle, L.G. ENSO and hydrologic extremes in the western United States*. *J. Clim.* **1999**, *12*, 2881–2893.

Anthropogenic-Induced Changes in the Mechanism of Drylands Ephemeral Stream Recharge, Western Saudi Arabia

Khan Z. Jadoon, Samir Al-Mashharawi, Sherif M. Hanafy, Gerard T. Schuster and Thomas M. Missimer

Abstract: Wadi aquifers in Saudi Arabia historically have been recharged primarily by channel loss (infiltration) during floods. Historically, seasonal groundwater levels fluctuated from land surface to about 3 m below the surface. Agricultural irrigation pumping has lowered the water table up to 35 m below the surface. The geology surrounding the fluvial system at Wadi Qidayd consists of pelitic Precambrian rocks that contribute sediments ranging in size from mud to boulders to the alluvium. Sediments within the wadi channel consist of fining upward, downstream-dipping beds, causing channel floodwaters to pass through several sediment sequences, including several mud layers, before it can reach the water table. Investigation of the wadi aquifer using field observation, geological characterization, water-level monitoring, geophysical profiles, and a hypothetical model suggests a critical water level has been reached that affects the recharge of the aquifer. The wetted front can no longer reach the water table due to the water uptake in the wetting process, downstream deflection by the clay layers, and re-emergence of water at the surface with subsequent direct and diffusive evaporative loss, and likely uptake by deep-rooted Acacia trees. In many areas of the wadi system, recharge can now occur only along the channel perimeter via fractured rocks that are in direct horizontal hydraulic connection to the permeable beds above and below the water table.

Reprinted from *Water*. Cite as: Jadoon, K.Z.; Al-Mashharawi, S.; Hanafy, S.M.; Schuster, G.T.; Missimer, T.M. Anthropogenic-Induced Changes in the Mechanism of Drylands Ephemeral Stream Recharge, Western Saudi Arabia. *Water* **2016**, *8*, 136.

1. Introduction

In drylands regions, most stream discharges are of an ephemeral nature with no perennial discharges of surface water that are locally produced [1]. In the Middle East region, ephemeral streams are termed wadis and the underlying alluvial sediments are commonly the only sources of fresh groundwater in many basins [2]. Wadis have been historically areas where oases have occurred in which wells containing freshwater and small farms were developed.

The aquifer underlying a wadi channel is recharged primarily by infiltration and percolation coupled with transmission loss (infiltration) in the channel during flood events [1–12]. A number of important factors dictate the quantity of recharge to these aquifers. They include: (1) duration of the channel flood event, (2) degree of surface water ponding within the wadi channel during an event, (3) vertical hydraulic conductivity of the alluvial sediments, (4) degree of horizontal stratification (e.g., presence of mud layers within the sequence), (5) slope and structure of the bedding within the aquifer, (6) position of the water level in the aquifer at the time of the flood, and (7) the occurrence of deep-rooted vegetation (phreatophytes).

Depending upon the geology of the basin in which a wadi channel lies, the alluvial sediments constituting the underlying aquifer can consist primarily of boulders, gravel, and sand, or can contain significant quantities of mud that forms a stacked pattern with mud layers creating separation between the high hydraulic conductivity gravel beds ([12], Figure 1). Where the aquifer consists of fining upwards beds associated with individual flood events, the saturated front of water percolating downward in the aquifer may be intercepted and diverted horizontally downstream or can be extracted by vegetation that has horizontal oriented root masses occurring atop mud layers [1]. In wadi segments where the channel slope is steep, the bedded sediments can be truncated, causing the percolated water to emerge downstream and be lost to direct evaporation or can be transported to a shallow depth at which diffusive evaporation loss can occur. This issue is particularly problematic at transition locations where slope inflections occur between channel-reach segments. In addition, water can be lost in channel segments that have eroded after a flood event (Figure 2, left panel). The steepest slope inflections tend to occur at transitional locations from the distal to middle, and middle to proximal channel locations.

In highly-stratified alluvial aquifers, a substantial lowering of the water table can cause a major reduction in the rate of recharge for individual storm events as well as a long-term reduction in overall recharge. The three causes of the reduction in recharge rate are: (1) the increased thickness of the unsaturated zone which creates the requirement of a greater amount of water to enter the aquifer to allow wetting of the unsaturated sediment; (2) the occurrence of the clay layers that inhibit vertical movement of the wetting front; and (3) the greater number of horizontal diversions that inhibit downward movement of the wetted front. The increased thickness of the unsaturated zone causes a rainfall event to be of larger magnitude and of a longer duration before recharge can occur. The larger number of barriers to vertical flow with diversion amplifies the problem of vertical flow to allow recharge as the water table declines. Enhanced growth of deep-rooted vegetation, such as *Acacia* sp., may also contribute to the change in hydrology. Based on local conditions, a critical position of the water table will be reached whereupon recharge from channel flood loss will no longer occur.

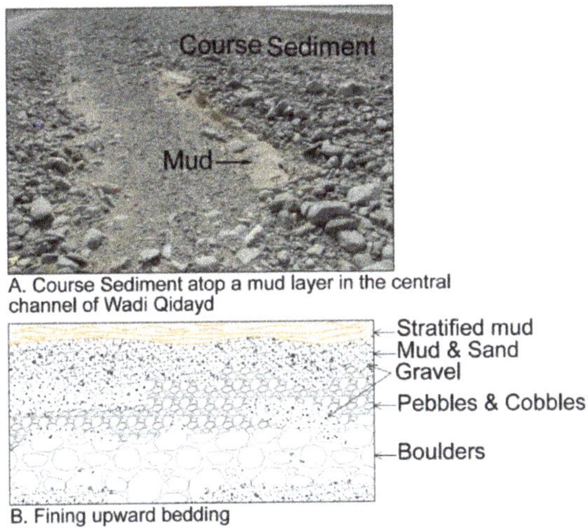

Figure 1. Channel sediments (**A**) and fining upward beds in the subsurface below the wadi channel (**B**).

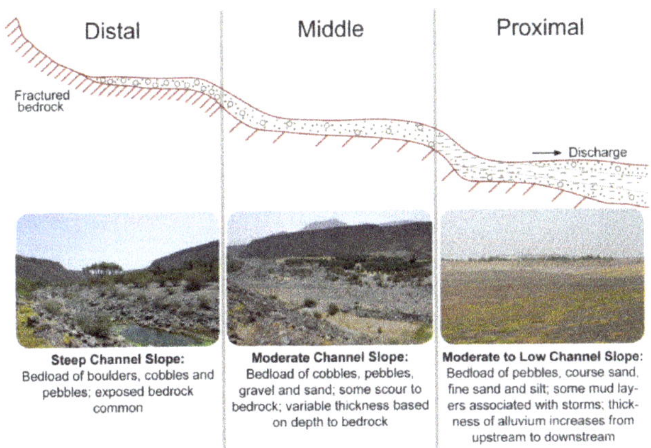

Figure 2. Characteristics of wadi channels in western Saudi Arabia showing the distal, middle, and lower segments (from Missimer *et al.* [12]).

It is the purpose of this research to explore the issue of anthropogenic lowering of water levels by pumping of wells in a wadi aquifer system that severely reduces or eliminates channel flood loss as a means of recharge. Loss of water from the alluvial aquifer system has been further exacerbated by the historical planting of *Acacia* trees which are phreatophytes. This can cause the only recharge mechanism to be vertical movement into fractured rocks occurring at the wadi perimeter or the permanent drying of the alluvial aquifer. The concept is investigated by field observation of the geology and hydrogeology of a wadi system occurring in a geologic setting that produces considerable quantities of mud during erosion. Water level measurements, geophysical measurements, and a theoretical assessment using one-dimensional unsaturated zone modeling were also made.

2. Background and Methods

2.1. Site Description

The western region of Saudi Arabia contains numerous wadi systems that drain into the Red Sea during flood events [13]. These wadis are very mature systems that have been formed as erosion channels incised into Precambrian-age rocks with diverse lithologies [14]. Wadis that occur in Precambrian rocks consisting of partially-metamorphosed shales, schists, metavolcanic, and meta-sediments tend to contain a greater quantity of mud within the alluvial sediment filling the channel and forming the underlying aquifer. At other locations, Precambrian granites produce a higher percentage of boulders, pebbles, gravel, and sand with lower mud content.

A series of field studies were conducted in Wadi Qidayd that is located about 120 km north of Jeddah in western Saudi Arabia (Figure 3). The geology of the Wadi Qidayd consists primary of schists, and meta-sedimentary rock types that produce a wide variety of sediments including boulders, cobbles, pebbles, gravel, sand, and mud [15]. The wadi channels are located in areas of enhanced faulting and fracturing of the Precambrian rocks and have been influenced by the influx of tertiary lava flows [16]. The Precambrian rocks surrounding the wadi channels are heavily fractured. An eroded tertiary lava field, eroded into boulders, commonly lies adjacent to the incised valleys formed by the wadi.

The composition of the sediments constituting the alluvial aquifer system that underlies the channel network varies depending on the geographic position. The geomorphic zones, as defined by Missimer *et al.* [12] (Figure 2), are the distal, middle, and proximal reaches. The aquifer in the distal channel reaches is 3 to 5 m thick and contains only a few mud layers which generally are located just above the fractured bedrock. In the middle segments, the aquifer thickness ranges from 5 to 50 m in thickness and it consists of a vertically stacked sequence of fining upward, bedded sediments. A mud layer does not always occur between layers of coarse-grained sediments due to removal by periodic erosion events. The proximal channel sediments range from 50 to 100 m in thickness and contain numerous fining upward flood deposited units along with 1 to 3 m thick units of muddy sand or sandy mud, some gravel beds, and some winnowed sand beds. Some of the bedding was directly observed in dry and deep wells with an exposed open-hole section (Figure 4). Local residents provided some descriptions of the sediments in the wells, which are large in diameter (2 to 3 m) and were dug by hand.

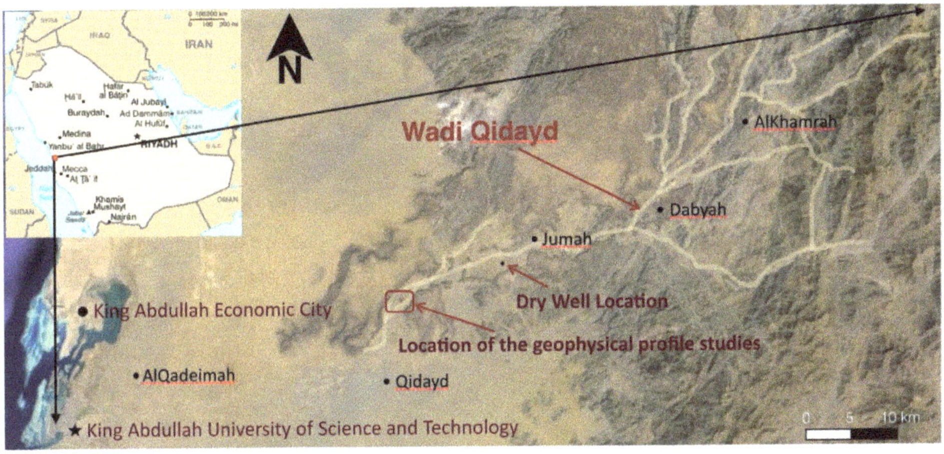

Figure 3. Location of Wadi Qidayd in west-central Saudi Arabia.

Figure 4. Dry well in Wadi Qidayd showing fining upward bedding in the lower part of the well.

A few aquifer tests were conducted in the area and showed a range in horizontal hydraulic conductivity values from 5 to 334 cm/h (1.4×10^{-3} to 9.3×10^{-2} cm/s) and vertical hydraulic conductivity values ranging from 4 to 34 cm/h (1.1×10^{-3} to 9.4×10^{-3} cm/s) [12]. Sediment samples were collected from cross sections of the channels at specific areas and the laboratory grain size distributions, porosity, and hydraulic conductivities were measured [17]. Estimates of the vertical hydraulic conductivity of the mud layer, when greater than 5 cm in thickness, are about 0.89 cm/h (2.5×10^{-4} cm/s) and at about 10 cm in thickness are less than 0.083 cm/h

(2.3×10^{-5} cm/s). The aquifer in the distal area is unconfined and in the middle and proximal is semi-unconfined to semi-confined in nature [18].

2.2. Field Observation Methods

Numerous field excursions and exercises were conducted at Wadi Qidayd over a period of three years. A large number of wells were inspected and discussions were held with the farmers in the area. The bedding was observed and described in some dry wells where it could be directly [4] recorded. Additionally, within the channel and particularly at the edges, the fining upward sediment flood deposits were observed. The channel was inspected and some areas of moisture emergence were observed (Figure 2, right panel). The occurrence of deep-rooted desert plants (*Acacia sp.*) was documented.

Water level recording devices (HOBO DataLoggers) were placed in several wells to document the depletion of the aquifer caused by pumping. The record from the well data is incomplete because during pumping cycles, the water level would commonly fall below the position of the transducer. Sometimes farmers would remove the transducers from the wells to dig below the depleted water table. The location of the well where water level data were recorded is shown in Figure 5.

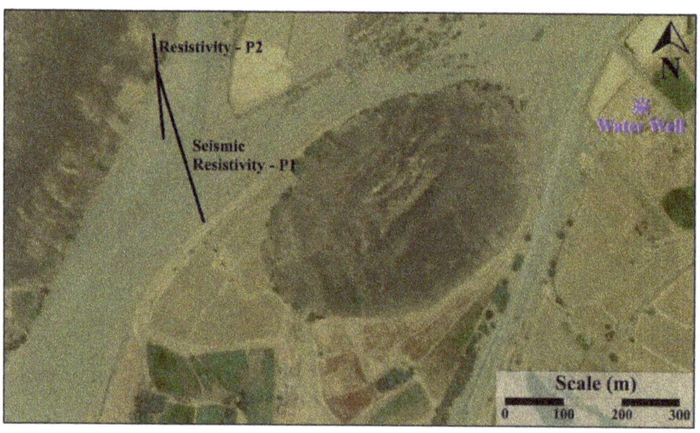

Figure 5. Locations of three geophysical profiles in Wadi Qidayd with both seismic refraction and resistivity profiles along tract P1 and only resistivity along tract P2.

2.3. Collection of Geophysical Profiles across Parts of Wadi Qidayd

Some geophysical profiles were collected across the wadi channel (Figure 5) after a rainfall event from the edge of the wadi to the central part of the channel. The target of the geophysical study was to find the thickness of the surface layer and the depth to the water table. In this regard, one seismic and two resistivity profiles were recorded at the selected study area (Figure 5).

The 2D seismic profile had a total of 117 vertical component receivers with 2.0 m spacing and one shot at every receiver position. Here, a 91 kg accelerated weight drop was used to generate the seismic energy with about 15 stacks used at each shot location.

The seismic refraction method was very sensitive to velocity/density variations in the subsurface, while the resistivity method is more sensitive to water content in the subsurface layer [19]. In this study two resistivity profiles were recorded (Figure 5) to identify the depth to the water table. Profile P-1 was recorded using 64 nodes with a 5 m node interval while profile P-2 was recorded using 32 nodes with node interval of 5 m. Both profiles were recorded using the Wenner-Schlumberger configuration.

2.4. Hydrodynamic Modeling

Vertical movement of the wetting front through the sediment sequence was assessed by using a simplified one-dimensional water flow model. The hydrological simulator Hydrus 1-D was utilized to numerically solve the Richards equation for the soil hydrodynamics, expressed here in terms of pressure head [20]:

$$C(h)\frac{\partial h}{\partial t} = \frac{\partial}{\partial Z}\left[K(\theta)\left(\frac{\partial h}{\partial Z}\right) - 1\right] \quad (1)$$

where the time and depth dependent pressure head (h) is given in cm, θ is the soil water content ($cm^3 \cdot cm^{-3}$), $C(h) = d\theta(h)/dh$ is the differential water capacity where $\theta(h)$ is the water content as a function of pressure head (cm^3/cm^3), t is the time (h), z is the depth which is considered positive downward (cm), and $K(\theta)$ is the hydraulic conductivity as a function of water content (cm/h).

The water retention and hydraulic conductivity functions of the subsurface system were described by the classical Mualem-van Genuchten model [21,22]. The water retention is given by:

$$\theta(h) = \theta_r + (\theta_s - \theta_r)\left[1 + |\alpha h|^n\right]^{-m} \quad (2)$$

where α (1/cm) and n(-) are shape parameters which are inversely related to the air entry value and the width of the pore size distribution, θ_s and θ_r are the residual and saturated water contents respectively, and m is defined as $m = 1 - 1/n$ with $n > 1$ [21]. The hydraulic conductivity is given by:

$$K(\theta) = K_s\left(\frac{\theta - \theta_r}{\theta_s - \theta_r}\right)^{\lambda}\left[1 - \left(1 - \left(\frac{\theta - \theta_r}{\theta_s - \theta_r}\right)^{\frac{1}{m}}\right)^{m}\right]^2 \quad (3)$$

where Ks is the saturated hydraulic conductivity (cm/h) and λ(-) represents pore tortuosity. The value of λ is 0.5 or it can be estimated from bulk density and hydraulic conductivity with which it is highly correlated [23].

Two scenarios were simulated to assess the infiltration of water in a 13-meter thick sediment subsurface section. The first scenario was considered with uniform coarse texture sediments having a porosity of 38% and saturated hydraulic conductivity of 24.81 cm/h. In the second scenario, thin layers of impervious fine texture soil were imbedded in the course texture sediment. Porosity of the fine texture soil was 42% and saturated hydraulic conductivity was 0.85/cm h. Parameters α and n were obtained with the Rosetta model using sand, silt, and clay contents for both the coarse and fine texture soil [24]. In total, fifteen soil samples were collected from the wadi channel and analyzed for grain size distribution having coarse to fine soil texture. The average of the grain sizes for the coarse texture samples have 93.39% sand, 4.93% silt, and 1.68% clay, whereas fine texture soil samples have 25.21% sand, 58.57% silt, and 16.2% clay. Parameters for the Mualem-van Genuchten model are shown in Table 1.

Table 1. Mualem-van Genuchten hydraulic parameters for the coarse and fine texture sediments.

Texture	θ_r(cm$^3\cdot$ cm^{-3})	θ_s(cm$^3\cdot$ cm^{-3})	α(cm^{-1})	n(-)	K_s(cm h^{-1})	λ(-)
Coarse	0.048	0.381	0.036	3.171	24.81	0.5
Fine	0.061	0.423	0.005	1.663	0.85	0.5

3. Results

3.1. Wadi Channel Geology

Data collected from dry wells was by visual observation. Eroded parts of the primary and tributary channels within the Wadi Qidayd drainage basin suggest that the alluvial aquifer is highly stratified with numerous beds occurring between the land surface and the base of the aquifer. The scale of these fining upwards beds ranges from 0.3 to 1.5 m. Flood events have removed the mud layer that separates the fining-upwards beds at some locations, thereby making the permeable bed thicker (up to 2 m). The overall aquifer ranges from less than 5 m in thickness in the upper (distal) reaches of minor tributaries to nearly 85 m in the primary channel in the proximal reach. These estimates are based on well depths that penetrate into or close to the base of the aquifer. The wells in the proximal reach may not yield an accurate estimate of the aquifer thickness because of saline-water occurrence which limits well depths. Based on field observations, it is estimated that there are at least four mud units occurring within each 5 to 30 m vertical segment of the aquifer. Three fining upward units were found in the dry well shown in Figure 4 within a 4 m open section.

The thickness of the mud part of a bed increases downstream (settling time during a flood event) with a corresponding overall reduction primarily in vertical hydraulic conductivity. The lateral continuity of each storm-deposited, fining upward bed cannot be predicted accurately. There are likely some discontinuities that allow deep penetration of the wetted front before reaching a deeper bed at some locations (e.g., bed slope inflection points).

The Precambrian rocks that occur along the perimeter walls of the wadi are highly fractured with the orientation of the fractures being vertical in many locations (Figure 6). Large-diameter, fractured-rock wells (2–3 m) are used to supply isolated dwellings that are built atop the bluffs surrounding the distal parts of the wadi system. These wells demonstrate that the rock has sufficient hydraulic conductivity and storage to supply enough fresh water to supply large houses (10–15 m^3/d). A drawing showing the conceptual connection between the fractured rock aquifer and the alluvial aquifer is shown in Figure 7. Commonly, during storms rainfall is trapped at the base of the heavily eroded "harrats" or lava fields that occur at the land surface along the upper rim of the wadis and flows between the boulders toward the edge of the wadi channels (Figure 8a). The soil occurring at the base of the eroded lava boulders consists of a thick layer of dense clay (Figure 8b). The surface water can then enter the fractured rock at the perimeter of the wadi channels and travel downward to enter the alluvium where the harrat system (eroded lava) comes in contact with the fractured rocks (Figure 9).

Figure 6. Highly fractured pelitic rocks consisting of schists and meta-shales along the Wadi Qidayd perimeter.

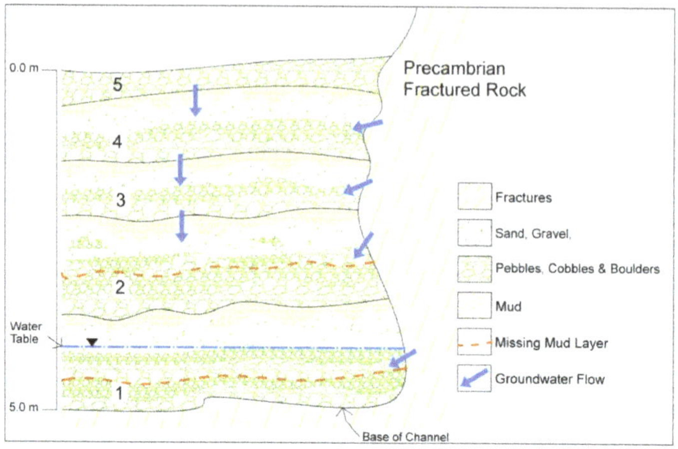

Figure 7. Stacked sequence of fining upward beds showing sediment composition and hydraulic connection to the perimeter fractured rock. Note that the vertical scale is what would be expected if all of the beds were preserved within a 5 m increment. It is expected that many mud layers have been removed in some reaches of the wadi, particularly the distal and middle sections, but in all cases there are at least three full beds preserved.

Figure 8. Eroded basaltic boulders of the harrat (**a**) trap water during rainfall events and it ponds atop the underlying clay (**b**). The water flows downhill on the clay surface to where it intersects fractured rocks along the perimeter walls of the wadi. The fractured rock conducts the water downward into the low hydraulic conductivity beds within the alluvial sequence, thereby providing perimeter recharge.

Figure 9. Harrat lava boulders draped over the edge of Wadi Qidayd allowing trapped rainfall to enter the perimeter fractured rock.

3.2. Water Level Fluctuations

The water levels throughout the alluvial aquifer system of Wadi Qidayd are declining, but not in a regular manner. Many wells in the distal area where the alluvium is 5 m or less in thickness are dry most of the year. Some wells do receive some recharge from channel loss during floods. So, many wells have about 1 m of water in storage above the basement rock with the water table position being 4 m below surface. Water levels in the middle to proximal reaches of the wadi are between 13 and 35 m below surface and are either dry or the water is unusable due to high salinity in many areas. Wells are still being developed and pumped for sale of water in some parts of the wadi system.

The water level fluctuations in a well were recorded for eight months as shown in Figure 10. A single precipitation event occurred near the end of December. The water level dropped from 14 to 15.6 m and the HOBO logger was in the free air after 15.6 m. The water level showed stable measurements during the month of September and October as the logger measured the pressure of free air. There was no data during November and the first three weeks of December as the farmer removed the logger and dug the well 2 m deeper. Overall, the system water level was dropping without significant recovery during flood events. There was a delay of 10 days or greater after a rainfall/flood event between a pattern of declining water levels and a short-term rise in the water level with a return to a declining state. The temporary recovery of the water level in the well was likely due to either a pressure effect or water reaching the well from the fractured rock source at the edge of the channel. The rapid decline occurs because of the continuing pumping of the aquifer in the nearby area.

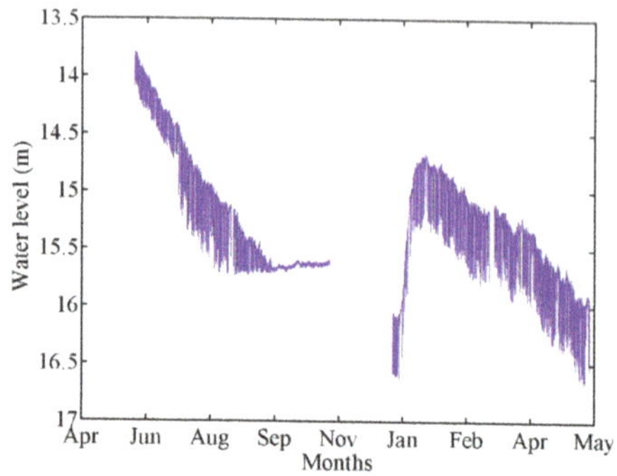

Figure 10. Water level record from a well at a farm (location in Figure 5). Note that one rainfall event was recorded in December of 2014, but water level recovery did not occur for 10 days after the event and began to rapidly recede within a few days. The missing record was caused by the removal of the transducer from the well by the farmer as it was deepened.

3.3. Hydrus Modeling Results

Figure 11 shows the precipitation and potential evaporation used for the hydrological simulator, HYDRUS 1-D (public domain free software) as a time-variable top boundary condition. The potential evaporation was calculated using the modified Penman-Monteith equation [24]. Hourly average values of various meteorological variables, including air temperature, relative humidity, wind speed, incoming shortwave radiation, and barometric pressure, were input into the model. Meteorological variables were recorded at a meteorological station located close to Wadi Qidayd. Two extreme precipitation events were simulated hypothetically. First, a precipitation event of 45 mm within four hours was chosen and second, an event equal to the precipitation of 90 mm within four hours was used which is the same as the devastating flood in Jeddah in 2009. The initial condition was considered as dry with 0.065 water content through the whole 13 m simulation domain.

Figure 12a,b shows that HYDRUS 1-D computed water content profiles based on the hydraulic parameters obtained for the first scenario with coarse texture sediment and second scenario with thin fine texture layers between the coarse texture sediment, respectively. In both scenarios, the effects of two extreme precipitation events are clearly visible in the water content profiles. In the time- and depth-dependent water content of scenario one, variations due to the first precipitation event can be observed in the top 5 m of the subsurface and after the second precipitation event the changes in water content remains in the top 10 m of sediment sequence (Figure 12a). Furthermore, the maximum water content simulated on the 350th day of the year was 0.084, which is 0.036 more than the residual water content. Hydrological simulations confirm that the thick unsaturated zone of 13 meter cannot be recharged even if the sediment sequence has an entirely coarse sediment sequence even with the heaviest rainfall event ever recorded in the region of Wadi Qidayd. In the second scenario, nine thin layers of fine texture soil were considered in the top 7.5 m within the coarse sediment sequence (Figure 12b). After the first precipitation event the first two thin layers of fine texture of soil represent the water infiltration and after second heavy precipitation event the water infiltration effect can be observed in the top four layers of fine texture of soil, which act as a mud layers (Figure 12b). Hydrological simulations were tested by placing the thin fine texture layers at different depths and in all cases the water did not infiltrate below the fourth layer. In the wadi, mostly *Acacia sp.* trees are grown, which have long roots and the water trapped in the mud layers is likely taken up by the root system. The effect of root water uptake was not incorporated in HYDRUS-1 modeling to keep the scenario simple and to simulate more subsurface infiltration.

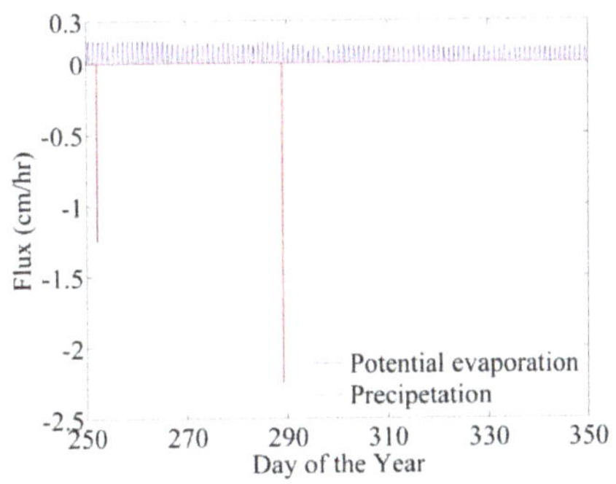

Figure 11. Precipitation and potential evaporation flux values recorded over a hundred-day period. Positive values are used for evaporation and negative for precipitation.

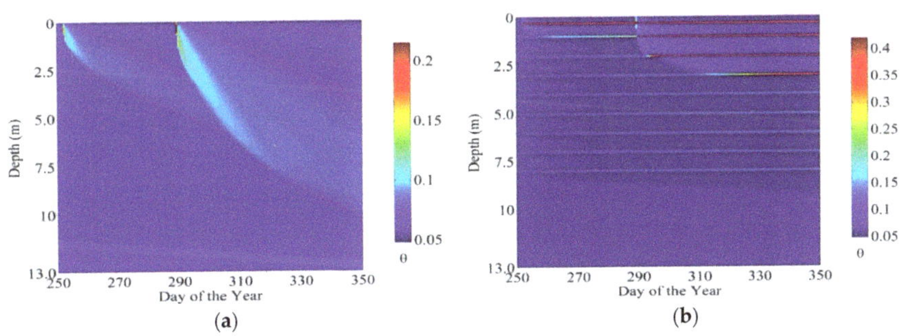

Figure 12. Simulated volumetric water content θ as a function of depth and time in days of the year, obtained for coarse texture sediments (**a**) and for nine think layers between coarse texture sediments (**b**).

3.4. Geophysical Profiles

A profile was made from the Precambrian rock located along the north wall of the main channel of the wadi toward the center of the channel. The first arrival travel times of the recorded data were picked and then inverted to find the subsurface velocity model shown in Figure 13 [25,26]. The travel time tomogram represents the variation in seismic wave propagation in both the horizontal and vertical directions. Figure 13 shows that three different layers are present. The first layer below the surface consists of loose sand and gravel. It has a seismic wave velocity ranging between 600 and 880 m/s. The second layer consists of clay which is partially saturated with water with a seismic wave velocity ranging between 1130 and 1240 m/s. The third layer consists of highly fractured bedrock with a seismic wave velocity of 1440 to 2000 m/s. The depth to the top of the second layer ranges between 13–16 m at the northern part and 4–7 m at the central and southern parts of the profile, while the depth to the top of the third layer is about 24 m at the northern part, and 7 to 14 m at the central and southern parts of the profile.

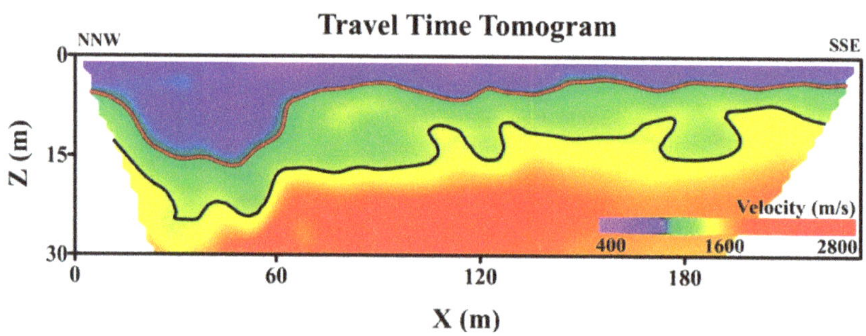

Figure 13. Seismic refraction tomogram (travel time *versus* depth) showing the velocities and the three layers.

Similar to the seismic refraction tomogram, the resistivity profiles [27] were analyzed using the travel-time tomography method described by Yu *et al.* [19] and show three subsurface layers (Figure 14). These layers include a surface unit with a thickness of 2 to 6 m, consisting of dry lose mud, sand, and gravel with a resistivity ranging between 800 and 1000 Ohm.m, a second layer, partially saturated with water with very low resistivity values (10–100 Ohm.m), because it consists mainly of clay and has a thickness of about 10 m, and a bottom layer consisting mainly of bedrock with high resistivity values of 700 and gradually increasing to 3000 Ohm.m with depth. The water table is consistent with the top of the third layer, and the saturation is the reason for the decreased the resistivity of this layer from 3000 Ohm.m to about 700 Ohm.m.

Resistivity profile P-2 shows a very similar subsurface model to that of profile P-1, except at the northern part of P-2 where the first layer disappears. This is because the northern part of P-2 is recorded along a rock cliff, where the clay layer is exposed at the surface, and no sand-gravel layer was deposited at this location.

Comparing the resistivity and seismic results shows some similarities, especially the depth to the top of the third layer at the central and southern part of the profile. Both methods should not give identical results since the seismic method is sensitive to the elastic properties of the rocks, while the resistivity method is sensitive to the water content and salinity of the rocks.

The profile shows that the water table within the wadi channel sloped downward from the fractured rock edge of the channel inward towards the center of the channel. There is approximately 3 m difference which is significant in that the water table in the aquifer within the central part of the channel is lower than at the sides. This suggests that little or no recharge is occurring in the channel, but only at the edges of the channel via the fractured rocks.

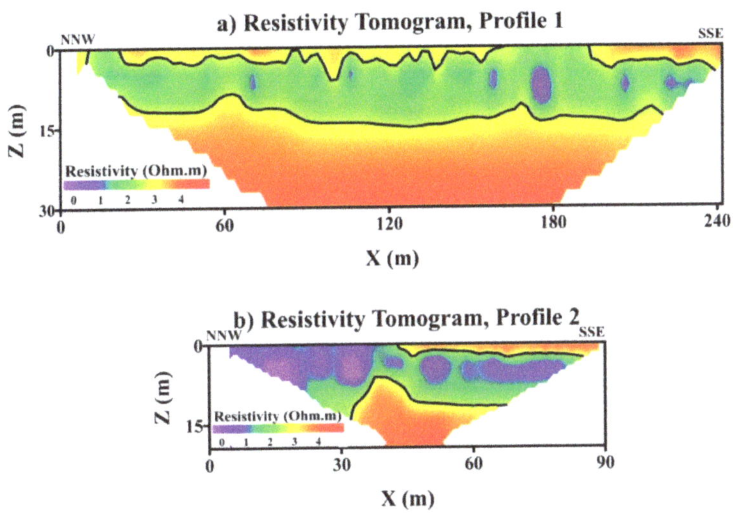

Figure 14. Resistivity profiles running from the channel edge on the left toward the middle of the channel.

4. Conclusions

Field observations of geology, water level measurements, geophysical profiles, and simple scenario-modeling using HYDRUS1-D, were conducted to investigate the current recharge mechanism within Wadi Qidayd in Western Saudi Arabia. The alluvial aquifer in this wadi has been heavily pumped and water levels are commonly greater than 25 m in depth in large areas of the system. The significant lowering of water levels by farm and commercial water sales wells is a relatively recent problem that did not occur historically in the wadi.

The data collected suggest that channel loss during floods is no longer the primary means of recharge in the wadi. This conclusion is based on the low water levels in the aquifer, resulting in a greater thickness of the unsaturated zone, the occurrence of naturally-occurring, fining upwards beds capped with low hydraulic conductivity clays, deflected water passing parallel to the mud layers downstream, higher water levels at the edge of the channel compared to the middle after a flood event, and the common occurrence of *Acacia* sp. which has long roots and removes water lying atop the mud layers. Simplified, scenario-modeling of the unsaturated zone using HYDRUS1-D suggests that the occurrence of four fining-upwards beds that contain a mud layer is sufficient to prevent the vertical recharge of the aquifer during flood events based on the actual range of rainfall and duration, as recorded at local weather stations.

Therefore, we conclude that most of the recharge to Wadi Qidayd currently occurs via the perimeter fractured rocks that have direct hydraulic connection to the horizontally-layered, high hydraulic conductivity gravel, and boulder beds to allow water collected in the harrat area to drain into the aquifer. There may be some areas of the channel that still may receive direct recharge via channel loss in the distal areas of the wadi where mud layers may not occur and at slope inflection points where erosion or non-deposition of the mud layers occur. The unsaturated zone modeling shows that there is a minimum water level that, once reached, will preclude recharge due to the and thickness of the unsaturated zone and number of preserved mud layers that define the top of flood deposits.

The importance of this research is that the activities of humans (over-pumping) can change the physical mechanisms of recharge within an ephemeral stream. Once the mechanisms have been changed, the quantity of water available for use from the aquifer will significantly reduce and exacerbate the depletion problem. Therefore, wadi aquifers, which are generally believed to provide renewable water resources, may not be recharged sufficiently to be viable water sources in the future and could, perhaps, be considered to be non-renewable resources within a short-term time context.

With regard to water management of the wadi aquifer system, the channel recharge mechanism could be restored by reducing the pumping rate, removal of the *Acacia sp.* trees (or reducing the density), and monitoring of the wadi aquifer water levels. If pumping is ceased and a "rest period" is allowed to occur, eventually the water table will rise to a level wherein channel loss or infiltration can, again, reach the water table. The likely range of water level fluctuations that should occur with Wadi Qidayd is 1 to 5 m below the surface to maintain the channel recharge function.

The usefulness of our research may be limited to wadi systems lying in geological formations that produce muddy sediments during erosion and the geometry of the observed flood deposits. However, nearly all wadi aquifers contain sufficient mud in storm deposits in the proximal reaches to exhibit similar behavior to the observed throughout Wadi Qidayd.

Acknowledgments: The authors thank the Water Desalination and Reuse Center and the Department of Earth Science and Engineering for the use of field and laboratory equipment. Funding for this research was provided by the Center and Department and from discretionary faculty research funding.

Author Contributions: KJ performed the modeling and assisted in writing the text. SM coordinated many of the field trips into the wadi, helped conduct field measurements, acted as a translator, and drafted many of the figures. SMH and JS worked on the seismic reflection tomograms using the stacking program. TM wrote most of the manuscript and coordinated the overall research effort.

Conflicts of Interest: The authors declare no conflicts of interest.

References

1. Maliva, R.G.; Missimer, T.M. *Arid Lands Water Evaluation and Management*; Springer-Heidelberg: Heidelberg, Germany, 2012.
2. Besbes, M.; Delhomme, J.P.; de Marsily, G. Estimated recharge from ephemeral streams in arid regions: A case study at Kairouan, Tunisia. *Water Res. Res.* **1978**, *14*, 281–290.
3. Abdulrazzak, M.J.; Morel-Seytoux, H.J. Recharge from an ephemeral stream following wetting front arrival to water table. *Water Resour. Res.* **1983**, *19*, 194–200.
4. Abdulrazzak, M.J.; Sorman, A.U.; Alhames, S.A. Techniques of artificial recharge from an ephemeral wadi channel under extreme arid conditions. In Proceedings of the International Symposium on the Artificial Recharge of Groundwater, Anaheim, CA, USA, 23–27 August 1988; Johnson, A.I., Finlayson, D.J., Eds.; American Society of Civil Engineers (ASCE): New York, NY, USA, 1989; pp. 602–611.
5. Sorman, A.U.; Abdulrazzak, M.I. Infiltration-recharge through wadi beds in arid regions. *Hydrol. Sci. J.* **1993**, *38*, 173–186.
6. Abdulrazzak, M.J.; Sorman, A.U. Transmission losses from ephemeral streams in arid regions. *J. Irrig. Drain. Eng.* **1994**, *120*, 669–675.
7. Abdulrazzak, M.J. Losses of flood water from alluvial channels. *Arid Soil Res. Rehabilit.* **1995**, *9*, 15–24.
8. Sorman, A.U.; Abdulrazzak, M.J. Estimation of wadi recharge from channel losses in Tabalah Basin, Saudi Arabia. In *Recharge of Phreatic Aquifers in (Semi-) Arid Areas*; Simmers, I., Ed.; International Association of Hydrogeologists: Reading, UK, 1997; Volume 19, pp. 187–200.
9. Sorman, A.U.; Abdulrazzak, M.J.; Morel-Seytoux, H.J. Groundwater recharge estimation from ephemeral streams, case study: Wadi Tabalah, Saudi Arabia. *Hydrol. Process.* **1997**, *11*, 1609–1619.

10. Bazuhair, A.S.; Nassief, M.O.; Al-Yamani, M.S.; Sharaf, M.A.; Batumi, T.H.; Ali, S. *Groundwater Recharge Estimation in Some Wadi Aquifers of Western Saudi Arabia: King Abdulaziz City for Science and Technology*; Project No. AT-17-63; CRC Press: Riyadh, Saudi Arabia, 2002.
11. Subyani, A.M. Use of chloride-mass balance and environmental isotopes for evaluation of groundwater recharge in the alluvial aquifer, Wadi Tharad, western Saudi Arabia. *Environ. Geol.* **2004**, *46*, 741–749.
12. Missimer, T.M.; Drewes, J.E.; Amy, G.; Maliva, R.G.; Keller, S. Restoration of wadi aquifers by artificial recharge with treated waste water. *Groundwater* **2012**, *50*, 514–527.
13. Vincent, P. *Saudi Arabia: An Environmental Overview*; Taylor & Francis: London, UK, 2008.
14. Grainger, D. *The Geological Evolution of Saudi Arabia: A Voyage through Space and Time*; Saudi Geological Survey: Jeddah, Saudi Arabia, 2007.
15. Roobol, M.J.; Kadi, K.K. *Cenozoic Faulting and Traces of Tertiary Gabbro Dikes in the Rabigh Area*; Saudi Geological Survey Technical Report SGS-TR-2008-6 (geological map); Saudi Geological Survey: Jeddah, Saudi Arabia, 2007.
16. Missimer, T.M.; Dagher, M.H.; Tao, W. Geology of Wadi Qidayd, western Saudi Arabia. In Proceedings of the Geological Society of America, Joint Annual Meeting Abstracts with Programs 124th Meeting, Minneapolis, MN, USA, 9–12 October 2011; Volume 43, p. 34.
17. Rosas, J.; Jadoon, K.Z.; Missimer, T.M. New empirical relationship between grain size distribution and hydraulic conductivity for ephemeral stream bed sediments. *Environ. Earth Sci.* **2014**, *73*, 1303–1315.
18. Kruseman, G.P.; De Ridder, N.A. *Analysis and Evaluation of Pumping Test Data*; International Institute for Land Reclamation and Improvement in the Netherlands: Wageningen, The Netherlands, 1970.
19. Yu, H.; Zhang, D.; Wang, X. Application of early-arrival waveform inversion to shallow land data. *J. Appl. Geophys.* **2014**.
20. Jury, W.A.; Horton, R. *Soil Physics*, 5th ed.; John Wiley & Sons: New York, NY, USA, 1996.
21. Mualem, Y. A new model for predicting the hydraulic conductivity of unsaturated porous media. *Water Resour. Res.* **1976**, *12*, 513–522.
22. Van Genuchten, M.T. A closed form equation for predicting the hydraulic conductivity of unsaturated soils. *Soil Sci. Soc. Am. J.* **1980**, *44*, 892–898.
23. Vereecken, H. Estimating the unsaturated hydraulic conductivity from theoretical models using simple soil properties. *Geodema* **1995**, *65*, 81–92.
24. Schaap, M.G.; Leij, F.J.; van Genuchten, M.T. ROSETTA: A computer program for estimating soil hydraulic parameters with hierarchical pedotransfer functions. *J. Hydrol.* **2001**, *251*, 163–176.
25. Aki, K.; Richards, P.G. *Quantitative Seismology*, 2nd ed.; University Science Books: Herndon, VA, USA, 2002.
26. Nemeth, T.; Normark, E.; Pacanowski, G. Dynamic smoothing in crosswell traveltime tomography. *Geophysics* **1997**, *62*, 168–176.

27. Ostrowski, S.; Lasocki, M.; Pacanowski, G. Electric Resistivity Tomography as a Tool in Geological Mapping. In Proceedings of the 72nd EAGE Conference & Exposition, Extended Abstracts, Barcelona, Spain, 14–17 June 2010.

Predicting Soil Infiltration and Horizon Thickness for a Large-Scale Water Balance Model in an Arid Environment

Tadaomi Saito, Hiroshi Yasuda, Hideki Suganuma, Koji Inosako, Yukuo Abe and Toshinori Kojima

Abstract: Prediction of soil characteristics over large areas is desirable for environmental modeling. In arid environments, soil characteristics often show strong ecological connectivity with natural vegetation, specifically biomass and/or canopy cover, suggesting that the soil characteristics may be predicted from vegetation data. The objective of this study was to predict soil infiltration characteristics and horizon (soil layer) thickness using vegetation data for a large-scale water balance model in an arid region. Double-ring infiltrometer tests (at 23 sites), horizon thickness measurements (58 sites) and vegetation surveys (35 sites) were conducted in a 30 km × 50 km area in Western Australia during 1999 to 2003. The relationships between soil parameters and vegetation data were evaluated quantitatively by simple linear regression. The parameters for initial-term infiltration had strong and positive correlations with biomass and canopy coverage ($R^2 = 0.64 - 0.81$). The horizon thickness also had strong positive correlations with vegetation properties ($R^2 = 0.53 - 0.67$). These results suggest that the soil infiltration parameters and horizon thickness can be spatially predicted by properties of vegetation using their linear regression based equations and vegetation maps. The background and reasons of the strong ecological connectivity between soil and vegetation in this region were also considered.

Reprinted from *Water*. Cite as: Saito, T.; Yasuda, H.; Suganuma, H.; Inosako, K.; Abe, Y.; Kojima, T. Predicting Soil Infiltration and Horizon Thickness for a Large-Scale Water Balance Model in an Arid Environment. *Water* **2016**, *8*, 96.

1. Introduction

Prediction of soil characteristics over large areas is desirable for environmental modeling, precision agriculture and natural resources management. In arid environments, soil permeability is one of the most important characteristics because it governs infiltration and runoff characteristics of the landscape. Horizon (soil layer) thickness above underground hardpans or basement rocks is also significant because hardpans and basement rocks are impermeable and constrain water movement. These soil characteristics are essential as input parameters or boundary conditions for water balance analysis by rainfall-runoff modeling. Rainfall-runoff models are important to predict the movement of precious water resources in arid landscapes. The accuracy of the input parameters used in such models has a significant effect on model performance [1]. A lack of data for the models is a major limitation for simulation of regional hydrologic processes [2]. However, it is well known that soil characteristics often show high spatial variability. Direct measurements of soil characteristics are time-consuming, labor-intensive and costly to characterize for areas >100 km^2.

Stemming from land system surveys of the 1950s to 1970s [3–5], two main approaches are widely used to predict the spatial variability of soil characteristics on large scales: geostatistics [6] and environmental correlation [7]. In this paper, we focus on the environmental correlation, specifically using correlations between soil and vegetation properties. An environmental correlation approach evaluates the spatial distribution of soil characteristics by quantifying the relationship between soil and environmental attributes. Terrain attributes derived from digital elevation models (DEM) are the most major predictive factor in this approach [8,9]. Explicit, quantitative, and usually simple empirical models are developed to describe the relationships between georeferenced soil samples and predictive factors (e.g., [10,11]). Multiple linear regression has been most widely used in the modeling processes (e.g., [12–15]). In addition to the topographical factors, climatic, organic, lithologic and temporal factors can be used as predictors [16].

Land cover is one of the useful indicators of soil properties and class because the natural vegetation and its dynamics should represent some kind of equilibrium relation with soil type [16–18]. Recently, vegetation data such as vegetation structure and canopy coverage can be obtained relatively easily even over large areas by remote sensing techniques. These data have been used as predictive factors in combination with terrain attributes and other factors in several studies [19–21]. For example, vegetation indices from Landsat Thematic Mapper and Enhanced Thematic Mapper Plus (Landsat TM and ETM+) were used to predict soil classes [22], soil drainage classes [23], soil hydraulic properties [2] and soil profile depth, total phosphorus and total carbon [24]. Zhu [25] used canopy coverage from Landsat TM to predict soil classes. Park and Vlek [26] used a vegetation map to predict soil properties. Vegetation indices from SPOT images were also used to predict soil organic matter [27] and soil organic carbon [28]. In these studies, vegetation data are just one category of the predictive factors, however, successful deployment of this data suggests that the vegetation attributes may play an important role in the environmental correlation approach.

To predict soil infiltration properties and horizon thickness, vegetation factors may be particularly useful because these soil characteristics are highly correlated to vegetation due to the co-evolution of landscapes [29]. A considerable number of studies have been reported on the strong relationship between infiltration parameters and vegetation data (e.g., [30–36]). Regarding the horizon thickness, Lamotte *et al.* [37] showed that the depth and distribution of hardpans were associated with differences in vegetation. Pracilio *et al.* [38] also reported that tree height was limited by the depth to hardpans in southwestern Australia. Therefore, these soil characteristics may be simply predicted from vegetation data only if quantitative and strong correlations can be found between the soil characteristics and vegetation attributes for large areas.

The objective of this study is to investigate the prediction of soil infiltration characteristics and soil horizon thicknesses using vegetation data for areas of the order of hundreds to thousands of km^2 in an arid environment. We conducted infiltration tests (at 23 sites), horizon thickness measurements (58 sites) and vegetation surveys (35 sites) in an established research area (30 km $\times$ 50 km) in Western Australia. The relationships between obtained soil parameters and vegetation data were evaluated quantitatively by simple linear regression analyses. The infiltration properties and horizon thickness in the whole research area were predicted from regression equations and vegetation maps derived by remote sensing.

2. Area Descriptions and Methods

2.1. Research Area and Measurement Sites

The Leonora research area (lat 28°38′13″ S, long 120°59′15″ E) was located about 600 km northeast from Perth, the capital city of Western Australia, covering approximately 1500 km^2 (30 km in an east-west direction and 50 km in a north-south direction) (Figure 1). This research area was originally established to assess arid zone afforestation and to estimate the effect of carbon fixation in arid lands [39]. A large-scale water balance model and a tree growth simulator have also been developed to estimate water and carbon flow in the research area [40,41]. Thus, information on soil characteristics for the whole research area was needed for these models and also to establish boundary conditions for the models [42].

The area is arid: average annual rainfall measured in the area (1917–1998) is 211 mm with a standard deviation of 109 mm [43]. Although the topographical gradient is less than 1% for almost the entire research area, there are many wadis (dry riverbeds containing water only during times of heavy rain). When intensive rainfall occurs, the wadis function as drainage paths for runoff to the large and shallow salt lake at the lowest elevation. The surface soil contains abundant quantities of ferric oxide and is colored red due to laterization [44]. The feature of the soil in this area is the presence of the Wiluna hardpan [45], which is extremely hard, impermeable and consists of laminated layers of silica with ferruginous and calcareous cement, and which is often up to 10 m thick [46]. According to a fluorescent X-ray analysis, these hardpans in the research area consist of 45% Al_2O_3, 39% SiO_2 and 12% Fe_2O_3.

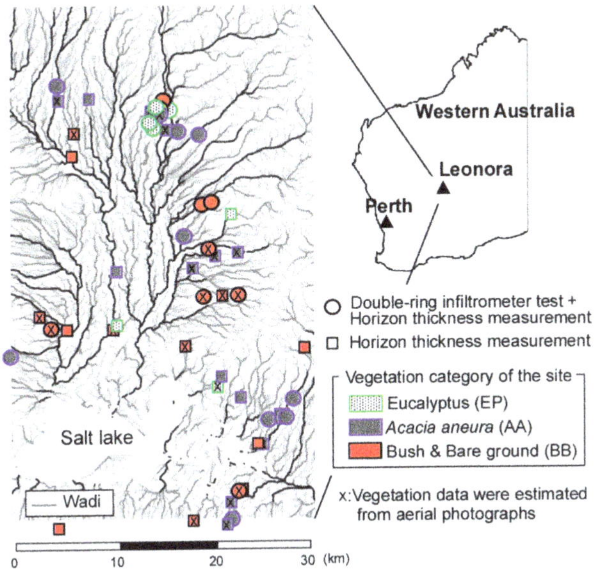

Figure 1. Map of the research area and the distribution of the measurement sites. The darker the line is, the larger the catchment area of the wadi is.

The research area has been used for extensive grazing by one farm family, but all of the forest and woodland in the research area is considered natural forest and woodland [47]. The vegetation of the area is a mixture of *Acacia* woodland, *Acacia* shrub land (*Acacia aneura* with small shrubs, e.g., *Eremophilla* sp.) and bare ground [47,48]. About half of the research area is bare ground or low open woodland, where canopy coverage is less than 10%, and the other half is woodland and open forest, where canopy coverage varies between 10% and 50% [49]. The dominant species is *Acacia aneura*; *Eucalyptus camaldulensis* open forest and woodland are adjacent to some wadis; and salt-tolerant shrubs and bushes are observed around the salt lake. Abe *et al.* [50] reported that the biomass distribution is related with the slope, position of the wadi and the distance from the wadi.

Measurements on soil infiltration, horizon thickness and vegetation were conducted at a maximum of 58 sites of size 50 m × 50 m. The soil surveys were intensively conducted in November and December in 1999 and 2002. The vegetation surveys were conducted from 1999 to 2003. All sites were classified into three vegetation categories by the dominant species at the sites: Eucalyptus (EP), *Acacia aneura* (AA), or bush or bare ground (BB) sites. The locations of sites on which measurements were taken were chosen by preliminary ground inspections and interpretation of aerial photographs in order to satisfy the following conditions: (i) the ecological and topographical conditions within each site were homogeneous:

and (ii) there was a wide range in the biomass (Figure 1). In addition, the distances of the sites from the nearest wadis were measured by visual interpretations of aerial photographs in order to categorize the sites into "around wadis" (within 100 m of the nearest wadis) and "others". The reason for the categorization was that the areas around wadis had a characteristic ecosystem. The distribution of wadis was able to be determined also from terrain analysis of a DEM [50].

2.2. *Analyzed Data Set*

2.2.1. Infiltration Data

Double-ring infiltrometer tests were performed at 23 sites (4 EP, 10 AA and 9 BB sites) (Figure 1). All other variables shown in Sections 2.2.2 and 2.2.3 were also recorded at these sites. At each site, one point was chosen that gave a satisfactory representation of the surface soil condition of the site. An inner ring, 0.30 m in diameter and 0.30 m deep, was carefully driven 0.10 m into the soil at the chosen point. Each infiltration run continued for 90 min by which time a steady state rate had been attained. Soil profile surveys to measure depth to hardpans were also conducted at all sites at which measurements were made. The horizons were generally dry during the measurement period and the variation of the volumetric water contents was small (0.02–0.08 $m^3 \cdot m^{-3}$). The effect of the difference in the soil moisture on infiltration rate was therefore small. At one site with little vegetation, infiltration tests were performed at three locations to evaluate the spatial variability in the soil permeability within the site.

Two commonly used infiltration equations were fitted to the infiltration data collected during tests: the Kostiakov [51] and Philip [52] equations. Cumulative infiltration (I, mm) is expressed by the Kostiakov equation as:

$$I = kt^m, \tag{1}$$

where t (min) is the time that water has been in contact with the soil, and k and m are constants. The Kostiakov equation is empirical; its parameters do not have clear physical meanings but it is simple and easy to define. Note that if $t = 1$ in Equation (1), then $I = k$. This means k is 1 minute of infiltration. In contrast, the Philip equation is based on physical parameters and given by:

$$I = St^{1/2} + At, \tag{2}$$

where S ($mm \cdot min^{-1/2}$) accounts for the sorption of water at the beginning of the infiltration process and A ($mm \cdot min^{-1}$) accounts for the long-term (steady-state) infiltration rate. We introduced a parameter I_t (mm), which is the measured cumulative infiltration for any period of time (t). As above, parameters k, m, S, A and I_t were obtained as soil infiltration characteristics at each of the sites.

According to visual interpretation of the aerial photographs, six of the 23 sites were within 100 m of the nearest wadis. Analyses excluding these six sites were also performed. A subscript _ew was added to the infiltration parameters when the six sites were excluded in the analyses (e.g., $k_{\_ew}$).

2.2.2. Horizon Thickness Data

Horizon thickness was measured at 58 sites (10 EP, 25 AA and 23 BB sites). The surface horizon above the underground hardpan was dug out by engine augers and the horizon thickness (T_h, cm) was determined by measuring the depths of the augered holes. The horizon thicknesses sometimes varied within a short distance thus they were measured at 10–15 locations within each site area, and the average horizon thickness was adopted as the T_h data at the site. The maximum measurement range of T_h was limited to 115 cm by the length of the drill. Therefore, at 10 sites (seven EP and three AA sites) where the horizon thickness was over 115 cm, T_h was tentatively assigned a value of 115 cm. Analyses excluding these 10 sites were also performed.

2.2.3. Vegetation Data

The stand biomass (B_m, kg$\cdot$m^{-2}) and canopy coverage (C_c, m$^2\cdot$m^{-2}) were measured or estimated at every site. The stand biomass was defined as the sum of the biomass of the individual trees above ground that comprised stand per the unit ground area. The canopy coverage was defined as the ratio of the vertical projection area of the forest and woodland canopy per unit area [47]. C_c was obtained relatively easily from both vegetation surveys and remote sensing analyses. In contrast, B_m was more difficult to estimate than C_c but we considered that it was a more efficacious measure of the amount of vegetation.

At 35 of the 58 sites, vegetation surveys were conducted originally for the remote sensing analyses. Various vegetation parameters were measured for all trees such as tree height, diameter at breast height and canopy silhouette area. Each tree biomass was calculated by the allometric equation for each tree species using a destructive sampling method [53]. From these results, the values of B_m and C_c were calculated for each site. To extrapolate the site based B_m and C_c values to the entire study area, satellite images (Landsat TM and ETM+; 16 scenes from June 1998–December 1999; 30 m resolution) and aerial photographs (on 10–11 November 1999; digitized ortho color photograph consisted of Red, Green and Blue bands, 0.25 m quadrate) were used. As results of analyses for both images, distribution maps of B_m and C_c were made for the entire area. The details of the estimation methods adopted are described by Taniguchi *et al.* [54] and Suganuma *et al.* [47,49]. An example of the B_m

distribution map developed by satellite image analyses is shown in Figure 2 [54]. At 23 sites at which vegetation surveys were not conducted (Figure 1), B_m and C_c were estimated from the aerial photographs that had higher resolution and accuracy than the satellite images. The C_c values were estimated by visual interpretation of the aerial photographs (e.g., R^2 = 0.99 [49]). The B_m values were estimated using high correlation between B_m and C_c for each vegetation type (e.g., R^2 = 0.92 [47]).

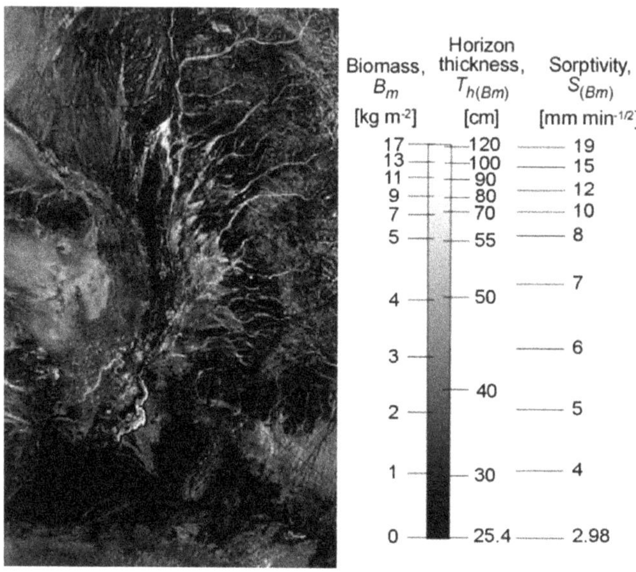

Figure 2. Distribution map of stand biomass and the distributions of sorptivity and horizon thickness predicted by the regression equations (Figures 4 and 7a). The stand biomass was estimated using SAVI (Soil Adjusted Vegetation Index) from satellite images (Landsat TM and ETM+; 16 scenes from June 1998–December 1999; 30 m quadrate) [54].

2.3. *Data Analysis and Development of Prediction Equation*

Liner regression analysis was used to quantify relationships between the infiltration parameters or horizon thickness and vegetation data. In this paper, soil characteristics were predicted from vegetation data, thus k, m, S, A, I_t and T_h were used as the dependent variables (y) and B_m and C_c were used as the independent variables (x). In addition to regression parameters, some statistical measures were provided to evaluate the model performance: mean (M) and standard deviation (SD) for each observed and predicted variable, mean bias error (MBE) and root mean square error (RMSE) [55].

The dependent variables highly correlated with the vegetation variables were predicted using their linear regression based equations. Prediction equations of time series of infiltration were also developed by substituting the regression equations for each infiltration constant into the original infiltration equations (Equations (1) and (2)). The detailed development is shown in the results section (Section 3.2).

3. Results

3.1. Relationships between Infiltration Parameters and Vegetation Data

The cumulative infiltrations varied greatly from site to site, suggesting a high spatial variability in the soil permeability in the research area (Figure 3). At a final cumulative infiltration (90 min), the maximum value was 685 mm, minimum 37 mm, mean 226 mm and standard deviation 155 mm. Both Equations (1) and (2) were in close agreement with the measured cumulative infiltration at every site; The coefficients of determination of the fittings were within a range of 0.99 to 1.00, and the RMSE between the measured and fitted cumulative infiltrations were within a range of 0.45 mm to 10.0 mm. At one site where the infiltration tests were performed at three locations, the final cumulative infiltrations (I_{90}) of three tests were 180, 193 and 214 mm and the standard deviation was 17 mm, suggesting the spatial variability in the soil permeability was small within the site.

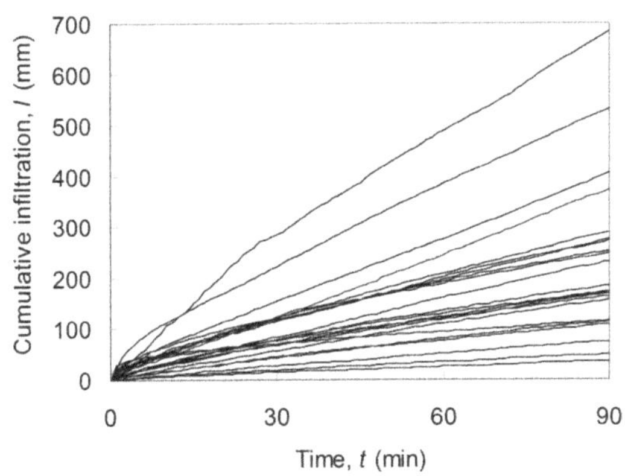

Figure 3. Cumulative infiltrations as a function of time at 23 sites.

Table 1 shows the output of the resulting linear regression analysis relating vegetation data to infiltration parameters and statistical measures. Several infiltration parameters correlated highly with vegetation data. The constants k in the Kostiakov equation and S in the Philip equation showed particularly high and positive correlations with B_m and C_c (Figure 4). Both k and S are measures of the infiltrability at the beginning of the infiltration process and results indicate that surface soil permeability was strongly correlated with the vegetation. With an increase in the t value of I_t, the correlation coefficients between I_t and B_m or C_c decreased (Table 1).

Thus, the rate of infiltration into shallower soil has a stronger correlation with vegetation. No significant correlation was obtained between the constant A of Philip's equation and B_m or C_c. This result suggests that permeability of horizons low in the profile is not correlated with the vegetation because the constant A accounts for the long-term infiltration rate.

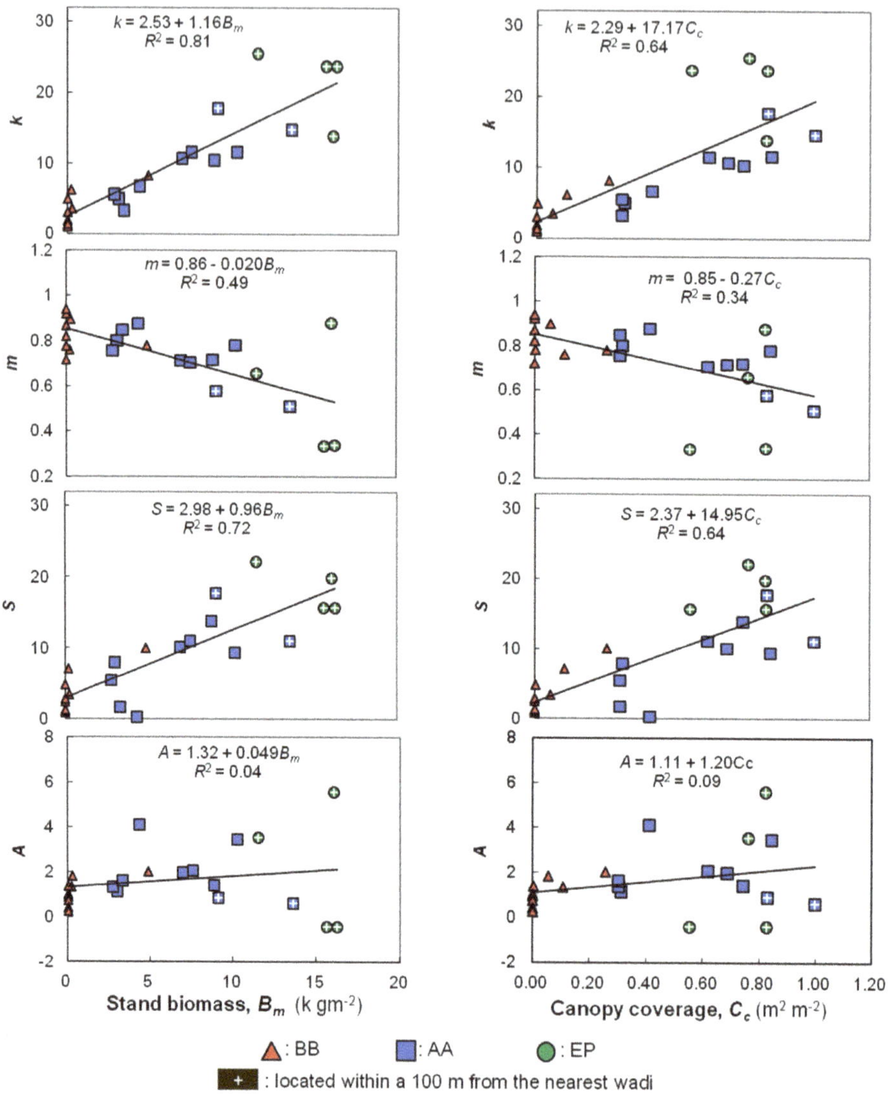

Figure 4. Relationships between infiltration parameters (k, m, S and A) and vegetation data (B_m and C_c). BB = bush or bare ground site; AA = *Acacia aneura* site; EP = Eucalyptus site.

Table 1. Linear regression equations relating the vegetation data to the infiltration parameters and statistical measures.

Vegetation Data (x)	Infiltration Parameters (y)	Regression Parameters				Statistical Measures						
		R^2	a	b	n	p	M_o	M_p	SD_o	SD_p	MBE	RMSE
Stand Biomass (B_m, kg· m^{-1})	k	0.81	2.53	1.16	23	<0.001	9.34	9.31	7.53	6.72	0.021	3.28
	m	0.49	0.86	−0.02	23	<0.001	0.74	0.74	0.16	0.11	0.048	0.11
	S	0.72	2.89	0.96	23	<0.001	8.51	8.51	6.57	5.56	−0.001	3.41
	A	0.04	1.32	0.049	23	0.360	1.60	1.61	1.43	0.28	−0.0038	1.37
	I_5	0.67	12.51	2.51	23	<0.001	27.43	27.20	18.04	14.54	3.50	9.83
	I_{10}	0.59	21.83	3.64	23	<0.001	43.12	43.13	27.43	21.09	−0.0039	17.19
	I_{30}	0.38	55.39	6.68	23	0.002	94.46	94.47	62.80	38.70	−0.010	48.38
	I_{60}	0.29	101.77	10.29	23	0.008	161.92	161.98	109.41	59.62	−0.055	89.76
	$k_{\_ew}$	0.83	2.69	0.95	17	<0.001	5.62	5.62	3.68	3.34	0.0039	1.49
	$m_{\_ew}$	0.27	0.84	−0.011	17	0.032	0.78	0.81	0.079	0.039	−0.030	0.051
	$S_{\_ew}$	0.58	2.69	0.91	17	<0.001	5.50	5.50	4.24	3.20	0.0083	2.68
	$A_{\_ew}$	0.40	1.05	0.17	17	0.007	1.59	1.57	0.98	0.59	0.017	0.74
	I_{60_ew}	0.71	84.34	17.5	17	<0.001	138.27	138.31	73.00	61.54	−0.031	38.06
Canopy coverage (C_c, m^2· m^{-2})	k	0.64	2.29	17.17	23	<0.001	9.34	9.34	7.53	6.02	−0.0044	4.41
	m	0.34	0.85	−0.27	23	0.004	0.74	0.74	0.16	0.095	0.133	0.13
	S	0.64	2.37	14.95	23	<0.001	8.51	8.51	6.57	5.25	−0.0055	3.86
	A	0.09	1.11	1.20	23	0.164	1.60	1.60	1.43	0.42	−0.0007	1.33
	I_5	0.67	10.19	41.36	23	<0.001	27.43	27.18	18.04	14.52	10.27	10.04
	I_{10}	0.62	17.72	61.86	23	<0.001	43.12	43.13	27.43	21.72	−0.0081	16.43
	I_{30}	0.42	46.4	117.00	23	<0.001	94.46	94.96	62.80	41.07	0.0017	46.46
	I_{60}	0.36	85.44	186.26	23	0.002	161.92	161.95	109.41	65.38	−0.033	85.81
	$k_{\_ew}$	0.81	2.52	11.33	17	<0.001	5.62	5.62	3.68	3.33	0.0043	1.52
	$m_{\_ew}$	0.28	0.84	−0.14	17	0.029	0.78	0.80	0.079	0.041	−0.026	0.05
	$S_{\_ew}$	0.58	2.59	10.64	17	<0.001	5.50	5.50	4.24	3.12	0.0041	2.78
	$A_{\_ew}$	0.40	1.02	2.10	17	0.007	1.59	1.59	0.98	0.61	−0.0030	0.066
	I_{60_ew}	0.69	81.56	207.4	17	<0.001	138.27	138.29	73.00	60.89	−0.017	39.02

R^2 = Correlation coefficient. a and b = Regression coefficients (intercept and slope, respectively). n = Number of sites. M = mean. SD = standard deviation. MBE = mean bias error. RMSE = root mean square error. I_5, I_{10}, I_{30} and I_{60} = Measured cumulative infiltration at 5, 10, 30 and 60 min, respectively. p = P-value for a correlation coefficient test. $_{\_ew}$ = Denotes the exclusion of the sites located within a 100 m of the nearest wadis. $_{\_o}$ and $_{\_p}$ = Denote the observed and predicted variables, respectively.

When the six sites within 100 m of the nearest wadis were excluded from the regression analyses, the R^2 values between I_{60_ew} or $A_{\_ew}$ and the vegetation data substantially increased compared with those of I_{60} or A (Table 1). The initial infiltrability, $k_{\_ew}$ and $S_{\_ew}$, also showed a strong correlation with vegetation data, suggesting that the soil permeability from the surface to depth is strongly correlated with the vegetation except close to wadis.

3.2. Prediction of Infiltration Parameters from Vegetation Data

The initial infiltrability (k, S and I_5) can be predicted with high accuracy from B_m and C_c over the whole area using the linear regression coefficients shown in Table 1. For example, S (sorptivity: mm· min$^{-1/2}$) can be expressed as a function of B_m using $S_{(Bm)} = 2.98 + 0.96B_m$ from Figure 4, where $S_{(Bm)}$ is the estimated sorptivity from B_m. Using this equation, Figure 2 is regarded as the distribution map of S for the whole research area. The long-term infiltrability (I_{60} and A), which was poorly correlated, was not predicted for the whole area, however, it can be predicted as well as the initial infiltrability if the area of within 100 m wadis is eliminated. I_{60} is an important long-term infiltrability parameter because it is used to model hourly runoff generation when hourly rainfall data are available. I_{60_ew} showed a strong correlation with B_m and C_c. Therefore, the prediction of hourly runoff becomes possible by the concurrent use of the hourly rainfall data, the regression equation for I_{60_ew}, and distribution maps of wadis and B_m or C_c.

An example of the development of the prediction equation of time series infiltration, using the relationship between k, m and B_m is given below. Both constants k and m of the Kostiakov equation are closely correlated with B_m. The regression equations between k and B_m, and m and B_m from Table 1 are $k = 2.53 + 1.16B_m$ and $m = 0.86 - 0.020B_m$. Substituting these into Equation (1) gives the estimated cumulative infiltration, $I_{K(Bm)}$ (mm), as:

$$I_{K(Bm)} = (2.53 + 1.16B_m)t^{-0.020Bm+0.86}. \quad (3)$$

Simultaneous use of Equation (3) and the B_m distribution map (Figure 2) enables the prediction of cumulative infiltration at any elapsed time and at any place in the area. The value of $I_{K(Bm)}$ increases with increases in t and B_m (Figure 5). However, at high values of t and B_m, $I_{K(Bm)}$ decreases with an increase in B_m. The decrease in $I_{K(Bm)}$ with the increase in B_m is incongruous with the observed positive correlation between infiltrability and vegetation.

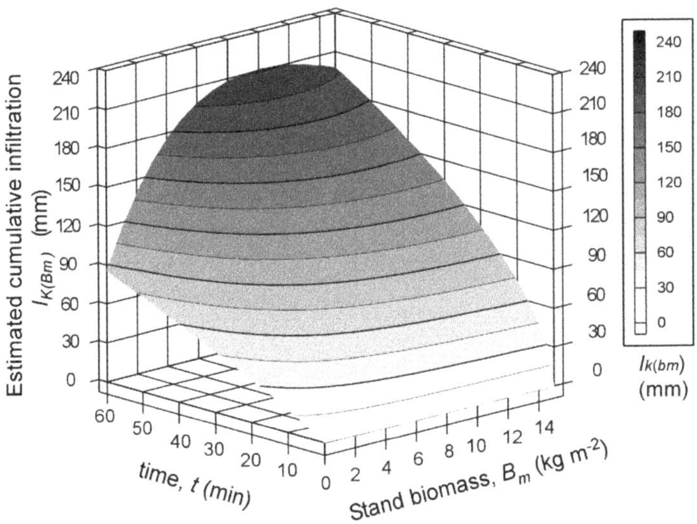

Figure 5. Estimated cumulative infiltration by Equation (3) ($I_{K(Bm)}$: mm) as a function of stand biomass (B_m: kg· m^{-2}) and time (t: min).

A similar procedure was used for the other infiltration parameters. The relationships between $k_{\_ew}$ and $m_{\_ew}$ for Equation (1), and $S_{\_ew}$ and $A_{\_ew}$ for Equation (2) are shown below. The prediction equations are presented as, respectively:

$$I_{K(Bm)\_ew} = (2.69 + 0.95B_m)t^{-0.011Bm+0.84}, \quad (4)$$

$$I_{P(Bm)\_ew} = (2.69 + 0.91B_m)t^{1/2} + (1.05 + 0.17B_m)t, \quad (5)$$

where $I_{K(Bm)\_ew}$ and $I_{P(Bm)\_ew}$ are the estimated cumulative infiltrations in mm. The subscript K and P denote that the bases of the equations are Kostiakov and Philip equations, respectively. These equations can be applied to the study area excluding the wadi surrounds.

The prediction accuracy of Equations (3)–(5) was evaluated by the coefficient of determination (R^2) between the estimated cumulative infiltration and I_t (measured cumulative infiltration) for all sites. Figure 6 presents the variations in the R^2 values with time. The R^2 value between $I_{K(Bm)}$ and I_t was initially high ($R^2 > 0.6$, $t < 10$ min), but decreased rapidly with time because of: (i) the low correlation between long-term infiltrability and vegetation; and (ii) the decrease in $I_{K(Bm)}$ with the increase in B_m in the high ranges of t and B_m (Figure 5), as previously mentioned. We propose that Equation (3) should be used only for the prediction of initial-term infiltrations, such as for the prediction of infiltrations and runoff generations for brief rainfall events ($t < 20$ min). In contrast, the R^2 value between $I_{K(Bm)\_ew}$ or $I_{P(Bm)\_ew}$ and I_t was more than 0.7 for all time periods, suggesting that even long-term infiltration can be predicted by Equations (4) or (5) with an acceptable degree of accuracy in the area outside the wadi surrounds. There was no significant difference between Equations (4) and (5) in prediction accuracy. The RMSE between the measured cumulative infiltration and $I_{K(Bm)}$, $I_{K(Bm)\_ew}$ and $I_{P(Bm)\_ew}$ at 60 min were 83.5, 35.6 and 36.0 mm, respectively. These values were slightly smaller than the RMSE values of B_m-I_{60} and B_m-I_{60_ew} (Table 1), suggesting the developed time series equations predicted the cumulative infiltration more accurately than the original regression equations.

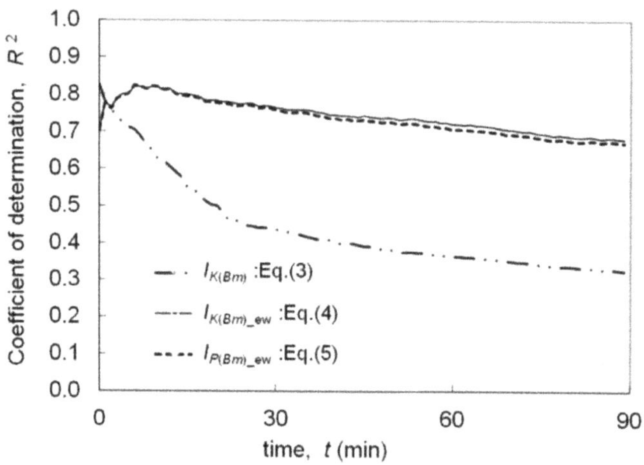

Figure 6. Variations in the coefficient of determination (R^2) values between measured and estimated cumulative infiltrations for all sites with time.

3.3. *Prediction of Horizon Thickness from Vegetation Data*

The horizon thickness had strong positive correlations with B_m ($R^2 = 0.67$; $n = 58$; $p < 0.001$) and C_c ($R^2 = 0.\ 56$; $n = 58$; $p < 0.001$) (Figure 7). T_h can be expressed as a function of B_m using $T_{h(Bm)} = 25.40 + 6.00B_m$ from Figure 7a, where $T_{h(Bm)}$ is the estimated horizon thickness from B_m in cm. Using this equation, Figure 2 is regarded as the distribution map of T_h for the whole research area. When 10 sites, the horizon thicknesses of which were greater than 115 cm were excluded in regression analyses, T_h showed strong correlation with B_m ($R^2 = 0.58$; $n = 48$; $p < 0.001$) and C_c ($R^2 = 0.53$; $n = 48$; $p < 0.001$), meaning the excluding these sites does not affect the correlation relationship. Eight of these 10 sites and another two sites were located within 100 m of the nearest wadi. If the total of 12 irregular sites, $T_h > 115$ cm or around wadis, were excluded in regression analyses, T_h still showed strong correlation with B_m ($R^2 = 0.55$; $n = 46$; $p < 0.001$) and C_c ($R^2 = 0.53$; $n = 46$; $p < 0.001$).

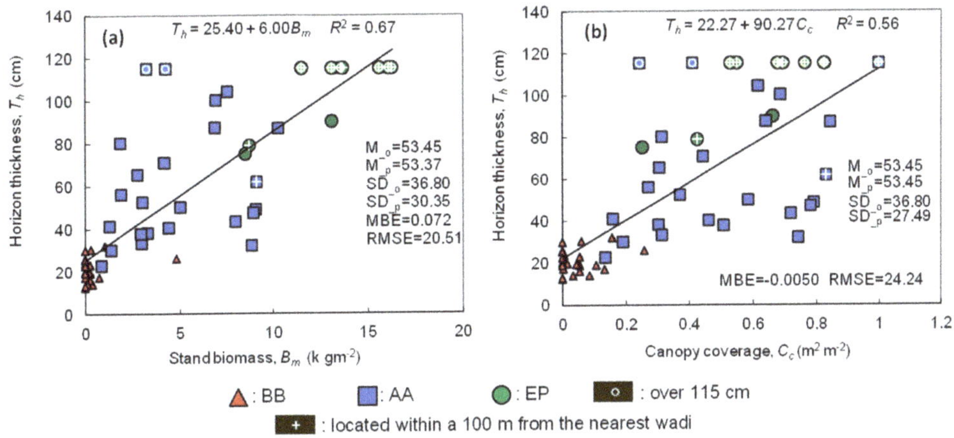

Figure 7. Relationships between (**a**) horizon thickness (T_h) and stand biomass (B_m) and (**b**) horizon thickness (T_h) and canopy coverage (C_c). BB = bush or bare ground site; AA = *Acacia aneura* site; EP = Eucalyptus site.

4. Discussion

The six sites within 100 m of the nearest wadis had relatively high B_m and C_c values, and showed high initial infiltration rates (Figure 4). From the result of soil profile surveys, horizons with low permeability were observed at three of these six sites. The depths of these horizons were about 20–50 cm from the soil surface. Although these horizons were not hardpans, they were extremely compacted and hard; the values for the long-term infiltration parameters were strongly affected by them. It seemed that these horizons of low permeability served as transient river beds around wadis during large runoff events. In contrast, there were no horizons of low permeability in the other three sites within 100 m of wadis and the soil permeability was high even at depth. These results indicate that the spatial variability of permeability was particularly high in lower horizons around wadis.

The constant m in the Kostiakov equation had a negative and strong correlation with B_m and C_c (Figure 4). This constant does not have clear physical meaning, but when the m value is small, the infiltration rate rapidly decreases with time (Equation (1)). Thus, m values were particularly small at the three sites around wadis where there are horizons of low permeability. These m values enhanced the negative slope of the regression line (b) between m and B_m or C_c (Table 1 and Figure 4). Therefore, the slope values of $m_{\_ew}$ for both B_m and C_c were smaller than the slope values of m for B_m and C_c. The prediction equation of time-series infiltration was also affected by the steep negative gradient due to the small m values; this was the reason why the predicted $I_{K(Bm)}$ by Equation (3) decreased with an increase in B_m at high values of t and B_m (Figure 5).

BB (bush or bare ground) sites were concentrated in the lower range of k and S, and the lower range of B_m and C_c (Figure 4). In contrast, EP (Eucalyptus) sites concentrated in the upper range of k and S, and B_m and C_c, respectively. AA (*Acacia aneura*) sites were distributed along regression lines, which means the stand biomass and canopy coverage of *Acacia aneura* spatially varied and correlated with the surface soil permeability.

It appeared that the main reason for the strong correlation between horizon thickness and vegetation was the restriction of the rooting depth for plants in the vicinity of hardpans [38,56]. Furthermore, vegetation may also affect the horizon thickness as a natural selection process and for its advantage [57–59]. For example, rich vegetation supplies surface soils through litter falls and micro-organisms, in contrast, poor vegetation cover often causes runoff and erosion which reduces the horizon thickness. BB sites were concentrated in the lower ranges of T_h and B_m or C_c, whereas EP sites were concentrated in higher ranges (Figure 7). These tendencies were similar to the distributions of BB and EP shown in the relationships between initial infiltrability and vegetation data (Figure 4). AA sites were dispersed widely in thin to deep soil areas; *Acacia aneura* may be able to grow regardless of the limitation of depth to the hardpan. This would be one of the reasons why *Acacia aneura* was the dominant species in this area where hardpans distribute in shallow underground.

It seemed that the strong correlation between soil infiltration, horizon thickness and vegetation was caused by their ecological connectivity through water movement. Such associations may be particularly pronounced in arid lands because water is the main driving force of vegetation distribution and hence ecosystems in arid environments (e.g., [60–62]). The regression equations presented in this paper are region-specific. We contend that the same concept of prediction of these soil characteristics would be applicable for other arid lands that have similar ecosystems.

In most cases, above ground stand biomass showed slightly higher correlations with soil characteristics than canopy coverage (Table 1). We propose that this was because the stand biomass obtained in this study included the allometric equation for each tree species in the calculation, and hence it was a more robust measure of the current condition and amount of vegetation than canopy coverage. However, canopy coverage can be obtained more easily than biomass. The choice of either stand biomass or canopy coverage as the predictor will depend on the required accuracy and the resources required to collect the field data.

Strong statistical relationships are proven between vegetation indices calculated from satellite imagery such as NDVI (Normalized Difference Vegetation Index) or SAVI (Soil Adjusted Vegetation Index) and biomass or canopy coverage. Vegetation indices can be obtained more easily for even larger areas than our current study site; they may be able to be effective soil characteristic predictors. Further studies are required to explore other predictors based on remote sensing. Improvement of field measurements is another task. To determine depths to deep hardpans, electric resistivity surveys may be useful [37]. We used the double-ring infiltrometer for the convenience of use and transportation; however, infiltrability is often overestimated under ponding conditions compared with dripping conditions (e.g., [63,64]). Quantification of spatial variability of infiltrability within sites may also be important, specifically at high vegetation sites. Conducting multiple infiltration tests within a site using rainfall simulators will contribute to increase in accuracy of runoff prediction.

5. Conclusions

Soil infiltration characteristics and horizon thickness were predicted using vegetation data for a 1500 km^2 site in an arid environment. The parameters for the initial-term infiltration and horizon thickness had strong and positive correlations with the biomass and canopy coverage over the entire area studied. The parameters for the long-term infiltration also had strong, positive correlations with vegetation when the data obtained around wadis was excluded. We conclude that the infiltration parameters and horizon thickness above hardpans can be predicted by use of the linear regression equations with vegetation data. The prediction equations of time series infiltration were developed by substituting the regression equations for each infiltration parameter into the original infiltration equations. Concurrent use of these equations and the distribution map of vegetation (and wadis) enabled the prediction of the cumulative infiltration at any time for place in the research area. We contend that the same concept of prediction of soil characteristics would be applicable for other arid lands that have similar ecosystems.

Acknowledgments: This work was conducted under the supports of the Global Environment Research Fund of The Ministry of Environment (GHG-SSCP Project), CREST of JST (Japan Science and Technology Agency) and International Platform for Dryland Research and Education, Tottori University. We thank Masahiko Taniguchi for providing data and helpful discussions.

Author Contributions: Tadaomi Saito, Hiroshi Yasuda, Yukuo Abe and Toshinori Kojima conceived and designed the experiments; Tadaomi Saito and Hideki Suganuma performed the experiments; Tadaomi Saito and Koji Inosako analyzed the data; Tadaomi Saito wrote the paper.

Conflicts of Interest: The authors declare no conflict of interest.

References

1. Guo, W.; Wang, C.; Zeng, X.; Ma, T.; Yang, H. Subgrid Parameterization of the Soil Moisture Storage Capacity for a Distributed Rainfall-Runoff Model. *Water* **2015**, *7*, 2691–2706.
2. Sanjay, K.S.; Mohanty, B.P.; Zhu, J. Including topography and vegetation attributes for developing pedotransfer functions. *Soil Sci. Soc. Am. J.* **2006**, *70*, 1430–1440.

3. Barshad, I. Factors affecting clay formation. In Proceedings of the 6th National Conference on Clays and Clay Mineralogy, Berkeley, CA, USA, 19–23 August 1957; pp. 110–132.
4. Nye, P.H.; Greenland, D.J. *The Soil under Shifting Cultivation*; Commonwealth Bureau of Soils: Harpenden, UK, 1960.
5. Giltrap, D.J. Mathematical Techniques for Soil Survey Design. Ph.D. Thesis, University of Oxford, Oxford, UK, 1977.
6. Chilès, J.P.; Delfiner, P. *Geostatistics: Modeling Spatial Uncertainty*; Wiley: New York, NY, USA, 1999.
7. McKenzie, N.J.; Austin, M.P. A quantitative Australian approach to medium and small scale surveys based on soil stratigraphy and environmental correlation. *Geoderma* **1993**, *57*, 329–355.
8. Gallant, J.C.; Wilson, J.P. Primary Topographic Attributes. In *Terrain Analysis*; Wilson, J.P., Gallant, J.C., Eds.; John Wiley & Sons, Inc.: New York, NY, USA, 2000; pp. 51–85.
9. Maynard, J.J.; Johnson, M.G. Scale-dependency of LiDAR derived terrain attributes in quantitative soil-landscape modeling: Effects of grid resolution *vs.* neighborhood extent. *Geoderma* **2014**, *230*, 29–40.
10. McSweeney, K.; Gessler, P.E.; Slater, B.; Hammer, R.D.; Bell, J.C.; Petersen, G.W. Towards a new framework for modeling the soil-landscape continuum. In *Factors of Soil Formation: A Fiftieth Anniversary Retrospective*; Amundson, R.G., Tandarich, J., Harden, J., Singer, M., Luxmoore, R.J., Bartels, J.M., Eds.; Soil Science Society of America, Inc.: Madison, WI, USA, 1994; pp. 127–145.
11. Thompson, J.A.; Pena-Yewtukhiw, E.M.; Grove, J.H. Soil-landscape modeling across a physiographic region: Topographic patterns and model transportability. *Geoderma* **2006**, *133*, 57–70.
12. Moore, I.D.; Gessler, P.E.; Nielsen, G.A.; Peterson, G.A. Soil attribute prediction using terrain analysis. *Soil Sci. Soc. Am. J.* **1993**, *57*, 443–452.
13. Gessler, P.E.; Moore, I.D.; McKenzie, N.J.; Ryan, P.J. Soil–landscape modeling and the spatial prediction of soil attributes. *Int. J. Geogr. Inf. Syst.* **1995**, *9*, 421–432.
14. Chaplot, V.; Walter, C.; Curmi, P. Improving soil hydromorphy prediction according to DEM resolution and available pedological data. *Geoderma* **2000**, *97*, 405–422.
15. Park, S.J.; McSweeney, K.; Lowery, B. Identification of the spatial distribution of soils using a process-based terrain characterization. *Geoderma* **2001**, *103*, 249–272.
16. McBratney, A.; Mendonça, S.M.; Minasny, B. On digital soil mapping. *Geoderma* **2003**, *117*, 3–52.
17. Donohue, R.J.; Roderick, M.L.; McVicar, T.R. On the importance of including vegetation dynamics in Budyko's hydrological model. *Hydrol. Earth Syst. Sci.* **2007**, *11*, 983–995.
18. Gasch, C.K.; Huzurbazar, S.V.; Stahl, P.D. Small-scale spatial heterogeneity of soil properties in undisturbed and reclaimed sagebrush steppe. *Soil Till Res.* **2015**, *153*, 42–47.
19. Feng, Q.; Zhao, W.; Qiu, Y.; Zhao, M.; Zhong, L. Spatial Heterogeneity of Soil Moisture and the Scale Variability of Its Influencing Factors: A Case Study in the Loess Plateau of China. *Water* **2013**, *5*, 1226–1242.

20. De Carvalho, W.; Chagas, C.D.; Lagacherie, P.; Calderano, B.; Bhering, S.B. Evaluation of statistical and geostatistical models of digital soil properties mapping in tropical mountain regions. *Rev. Bras. Cienc. Solo* **2014**, *38*, 706–717.
21. Fu, T.; Chen, H.; Zhang, W.; Nie, Y.; Gao, P.; Wang, K. Spatial variability of surface soil saturated hydraulic conductivity in a small karst catchment of southwest China. *Environ. Earth Sci.* **2015**, *74*, 6847–6858.
22. Grinand, C.; Arrouays, D.; Laroche, B.; Martin, M.P. Extrapolating regional soil landscapes from an existing soil map: Sampling intensity, validation procedures, and integration of spatial context. *Geoderma* **2008**, *143*, 180–190.
23. Campling, P.; Gobin, A.; Feyen, J. Logistic modeling to spatially predict the probability of soil drainage classes. *Soil Sci. Soc. Am. J.* **2002**, *66*, 1390–1401.
24. McKenzie, N.J.; Ryan, P.J. Spatial prediction of soil properties using environmental correlation. *Geoderma* **1999**, *89*, 67–94.
25. Zhu, A.X. Mapping soil landscape as spatial continua: The neural network approach. *Water Resour. Res.* **2000**, *36*, 663–677.
26. Park, S.J.; Vlek, L.G. Prediction of three-dimensional soil spatial variability: A comparison of three environmental correlation techniques. *Geoderma* **2002**, *109*, 117–140.
27. Liu, S.; An, N.; Yang, J.; Dong, S.; Wang, C.; Yin, Y. Prediction of soil organic matter variability associated with different land use types in mountainous landscape in southwestern Yunnan province, China. *Catena* **2015**, *133*, 137–144.
28. Oueslati, I.; Allamano, P.; Bonifacio, E.; Claps, P. Vegetation and Topographic Control on Spatial Variability of Soil Organic Carbon. *Pedosphere* **2013**, *23*, 48–58.
29. Berry, S.L.; Farquhar, G.D.; Roderick, M.L. Co-evolution of climate, soil and vegetation. In *Encyclopedia of Hydrological Sciences*; Anderson, M., Ed.; John Wiley: Indianapolis, IN, USA, 2005; Volume 1.
30. Berndtsson, R.; Larson, M. Spatial variability of infiltration in a semi-arid environment. *J. Hydrol.* **1987**, *90*, 117–133.
31. Bruce, R.R.; Langdale, G.W.; West, L.T.; Miller, W.P. Soil surface modification by biomass inputs affecting rainfall infiltration. *Soil Sci. Soc. Am. J.* **1992**, *56*, 1614–1619.
32. Stroosnijder, L. Modelling the effect of grazing on infiltration, runoff and primary production in the Sahel. *Ecol. Model.* **1996**, *92*, 79–88.
33. Weltz, L.; Frasier, G.; Weltz, M. Hydrologic responses of shortgrass prairie ecosystems. *J. Range Manag.* **2000**, *53*, 403–409.
34. Wainwright, J.; Parsons, A.J.; Schlesinger, W.H.; Abrahams, A.D. Hydrology-vegetation interactions in areas of discontinuous flow on a semi-arid bajada, Southern New Mexico. *J. Arid Environ.* **2002**, *51*, 319–338.
35. Rietkerk, M.; Ouedraogo, T.; Kumar, L.; Sanou, S.; Van Langevelde, F.; Kiema, A.; Van de Koppel, J.; Van Andel, J.; Hearne, J.; Skidmore, A.K.; De Ridder, N.; Stroosnijder, L.; Prins, H.H.T. Fine-scale spatial distribution of plants and resources on a sandy soil in the Sahel. *Plant Soil* **2002**, *239*, 69–77.

36. Kelishadi, H.; Mosaddeghi, M.R.; Hajabbasi, M.A.; Ayoubi, S. Nearsaturated soil hydraulic properties as influenced by land use management systems in Koohrang region of central Zagros, Iran. *Geoderma* **2014**, *213*, 426–434.
37. Lamotte, M.; Bruand, A.; Dabas, M.; Donfack, P.; Gabalda, G.; Hesse, A.; Humbel, F.X.; Robain, H. Distribution of hardpan in soil cover of arid zones—Data from a geoelectrical survey in northern Cameroon. *Comptes Rendus de l'Academie des Sci. Serie II* **1994**, *318*, 961–968. (In French).
38. Pracilio, G.; Smettem, K.R.J.; Bennett, D.; Harper, R.J.; Adams, M.L. Site assessment of a woody crop where a shallow hardpan soil layer constrained plant growth. *Plant Soil* **2006**, *288*, 113–125.
39. Yamada, K. Establishment of carbon fixation system by afforestation of arid land. In *Final Report of Heisei 10th Adoption Project—Resource Circulation and Energy Minimized System Engineering*; Japan Science Technology Agency, Ed.; CREST, JST: Tokyo, Japan, 2004; pp. 355–455. (In Japanese).
40. Kojima, T.; Asaka, N.; Ishida, J.; Hamano, H.; Yamada, K. Development of a model for large scale water balance in arid land. *J. Arid Land Stud.* **2004**, *14*, 223–226.
41. Kojima, T.; Yokohagi, O.; Suganuma, H.; Ito, T.; Suzuki, S. Site selection and environmental effect evaluation of large scale plantation using arid area runoff model. *J. Arid Land Stud.* **2015**, *25*, 101–104.
42. Tabuchi, H.; Tanaka, N.; Koyanagi, S.; Kurosawa, K.; Hamano, H.; Suganuma, H.; Kojima, T. Runoff Model Validation for Large-Scale Afforestation in Arid Land. *Int. J. Global Environ. Issues* **2012**, *12*, 282–292.
43. Yasuda, H.; Abe, Y.; Yamada, K. Periodic fluctuation of the annual rainfall time series at Sturt Meadows, the Western Australia. *J. Arid Land Stud.* **2001**, *11*, 71–74. (In Japanese).
44. Vreeswyk, A.M.E. *An Inventory and Condition Survey of the North-eastern Goldfields, Western Australia*; Technical Bulletin No. 87; Department of Agriculture: Perth, Australia, 1994; pp. 98–117.
45. Bettenary, E.; Churchward, H.M. Morphology and stratigraphic relationships of the Wiluna hardpan in arid Western Australia. *J. Geol. Soc. Aust.* **1974**, *21*, 73–80.
46. Teakle, L.J.H. Red and brown hardpan soils of Western Australia. *J. Aust. Inst. Agric. Sci.* **1950**, *16*, 15–17.
47. Suganuma, H.; Abe, Y.; Taniguchi, M.; Tanouchi, H.; Utsugi, H.; Kojima, T.; Yamada, K. Stand biomass estimation method by canopy coverage for application to remote sensing in an arid area of Western Australia. *For. Ecol. Manag.* **2006**, *222*, 75–87.
48. National Land and Water Resources Audit. *Australia's Native Vegetation—A Summary of the National Land and Water Resources Audit's Australian Native Vegetation Assessment 2001*; Goanna Print: Canberra, Australia, 2002; Volume 27.
49. Suganuma, H.; Abe, Y.; Taniguchi, M.; Yamada, K. Evaluation of the stand biomass estimation method by digitized aerial photographs in an arid area of Western Australia. *J. Jpn. Soc. Photogramm. Remote Sens.* **2006**, *45*, 12–23. (In Japanese).

50. Abe, Y.; Taniguchi, M.; Suganuma, H.; Saito, M.; Kojima, T.; Egashira, Y.; Yamamoto, Y.; Yamada, K. Comparative analysis between biomass and topographic features in an arid land, Western Australia. *J. Chem. Eng. Jpn.* **2003**, *36*, 376–382.
51. Kostiakov, A.N. On the dynamics of the coefficient of water-percolation in soils and on the necessity for studying it from a dynamic point of view for purpose of amelioration. *Trans. Int. Congr. Soil Sci.* **1932**, *6*, 17–21.
52. Philip, J.R. The theory of infiltration. Sorptivity and algebraic infiltration. *Soil Sci.* **1957**, *84*, 257–264.
53. Taniguchi, M.; Abe, Y.; Saito, M.; Owada, M.; Yamada, K. Biomass estimation of representative plant communities in arid area of southwestern Australia. *Jpn. J. For. Environ.* **2002**, *44*, 21–29. (In Japanese).
54. Taniguchi, M.; Abe, Y.; Suganuma, H.; Saito, M.; Yamada, K. Estimation of biomass using Landsat in arid area, Western Australia. *J. Arid Land Stud.* **2002**, *12*, 55–66. (In Japanese).
55. Willmott, C.J. Some comments on the evaluation of model performance. *Bull. Am. Meteorol. Soc.* **1982**, *63*, 1309–1313.
56. Hingston, F.J.; Galbraith, J.H.; Dimmock, G.M. Application of the process-based model BIOMASS to Eucalyptus globulus subsp. globulus plantations on exfarmland in south western Australia I. Water use by trees and assessing risk of losses due to drought. *For. Ecol. Manag.* **1998**, *106*, 141–156.
57. Mclvor, J.G.; Williams, J.; Gardener, C.J. Pasture management influences runoff and soil movement in the semi-arid tropics. *Aust. J. Exp. Agric.* **1995**, *35*, 55–65.
58. Loch, R.J. Effects of vegetation cover on runoff and erosion under simulated rain and overland flow on a rehabilitated site on the Meandu Mine, Tarong, Queensland. *Aust. J. Soil Res.* **2000**, *38*, 299–312.
59. Verboom, W.H.; Pate, J.S. Bioengineering of soil profiles in semiarid ecosystems: The "phytotarium" concept. A review. *Plant Soil* **2006**, *289*, 71–102.
60. Snyman, H.A.; Fouché, H.J. Production and water-use efficiency of semi-arid grasslands of South Africa as affected by veld condition and rainfall. *Water South Afr.* **1991**, *17*, 263–268.
61. Palmer, A.R.; Van Staden, J.M. Predicting the distribution of plant communities using annual rainfall and elevation: An example from southern Africa. *J. Veg. Sci.* **1992**, *3*, 261–266.
62. Smit, G.N.; Rethman, N.F.G. The influence of tree thinning on the soil water in a semi-arid savanna of southern Africa. *J. Arid Environ.* **2000**, *44*, 41–59.
63. Free, G.R.; Browning, G.M.; Musgrave, G.W. *Relative Infiltration and Related Physical Characteristics of Certain Soils*; Technical Bulletins No 729; United States Department of Agriculture, Economic Research Service: Washington, DC, USA, 1940; pp. 7–9.
64. Rao, K.P.C.; Steenhuis, T.S.; Cogle, A.L.; Srinivasan, S.T.; Yule, D.F.; Smith, G.D. Rainfall infiltration and runoff from an Alfisol in semi-arid tropical India. I. No-till Systems. *Soil Tillage Res.* **1998**, *48*, 51–59.

Simulation of Water Use Dynamics by *Salix* Bush in a Semiarid Shallow Groundwater Area of the Chinese Erdos Plateau

Jinting Huang, Yangxiao Zhou, Rongze Hou and Jochen Wenninger

Abstract: This study analyzed the water use of the *Salix psammophila* bush in a semi-arid area in northwest China using a Hydrus-1D model. The model incorporated the effect of thermally driven water flow coupling liquid water, water vapor and heat transport. The model was calibrated and validated using hourly field measurements of soil water content and temperature at different depths for a growing season of 154 days. Furthermore, another Hydrus-1D model was established to simulate environments with decreased heat, rainfall or temperature and an increased leaf area index using calibrated and validated parameters. Our results show that upward and downward thermally driven water vapor fluxes account for 0.11% and 0.28%, respectively, of the corresponding direction of total water flux during the bush's growing season. Although the vapor flux is very small, simulations incorporating heat flow revealed alterations in the temperature and pressure head gradients over the root zone, especially during dry periods. Consequently, the cumulative contributions of groundwater to evapotranspiration (*ETg*) with heat flow and without heat flow were 26.9% and 40.6%, respectively, during the simulation period. Therefore, the cumulative contribution of groundwater to *ETg* is overestimated when heat flow is excluded. Thus, we recommended that heat transport be incorporated when evaluating *ETg* in arid and semi-arid areas.

Reprinted from *Water*. Cite as: Huang, J.; Zhou, Y.; Hou, R.; Wenninger, J. Simulation of Water Use Dynamics by *Salix* Bush in a Semiarid Shallow Groundwater Area of the Chinese Erdos Plateau. *Water* **2015**, *7*, 6999–7021.

1. Introduction

Arid and semi-arid regions compose over half of the global land area [1]. These areas often suffer from frequent sandstorms, and farmland is threatened by moving sands, particularly in China [2]. In the early 1980s, the Chinese government initiated a reforestation project called the "Three North Forest Shelterbelts" [3], and in 2000, the "Return Farmland to Forest and Grassland" project was initiated [4]. These reforestation projects have prevented further deterioration of arid regions and have rehabilitated the eco-environment. However, recent research has revealed side effects stemming from the improper selection of plants [5], such as the decay of re-vegetation and a decrease in groundwater levels [6]. In arid areas, water resources are extremely scarce and the environment is very fragile. Surface water is usually limited, and groundwater sustains the co-existence of social-economic development and the natural environment. Thus, understanding the relationship between water use by vegetation and groundwater dynamics is crucial to managing and maintaining healthy ecosystems while providing water for human needs [7].

In desert regions, rainfall events can be characterized as rainfall pulses with a discontinuous, highly variable, and largely unpredictable frequency and intensity [8,9]. This small amount of rainfall only recharges the shallow soil layer; in contrast, heavy rainfall can permeate into the deep soil layer and groundwater [2]. Many perennial plants in arid and semiarid zones have adapted to these sporadic rainfall pulses by adopting a dimorphic root system composed of branched surface roots that acquire water from shallow soil layers and deeper roots that can access perennially available groundwater [10,11]. Plants able to extract groundwater from shallow aquifers are commonly referred to as phreatophytes. Some phreatophytes appear to be largely dependent on groundwater, while others show only a slight dependency. Thus, phreatophytes exhibit an obligate and facultative groundwater dependency [12]. However, the identification of obligate and facultative phreatophytes is complicated by species characteristics [13,14] and shifts between wetter and drier soil conditions [11].

Temperature-driven soil water flow is another process that can affect water balance and movement in arid and semi-arid areas [15]. The dynamics of liquid water and water vapor movement have been determined based on the mathematical description of liquid water and vapor flux by Philip and de Vries [16] for unsaturated soils under a soil water pressure head and a soil temperature gradient. This study produced a mathematical model [1,17] based on the influence of air flow [18], meteorological conditions [19], and vegetation cover [15,20]. However, there have been limited modeling studies evaluating the role of coupled fluxes and their temporal variation in desert bush-dominated areas under shallow groundwater level conditions. In particular, information about the movement of liquid water, water vapor, and heat in bush-dominated sandy soil as well as the contribution of water vapor flux to total water flux remains limited for areas such as the Mu Us Desert.

The objective of this study was to assess the water use of *Salix psammophila* (*S. psammophila*) in a semi-arid area of the Mu Us Desert of northwest China using a coupled liquid water, water vapor, and heat transport model. Systematic field measurements were conducted to investigate the relationships between *S. psammophila* transpiration and groundwater level changes. First, we calibrated the HYDRUS-1D model [21] using soil water and temperature measurements obtained in an *S. psammophila* bush plot. We then used the model to compute the contributions of groundwater to root water uptake to quantify the dependency of the *S. psammophila* bush on groundwater. Third, we investigated the movement of liquid water and water vapor in the absence of heat transport and under decreased rainfall, increased temperature and *LAI*.

2. Materials and Methods

2.1. Field Site Description

The experimental site was located in the Hailiutu River catchment (between 38°06′ and 38°52′ N, 108°47′ and 109°23′ E) in the Erdos Plateau of northwest China (Figure 1). The Hailiutu River is a branch of the Wuding River, which is a major tributary of the Yellow River. The total area of the Hailiutu River catchment is approximately 2645 km^2. A hydrological station located at the outlet of the Hailiutu catchment near Hanjiamao village registered a mean annual discharge of 2.64 m^3/s for the period 1957–2007. Studies show that the flow regime has changed dramatically over the last 51 years. Four major shifts were detected in 1968, 1986, 1992 and 2001, reflecting quasi-natural conditions, reservoir construction, combined river diversion and groundwater extraction, and increased crop area [22]. Based on 50 years of meteorological data (from 1957–2007) obtained from a station approximately 40 km northwest of the study site, the long-term average annual air temperature is 8.1 °C, and the minimum and maximum monthly average air temperatures are −8.6 °C in January and 23.9 °C in July. The average annual precipitation is 340 mm, approximately 70% of which occurs between July and September. The average annual potential evaporation is estimated to be 2180 mm [23]. The area's main crop is maize, which is sown in mid-April and harvested in mid-October, with slight variation depending on seasonal weather conditions. Irrigation depends on local rainfall [23].

The experimental site is located approximately 11 km northeast of the Hailutu River. The dominant natural vegetation around the research site is *S. psammophila*, which covers approximately 30% of the area, along with other sporadically distributed herbs, primarily *Artemisia ordosica*, *Korshinsk peashrub* and *Hedysarum laeve* Maxim.

2.2. Field Measurements

A set of instruments was installed to measure environmental variables, including the sap flow of *S. psammophila*, soil water, soil temperature, and groundwater level (Figure 2). The instruments used in the study are listed in Table 1. Detailed information regarding river catchment and groundwater level measurements can be found in a previous study [23]. Here, the results of a root survey and measurements of soil water content and soil temperature are described.

2.2.1. Measurement of Soil Water Content

Soil water content (SWC, cm^3/cm^3) was measured at eight depths using time-domain reflectometry (TDR) technology. All sensors were inserted horizontally into the soil. All sensors installed at 10, 20, 40, 60, 80, 100, 120 and 140 cm depths were 6005CL2 (Minitrase SEC Co. Ltd., Goleta, CA, USA, 2% resolution). Data were recorded every 10 s and stored every 1 hour by a data-logger (Minitrase SEC Co. Ltd., Goleta, CA, USA).

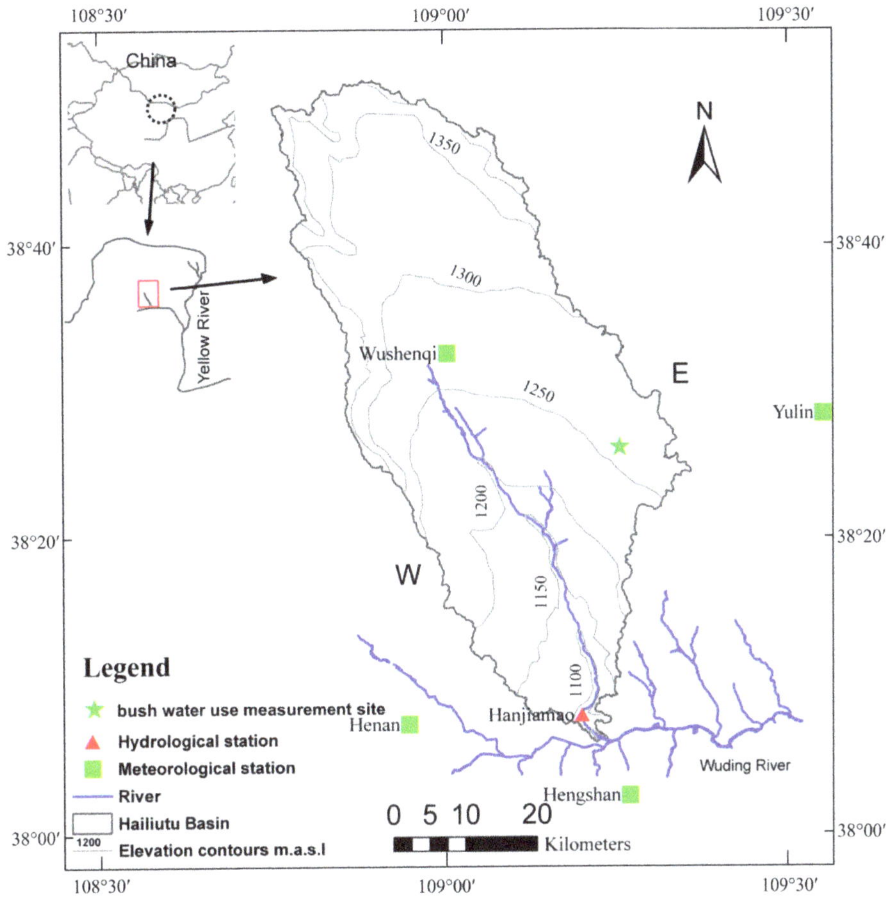

Figure 1. Location of the Hailiutu River catchment and the research site.

2.2.2. Measurement of Soil Temperature

The soil temperature was measured at 10 different depths using TCAV thermocouple sensors (Campbell Scientific, Logan, UT, USA). The thermocouple sensors were installed at 2, 5, 10, 20, 30, 40, 50, 60, 70 and 80 cm depths. All sensors were inserted horizontally into the soil. The data were recordedat 10-s intervals and were stored as 1-h averages using a CR1000 data-logger (Campbell Scientific, Logan, UT, USA).

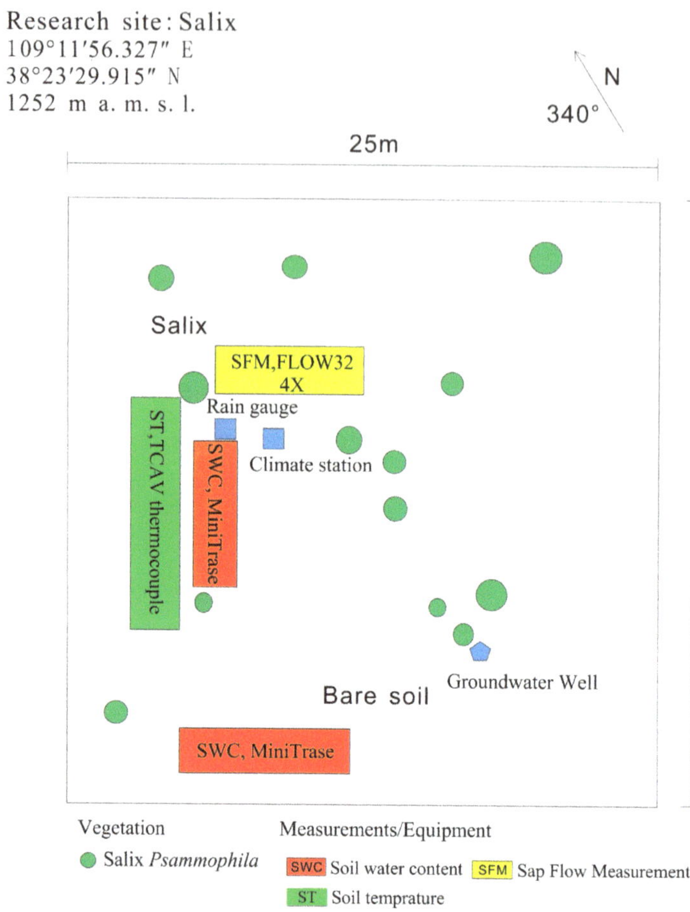

Figure 2. Schematic plot of instrument locations.

2.2.3. Root Survey

Root distribution reflects the location of the water source used by the plants. Because root distribution varies with depth and radial distance, soil cores were collected around *S. psammophila* plants using a root auger (Eijkelkamp, Giesbeek, The Netherlands) after defoliation. Undisturbed uniform soil samples 10 cm in diameter and 15 cm in height were obtained, and the position of each sample was recorded at the horizontal radial distance from the stem and the depth to the point of each sample. Four symmetrical transects centered at the stem with a radius of 4 m were investigated. The total area of the root distribution was determined to be 50.24 m^2. Roots were sieved out of the core sample and washed on the day of sampling. The cleaned roots from each core sample were weighed using an electric balance (readout 0.01 g; resolution limit of 1%), and photos were taken on scaled paper (resolution:

1 mm). The total root length of fine roots (diameter < 2 mm) was measured using gvSIG (a geographic information system) open source software [24]. The root length density (cm/cm^3) of each sample was determined by dividing the total root length by the core volume. The vertical distribution of fine roots was determined from the average of the root-length density at the same sample depth and the radial distance within the root zone of a *S. psammophila* plant. Two maxima of root length density were observed: one between 0 and 30 cm and another between the 60 and 120 cm (with a peak at approximately 80–90 cm) soil layers.

Table 1. Instruments used at the research site.

Variable	Height (m)	Horizontal Distance to the Bush Stem (m)	Instrument or Sensor Type
Wind speed (W· m^{-2})	5.25	3.0	05130-5 RM Yong wind monitor
Net radiation (W· m^{-2})	3.75	3.0	NR-LITE sensor
Rainfall (mm· hr^{-1})	–	3.0	52203 RM Young rain gauge
Air temperature (K) and	3.25	3.0	HMP45C temperature-humidity
relative humidity (%)	4.25	3.0	sensors
Soil heat flux (W· m^{-2})	−0.08	3.0	HFP01 heat flow plates
Soil temperature (°C)	−0.02, −0.05, −0.1, −0.2, −0.3, −0.4, −0.5, −0.6, −0.7, −0.8	1.5	TCAV thermocouple
Sap flow	0.7	–	Flow 32 1K
Soil water content	−0.1, −0.2, −0.4, −0.6, −0.8, −1.0, −1.2, −1.4	1.5	6005CL2
Groundwater level (cm) and Temperature (°C)	−2.0	1.5	DI501
Root auger	−0.15, −0.3, −0.45, −0.6, −0.75, −0.9, −1.05, −1.2, −1.35, −1.5	from 0.1 to 4.0 with 0.1 interval	–

2.3. *Simulations of Soil Water and Heat Transport*

HYDRUS-1D software [21] was used to simulate water flow, water vapor and heat transport under root water uptake conditions. The model simulation followed a 3-step approach similar to that previously published [25]. The first step involves the calibration of soil parameters using hourly measurements from 26 April to 26 May 2012. The second step is the validation of the model using hourly measurements from 27 May to 27 September 2012. The third step is a complete simulation from April to September 2012 to investigate the groundwater dependency of the *Salix* bush.

2.3.1. Governing Equation

The governing equation for the one-dimensional vertical flow of liquid water and water vapor in variably saturated media is given by the following mass conservation equation [17]:

$$\frac{\partial \theta_T(h)}{\partial t} = -\frac{\partial q_L}{\partial z} - \frac{\partial q_v}{\partial z} + S(h) \tag{1}$$

where θ_T is the total volumetric water content (cm^3/cm^3), which is the sum of the volumetric liquid and vapor water content ($\theta_T = \theta_L + \theta_v$); q_L and q_v are the flux densities of liquid water and water vapor (cm/h), respectively; t is time (h); z is the vertical axis position (cm); and $S(h)$ is the sink term ($cm^3/cm^3/h$).

The flux density of liquid water, q_L, has been defined by Philip, J.R., *et al.* [16]:

$$q_L = q_{Lh} + q_{LT} = -K_{Lh}(\frac{\partial h}{\partial z} + 1) - K_{LT}\frac{\partial T}{\partial z} \tag{2}$$

where q_{Lh} and q_{LT} are the isothermal and thermal liquid water flux densities (cm/h); h is the matrix potential head (cm); T is the temperature (°C); and K_{Lh} (cm/h) and K_{LT} (cm^2/°C/h) are the isothermal and thermal hydraulic conductivities for liquid-phase fluxes due to gradients in h and T, respectively.

Using the product rule for differentiation and assuming that the relative humidity in soil pores is constant at different temperatures, the flux density of water vapor, q_v, can be written as [17]:

$$q_v = q_{vh} + q_{vT} = -K_{vh}\frac{\partial h}{\partial z} - K_{vT}\frac{\partial T}{\partial z} \tag{3}$$

where q_{vh} and q_{vT} are, respectively, the isothermal and thermal water vapor flux densities (cm/h); and K_{vh} (cm/h) and K_{vT} (cm^2/°C/h) are the isothermal and thermal vapor hydraulic conductivities of water vapor, respectively.

The total heat flux density is defined as the sum of the conduction of sensible heat, as described by Fourier's law (the first term on the right side), the convection of sensible heat by liquid water (the second term) and water vapor (the third term), and the convection of latent heat by vapor flow (the fourth term) [17]:

$$(C_p\frac{\partial T}{\partial t}) + (L_0\frac{\partial \theta_v}{\partial t}) = \frac{\partial}{\partial z}(\lambda\frac{\partial T}{\partial z}) - (C_w q_L\frac{\partial T}{\partial z}) - (Cvq_V\frac{\partial T}{\partial z}) - (L_0\frac{\partial q_V}{\partial z}) \tag{4}$$

where λ is the apparent thermal conductivity of the soil (J/cm/h/°C); C_p, C_w, and C_v are the volumetric heat capacities (J/cm^3/°C) of the soil, liquid, and vapor phases, respectively; q_L is the sum of the isothermal and thermal liquid water flux densities (cm/h); L_0 is the latent heat of vaporization of liquid water (J/cm^3); and q_v is the sum of the isothermal and thermal vapor flux densities (cm/h). Details regarding the HYDRUS-1D model can be found in previously published work [17,21].

The sink term in Equation (1), $S(h)$, is defined as the volume of water removed from a unit volume of soil per unit time due to plant water uptake. Feddes, R.A., *et al.* [26] defined $S(h)$ as:

$$S(h) = \alpha(h)b(x)T_p \tag{5}$$

where the root-water uptake reduction factor *α(h)* is a prescribed dimensionless function of the soil water pressure head, and *b(x)* is a non-uniform distribution of the potential water uptake rate over a root zone normalized by the field root survey value determined by Huang, J.T. *et al.* [23]. T_p is the potential transpiration.

2.3.2. Soil Characteristics

The water retention curve is one of the most fundamental hydraulic characteristics in the soil water flow equation. The soil water retention equation is given by van Genuchten [27]:

$$\theta(h) = \begin{cases} \theta_r + \dfrac{\theta_s - \theta_r}{[1 + |\alpha h|^n]^m} & h \leqslant 0 \\ \theta_s & h > 0 \end{cases} \tag{6}$$

where θ is the volumetric water content (cm^3/cm^3) at pressure head h (cm); θr and θs are the residual and saturated water contents (cm^3/cm^3), respectively; α (> 0, in 1/cm) is related to the inverse of the air-entry pressure; and n (>1) is a measure of effect of the pore-size distribution on the slope of the retention function ($m = 1 - 1/n$).

The hydraulic conductivity of soil $K(h)$ is described as:

$$K(h) = K_s S_e^l \left\{ 1 - [1 - S_e{}^{n/(n-1)}]^{1-1/n} \right\}^2 \tag{7}$$

where K_s is the saturated hydraulic conductivity. S_e is the effective saturation:

$$S_e = \frac{\theta(h) - \theta_r}{\theta_s - \theta_r} \tag{8}$$

The thermal conductivity was defined by de Marsily [28] as:

$$\lambda(\theta) = \lambda_0 + \beta_t C_w |q| \tag{9}$$

where C_w is the volumetric heat capacity of the water and q is water flux; β_t is the thermal dispersivity (cm); and λ_0 is the baseline thermal conductivity, which has been described by Chung, S.O., *et al.* [29] as:

$$\lambda_0 = b_1 + b_2\theta + b_3\theta^{0.5} \tag{10}$$

where b_1, b_2 and b_3 are empirically determined parameters (W/cm/°C).

2.3.3. Root Water Uptake Modeling

In equation (5), the root distribution $b(x)$ was normalized according to the field root survey data. For $\alpha(h)$, we used the functional formula introduced by van Genuchten [30]:

$$\alpha(h) = \frac{1}{1 + (h/h_{50})^{p}} \tag{11}$$

where h_{50} is the pressure head at which transpiration is halved and p is an adjustable constant that determines the steepness of the transition from potential to reduced uptake rates as h decreases.

2.3.4. Boundary and Initial Conditions

An atmospheric boundary condition was implemented at the soil surface, while a variable pressure head condition was used at depth. The evaporation from the soil surface and transpiration by plants were simulated using the HYDRUS-1D model. To specify the potential transpiration (T_p) and potential evaporation (E_p), we calculated the potential evapotranspiration (ET_p) using the Penman-Monteith equation [31]. Potential evaporation (E_p) and potential transpiration were then calculated according to the Beer's law method:

$$\begin{aligned} T_p &= ET_p(1 - e^{-k \cdot LAI}) \\ E_p &= ET_p e^{-k \cdot LAI} \end{aligned} \tag{12}$$

where LAI is the leaf area index [−] and k is an extinction coefficient set to 0.463.

In our study, LAI was derived from MODIS imagery (MOD15A2) at a temporal resolution of 8 days, with LAI reaching a peak of ~1.69. Thus, the computed potential evaporation was used as an input to calculate the actual evaporation fluxes based on a reduction of van Genuchten's equation for transpiration and the h_{Crit} limit for soil evaporation. In our simulations, h_{Crit} was determined from the equilibrium conditions between soil water and atmospheric water vapor. During the experiments, the observed pressure heads at the research site were used for model calibration and validation. Figure 3 summarizes the imposed surface and bottom conditions, showing the hourly values of precipitation, potential evapotranspiration, and depth to groundwater table.

The upper and lower boundary conditions for heat transport were specified as temperature boundary conditions. The upper boundary values were given as the observed temperatures at 0.02 cm, and the lower boundary values were given as the groundwater temperature. The initial soil water content and soil temperature were determined from the measured values on April 26 at 0:00 h by interpolating the measured values between different depths.

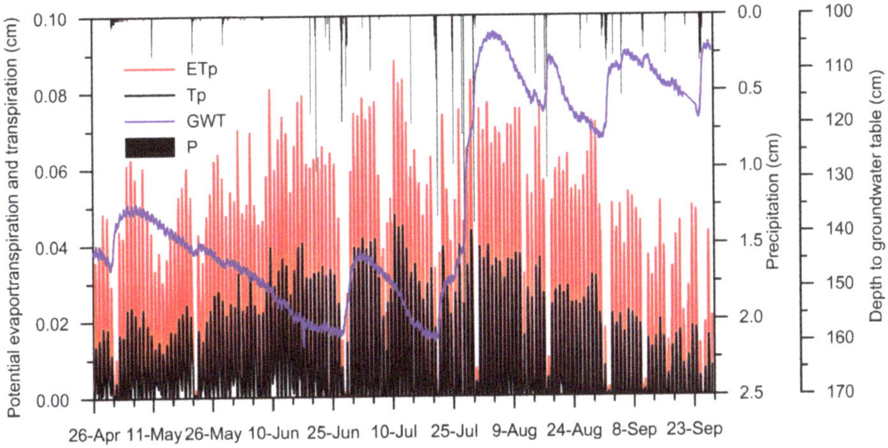

Figure 3. Summary of the modeled soil column boundary conditions (P = precipitation; ETp = potential evapotranspiration; Tp = transpiration; GWT = depth to groundwater table).

2.3.5. Criterion for Model Calibration

Model calibration was evaluated using root mean square error (*RMSE*), systematic *RMSE* ($RMSE_s$), unsystematic *RMSE* ($RMSE_u$), and index of agreement (d) [32,33], defined as follows:

$$RMSE = \sqrt{\sum_{i=1}^{N} \frac{(P_i - O_i)^2}{N}} \tag{13}$$

$$RMSE_s = \sqrt{\sum_{i=1}^{N} \frac{(\hat{P}_i - O_i)^2}{N}} \tag{14}$$

$$RMSE_u = \sqrt{\sum_{i=1}^{N} \frac{(P_i - \hat{P}_i)^2}{N}} \tag{15}$$

$$d = 1 - \left[\frac{\sum_{i=1}^{N} (P_i - O_i)^2}{\sum_{i=1}^{N} \left(|P_i - \overline{O}| + |O_i - \overline{O}|\right)^2} \right], \quad 0 \leqslant \mathrm{d} \leqslant 1 \tag{16}$$

where the $RMSE_s$ indicates errors due to under- or over prediction by showing how far the data fluctuate from the 1:1 line in a scatter plot of computed *versus* measured values. When the $RMSE_u$ is minimized and the $RMSE_s$ approaches *RMSE*, the model performs with maximum accuracy. The value of d ranges from 0 (no agreement) to 1 (a perfect fit between simulated and measured values). N is the number of paired observations, and P_i and O_i are the computed and measured values, respectively. $\overline{O}$ is the mean of measured values, and $\hat{P}_i$ is defined as:

$$\hat{P}_i = a + bO_i \quad (17)$$

where a and b are the intercept and slope of the least-squares linear regression between P_i and O_i.

2.3.6. Investigation of Liquid Water and Water Vapor Movement

To quantify the effect of water movement on the water use of the *Salix* bush, the calibrated model was used to investigate liquid water and water vapor movement under two scenarios. The first scenario quantified the contribution of soil water to *S. psammophila* under conditions with or without heat using the measured data. The second scenario evaluated liquid water and water vapor movement in the case of decreased rainfall (*i.e.*, half the amount of rainfall during the experimental period, in agreement with historical values during periods of low rainfall) while increasing temperature and LAI (5 °C temperature and 1.5 *LAI* increases [34] were assumed, respectively) using the calibrated model.

3. Results

3.1. Model Calibration and Validation

HYDRUS-1D software was used to simulate water movement at the experimental site. The soil hydraulic parameters and soil heat parameters were calibrated and validated using the soil water content and temperature data while varying the drought stress parameters h_{50} and p. During the model calibration period (from April 26 to July 11, 2012), in general, the soil water content gradually increased from the ground surface to the deep layers, unless rainfall occurred. Moreover, the soil water content in the shallow layers was more sensitive to rainfall than that in the deep layers. For example, a daily rainfall of 30.3 mm occurred on June 27 and triggered an increase in the soil water content at depths up to 100 cm, while after smaller rainfall events, the increase in the soil water content only reached 40 cm. During the validation period (July 12 to September 28, 2012), the soil water content increased with soil depth to groundwater. However, a large rainfall event of 41.1 mm on July 20 and subsequent rainfall resulted in an increase in the groundwater level, which thereafter remained high. The soil water content at a depth of 100 cm remained saturated until the conclusion of our experiments.

Six parameters in the van Genuchten model [27] were calibrated using field measurements (*i.e.*, θ_r, θ_s, α, n, l, and Ks). Calibration was performed by fitting the observed and modeled soil water contents using the Marquardt-Levenberg optimization algorithm. HYDRUS-1D ran the optimization process until it found the highest R^2 values [21] between observed and computed soil water content.

The soil column was schematized as six soil layers based on our *in situ* investigations. The parameter Ks was determined using the inverse auger method [35] for each layer. The remaining parameters were estimated using Rosetta [36], a pedotransfer function that predicts hydraulic parameters from soil texture data (Table 2).

Table 2. Soil texture at the observation site.

Depth (cm)	Sand (%)	Silt (%)	Clay (%)
0–40	97.3	2.7	0
41–55	97.9	2.1	0
56–70	98.5	1.6	0
71–90	98.7	1.3	0
91–110	99.1	0.9	0
111–200	98.8	1.2	0

Running Hydrus-1D with Rosetta hydraulic parameter estimates and empirically determined values for saturated hydraulic conductivity in simulations produced poor agreement for the index. We therefore attempted to calibrate the soil hydraulic property model using the measured soil water content. In Equation (6), there are five unknown parameters for each layer. When using the inverse model in Hydrus-1D software, we found that fitting all five parameters in six layers at the same time tended to cause the inverse algorithm to fail. Thus, the five parameters were fitted layer by layer. This fitting method has been reported previously by other researchers [37]. When the simulated and measured soil water contents exhibited a good agreement index, the hydraulic parameters were fixed and the drought stress parameters h_{50} and p were fitted via the measured sap flow of *S. psammophila*. Finally, the hydraulic and drought stress parameters were fixed, and the thermal parameters were fitted using the soil temperature.

The temporal variation of the computed soil water content at the eight measurement depths during the calibration and validation periods was reasonably consistent with the field measurements at each depth (Figure 4). We found that the parameters α and n were more sensitive than the other parameters. Table 3 shows the calibrated hydraulic parameters. The statistical criteria for model calibration (Equations (13)–(16)) are summarized in Table 4. The RMSE values were very low for both the calibration and validation periods at all measured depths. The index of agreement (d) was very high, ranging from 0.65 to 0.96 and from 0.86 to 1.00, respectively, for the model calibration and validation periods at all depths. Generally, the calibration and validation results were acceptable. The soil water contents computed using the model captured the sharp increase in the soil water content after heavy rainfall events.

Table 3. The calibrated parameters for the HYDRUS-1D model.

Depth (cm)	θ_r (cm^3· cm^{-3})	θ_s (cm^3· cm^{-3})	α (cm^{-1})	n	Ks (cm· h^{-1})	l
0–40	0.045	0.39	0.0550	2.180	115.70	0.5
41–55	0.075	0.40	0.0245	3.812	118.22	0.5
56–70	0.07	0.39	0.0198	4.873	113.42	0.5
71–90	0.072	0.29	0.0189	5.716	110.42	0.5
91–110	0.124	0.39	0.0295	2.973	103.02	0.5
111–200	0.065	0.39	0.01618	5.654	110.02	0.5

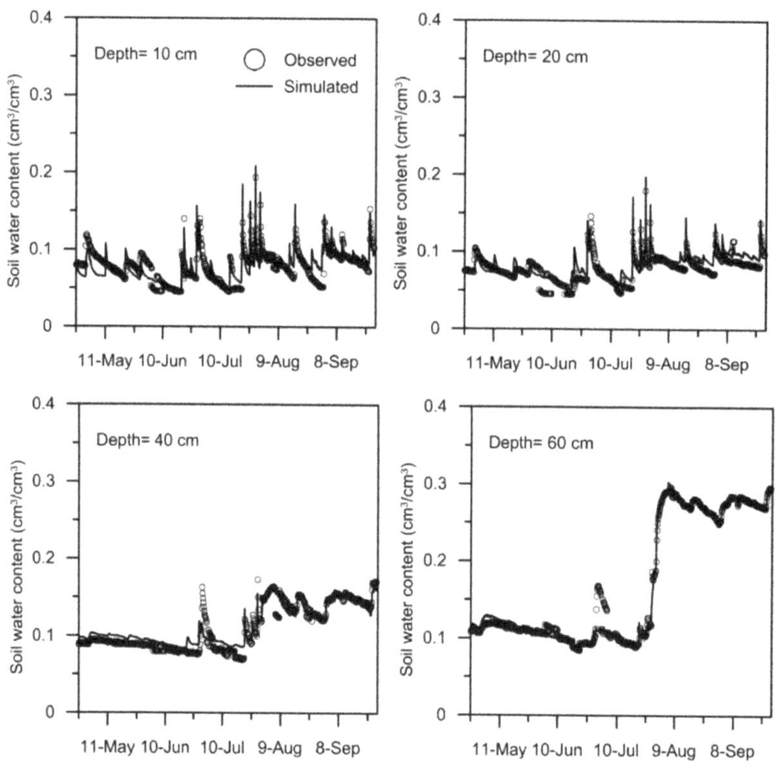

Figure 4. *Cont.*

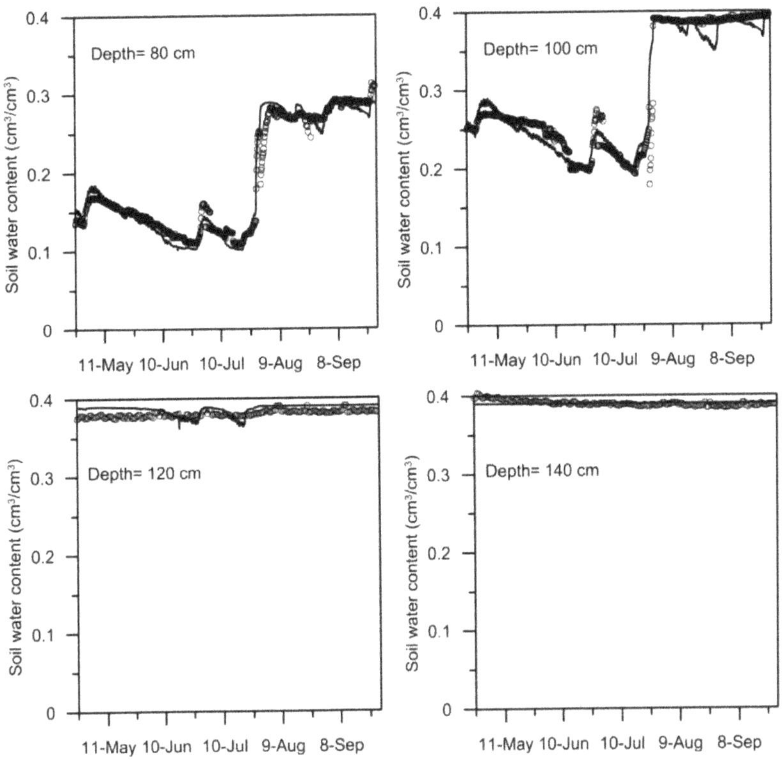

Figure 4. The fit of measured soil water content at eight different depths based on soil profile (points) using model-computed values (solid lines) for the calibration period (26 April to 12 July) and the validation period (13 July to 27 September).

Reported parameter values for root-water uptake reduction by specific plants and soils range from approximately −1000 cm to −5000 cm for h_{50} and from 1.5 to 3 for p [38]. However, especially coarse soil, such as sand, is almost completely drained of water at a fairly modest pressure head (*i.e.*, −300 or −400 cm). Using values for h_{50} and p that are similar to those reported in the literature caused the uptake reduction to perform essentially as a step function [38]. On this basis, we employed an h_{50} value that was considerably lower than that reported in the literature. Likewise, a larger value of p was required to account for the steepness of the soil water retention curve. Similar to the approach reported by Zhu, Y., *et al.* [39], we performed simulations using a range of values for h_{50} and p in an effort to calibrate the HYDRUS-1D model. As expected for Aeolian sand, the simulated water contents were not very sensitive to h_{50} and p. This finding agrees with the results of Zhu, Y., *et al.* [39], which simulated *Populus euphratica* root uptake in coarse sand soil. A comparison between simulated and measured soil water contents is presented in Figure 4. For our research plants, the values of h_{50} and p were found to be −630 cm and 3, respectively, from the results of fitting the predicted and measured values for transpiration (Figure 5). It can be seen that the observed and simulated transpiration of *S. psammophila* fitted well (with R^2 = 0.99), indicating that the calibration values of h_{50} and p were acceptable.

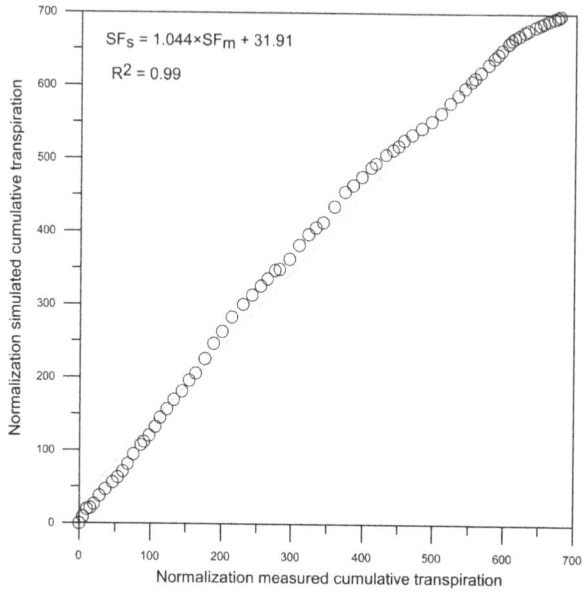

Figure 5. The fit of the measured (SF_m) and computed (SFs) cumulative transpiration rates of the *S. psammophila* bush.

Hourly measurements of soil temperature at nine different depths reveal clear diurnal fluctuations (Figure 6). However, both the temperature and the amplitude of the diurnal fluctuations decrease with the increase of depth because the air temperature is higher than the groundwater temperature during the measurement period spanning from 26 April to 27 September 2011, indicating downward heat transport. The temperature also exhibited seasonal variation, increasing beginning in late April, reaching the highest temperature in July and August and decreasing in September. The calibrated HYDRUS-1D model is able to simulate both seasonal and diurnal variations in soil temperature at nine different depths (Figure 6). The statistical measures used for temperature calibration and validation are shown in Table 5. The RMSE ranges from 0.446 °C to 2.396 °C and from 0.784 °C to 2.175 °C during the calibration and validation periods, respectively. The index of agreement (d) is larger than 0.88, during both the calibration period and the validation period, indicating a good agreement between the computed and measured soil temperatures (Table 5).

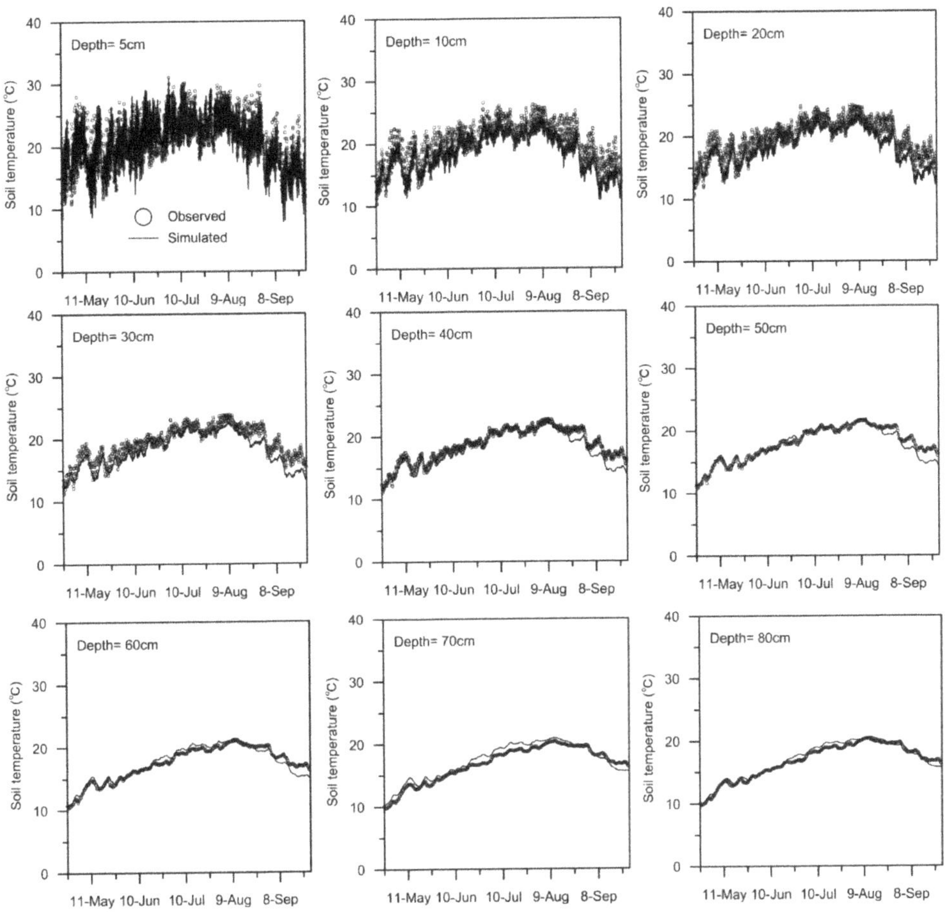

Figure 6. The fit of computed soil temperatures (solid lines) to measured temperatures (points) at nine depths in the soil profile used for calibration (26 April to 12 July) and validation (13 July to 27 September).

Table 4. Statistical measures of HYDRUS-1D model performance for simulations of volumetric water content.

	Simulated *vs.* Measured Water Content ($cm^3 \cdot cm^{-3}$)															
	Calibration (cm)								Validation (cm)							
depth	10	20	40	60	80	100	120	140	10	20	40	60	80	100	120	140
N	1848	1848	1848	1848	1848	1848	1848	1848	1859	1859	1859	1859	1859	1859	1859	1859
a	0.03538	0.019486	0.046836	−0.01887	−0.01928	0.019124	0.28803	0.386439	0.03538	0.019767	0.025305	−0.00255	0.00503	0.048224	−0.06376	0.390254
b	0.617928	0.740785	0.546443	1.194489	1.132598	0.915269	0.256591	0.008968	0.617928	0.868301	0.801304	1.009043	0.981768	0.855417	1.180737	−0.00073
RMSE	0.009	0.008	0.009	0.005	0.00	0.012	0.009	0.005	0.009	0.008	0.003	0.000	0.015	0.036	0.007	0.003
RMSEs	0.006573	0.003	0.008	0.003	0.003	0.002	0.007	0.004533	0.006573	0.004841	0.004788	7.18×10^{-5}	0.001354	0.017426	0.00535	0.003289
RMSEu	0.006573	0.007	0.004	0.005	0.007	0.011	0.006	8.14×10^{-5}	0.006573	0.004841	0.004788	7.18×10^{-5}	0.014765	0.031006	0.00496	8.65×10^{-5}
d	0.86	0.88	0.65	0.93	0.96	0.94	0.89	0.85	0.86	0.86	0.97	1.00	0.99	0.94	0.87	0.89

Table 5. Statistical measures of HYDRUS-1D model performance for simulations of soil temperature at different soil depths.

	Simulated *vs.* Measured Soil Temperature (°C)																	
	Calibration (cm)									Validation (cm)								
depth	5	10	20	30	40	50	60	70	80	5	10	20	30	40	50	60	70	80
N	1848	1848	1848	1848	1848	1848	1848	1848	1848	1859	1859	1859	1859	1859	1859	1859	1859	1859
a	0.379587	1.209036	0.883048	0.470773	−0.32943	−0.20278	−0.08195	0.151507	0.477904	−3.406	−3.516	−4.168	−4.359	−6.509	−8.936	−9.418	−8.413	−7.49616
b	0.909248	0.879813	0.906546	0.928757	0.995939	1.020258	1.029854	1.04017	1.048278	1.042	1.075	1.121	1.144	1.274	1.427	1.477	1.455	1.437842
RMSE	2.396	1.543	1.282	1.201	0.735	0.446	0.524	0.807	1.185	2.175	1.741	1.588	1.426	1.341	1.645	0.865	0.784	0.951
RMSEs	1.882521	1.145	0.892	0.837	0.399	0.134	0.388	0.748926	1.153619	0.781	0.738	0.730	1.111	1.128	1.356	0.665	0.540	0.705828
RMSEu	1.882521	1.033	0.921	0.861	0.617	0.426	0.353	0.301853	0.268979	0.781	0.738	0.730	0.644	0.725	0.931	0.553	0.569	0.637423
d	0.91	0.93	0.94	0.94	0.97	0.99	0.99	0.97	0.94	0.88	0.89	0.90	0.90	0.91	0.82	0.93	0.92	0.85

3.2. *Pressure Head over the Root Zone*

The mean values of the pressure head over the root zone with and without heat flow are shown in Figure 7. During the simulated period, the mean value of the pressure head ranged from −9408.9 to −31.11 cm with heat flow and from −244.3 to −32.5 cm without heat flow. In the simulation without heat, liquid connectivity breaks down (via Equation 7) and the water content cannot decrease further, whereas in the simulation with heat transport, very small amounts of water continue to be transported. These small changes have a large effect on the simulated matrix potential because the simulation approaches the residual water content. The lowest value of the pressure head occurred mainly during periods of no precipitation and a deep groundwater table (Figure 3).

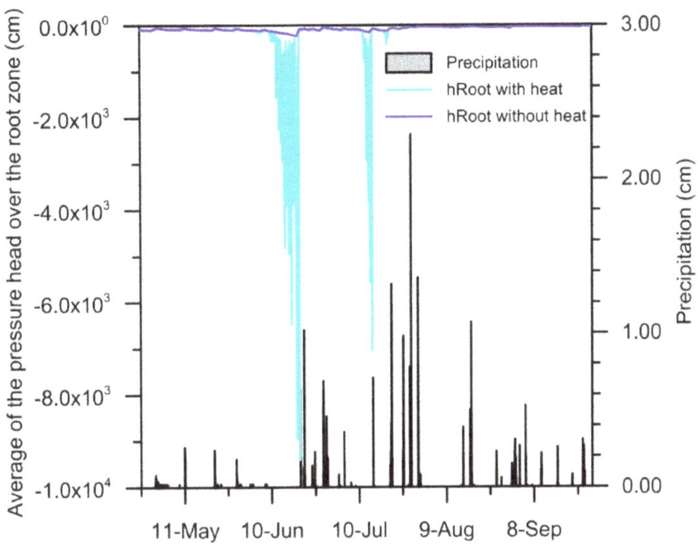

Figure 7. Comparison of the mean pressure head values over the root zone with heat flow (*hRoot* with heat) and without heat flow (*hRoot* without heat).

3.3. *Evapotranspiration*

The calculated evapotranspiration is shown in Figure 8, both with and without heat flow. During the simulated period, the total transpiration rates were 27.9 cm and 28.1 cm with and without heat flow, respectively. The total evaporation rates were 15.9 cm and 16.7 cm with and without heat flow, respectively. The differences were 0.2 cm and 0.8 cm for transpiration and evaporation, respectively (Figure 8a). In total, the simulated value of *ET* with heat flow is lower than that without heat flow; in other words, the value of *ET* is overestimated without heat flow.

To further clarify the effect of heat flow on transpiration, root water uptake values for two zones are shown in Figure 8b. It can be seen that the root water uptake is affected by heat flow mainly from 0 cm to 30 cm, and the value was 10.4 cm and 10.6 cm with and without heat flow, respectively. The value of root water uptake in the zone from 60 cm to 90 cm is the same—6.1 cm—both with and without heat flow.

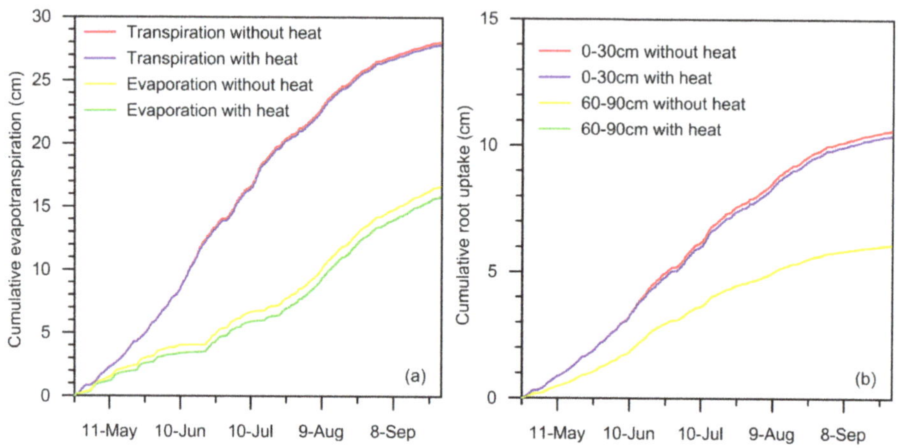

Figure 8. (**a**) Cumulative evapotranspiration; and (**b**) root water uptake for selected depth intervals with heat flow and without heat flow.

3.4. Water Storage and Bottom Flux

The change in soil water storage and bottom flux during the simulated period is shown in Figure 9. The cumulative changes in soil water storage were 11.7 cm and 11.4 cm with and without heat flow, respectively. The heat flow thus has little influence on the change in soil water storage.

During the simulated period, the cumulative bottom flux was positive, indicating the net groundwater inflow to the soil column. The total bottom fluxes were 14.4 cm and 15.1 cm with heat flow and without heat flow, respectively. Furthermore, groundwater inflow occurs during dry days, indicating the dependency of this process on groundwater. During heavy rain, the cumulative bottom fluxes decreased, indicating groundwater recharge. For example, after 4.87 cm rain fall occurred on the June 27 and June 28, value of cumulative bottom fluxes decreased from 11.18 cm to 9.55 cm without heat and from 9.42 cm to 7.89 cm with heat, respectively. This indicated value of recharge was 1.63 cm without heat and 1.53 cm with heat, respectively.

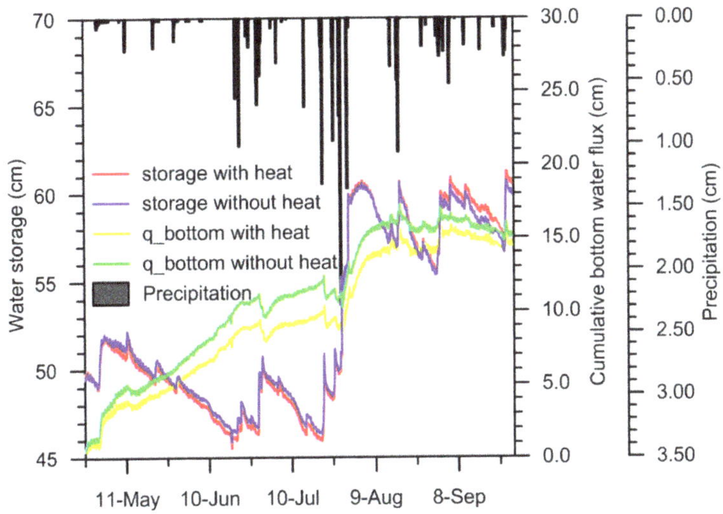

Figure 9. A comparison of the cumulative change in soil water storage and cumulative bottom water flux, both with and without heat flow.

3.5. Movement of Liquid Water and Water Vapor

3.5.1. Daily Movement of Liquid Water and Vapor Water in Summer

Figure 10a shows the liquid water movement temporal-space character before and after a heavy rainfall event (48.7 mm, occurred from 11:00 27 June to 4:00 29 June). Before the rainfall, the liquid water moved upward because of evapotranspiration, and the value ranged from 0 to 0.038 cm/h during the daytime; during the same period, transpiration ceased, a zero gradient zone formed at a depth of approximately 80 cm, and the liquid water below the gradient zone moved downward with a value ranging from 0 cm/h to 0.08 cm/h. During the rainfall, the liquid water moved downward, and the maximum water flux reached 0.69 cm/h. Following the rainfall, because of *ET*, the liquid water moved upward at the upper soil profile and moved downward continually over a 54-h period. At the end of the infiltration process, the change in the liquid water showed the same characteristics as before the rainfall.

Figure 10b shows the water vapor characteristics in the temporal-space before and after the same heavy rainfall event. It can be seen that the rainfall event only influenced the shallow soil layer (around a depth of 0 cm to 40 cm) water vapor movement, and the direction was upward. During the other days, the movement direction of water vapor was different during the day (moved upward) and night (moved downward) at a depth of 0 to 40 cm. At a depth of 40 cm to the groundwater table, the water vapor moved downward. This phenomenon indicates that the vapor circulation-condensation process supplied additional moisture for root water uptake at the shallow soil profile.

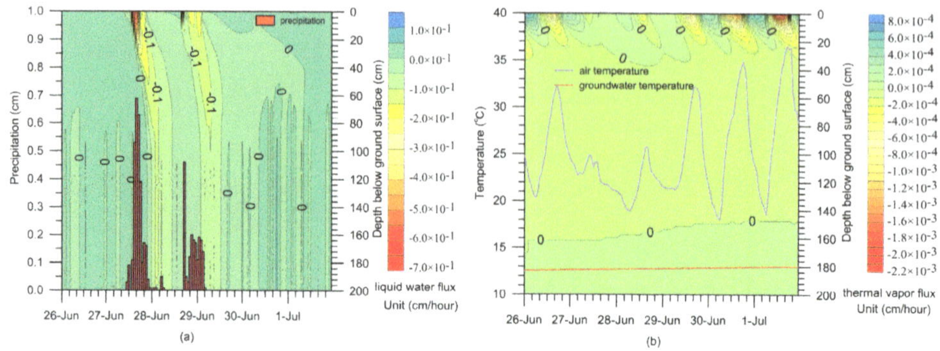

Figure 10. Simulated vertical distribution of the isothermal and thermal liquid water and water vapor fluxes during a typical period from the experimental period before and after a rainfall event (0:00 26 June to 23:00 1 July); (**a**) liquid water fluxes; (**b**) thermal vapor water fluxes.

3.5.2. Movement of Liquid Water and Water Vapor in Autumn

Figure 11 shows the liquid water and water vapor movement in the beginning of autumn. The flux of the liquid water movement in the autumn is much smaller than that in the summer (Figure 11a). However, in arid and semi-arid areas, the air temperature dropped lower than the soil temperature at night during autumn. Consequently, the movement of water vapor shows slightly different characteristics compared with that in the summer (Figure 11b). At a depth of 0 cm to 40 cm, the movement of water vapor shows the same characteristics as that in the summer. However, when the air temperature is lower than the groundwater temperature, the water vapor zero flux interface moves to a depth of approximately 80 cm and the water vapor moves upward at a depth of 40 cm to 80 cm, indicating that the water vapor movement is upward. In contrast, when the air temperature is higher than the groundwater temperature, the water vapor moves downward at a depth of 40 cm to groundwater table.

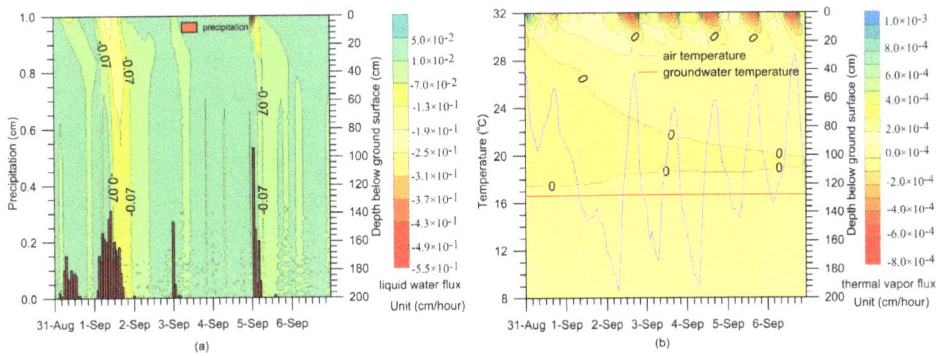

Figure 11. The movement of (**a**) liquid water; and (**b**) water vapor in autumn.

3.6. Contributions of Liquid Water and Water Vapor to Water Movement

The cumulative change of the average node flux with time is shown in Figure 12. The cumulative upward liquid water and water vapor flux are 26.49 cm and 0.03 cm, accounting for 99.89% and 0.11% of the total upward water flux, respectively. Additionally, the cumulative downward liquid water and water vapor flux are 26.75 cm and 0.28 cm, accounting for 99.72% and 0.28% of the total downward water flux, respectively. The liquid water flux is three orders of magnitude higher than the water vapor flux. The water vapor flux is negligible.

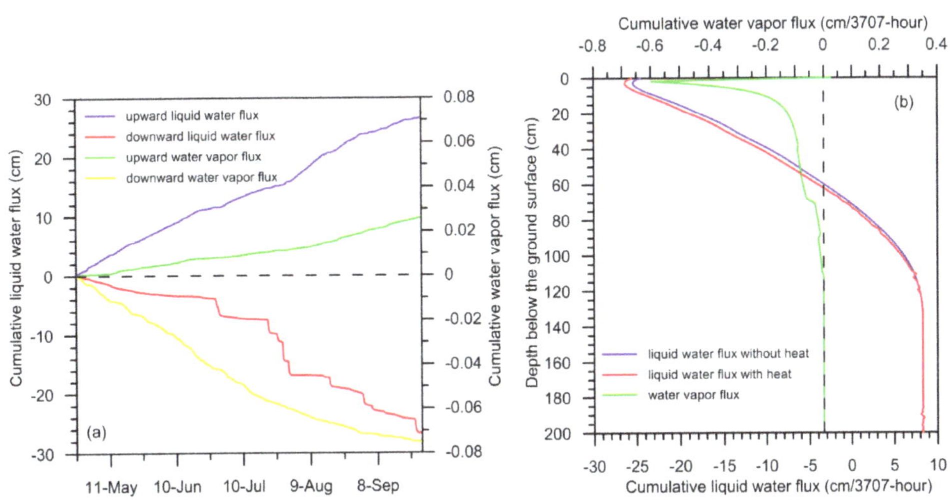

Figure 12. Changes in the cumulative liquid water and water vapor flux with time (**a**) and with depth; (**b**) during the investigation period.

Figure 12b shows the change in the water flux with depth. It can be seen that the zero flux interface of liquid water is formed at depths of 73 cm and 70 cm with heat flow and without heat flow. This indicates that the liquid water moves farther downward with the effect of heat flow during the simulated period. Moreover, the cumulative upward water vapor is lower than the downward water vapor (Figure 12a), so the cumulative water vapor was directed to the deep soil layer at a depth of approximately 158 cm.

3.7. Soil Water Contributions to Evapotranspiration

According to the previous calculations under the influence of heat flow, the soil water storage is increased by 11.7 cm. The precipitation is 41.1 cm, and the *ET* is 43.8 cm, including 27.9 cm of transpiration and 15.9 cm of evaporation. The total bottom water flux is 14.4 cm (upward), indicating the contribution of groundwater to the soil water balance.

To calculate the soil water contributions to evapotranspiration, the difference in the soil distribution between hydrostatic and actual conditions was used as described by Shah, N., *et al.* [40]. When distribution of the pressure and the soil moisture reached the hydrostatic equilibrium condition, the soil acted as a vessel and the ET was supplied entirely by the groundwater without a vadose zone contribution (VZC). As the depth of the groundwater table (DWT) increased, the hydraulic connections weakened. The hydraulic connections between groundwater are lost at a rate that exceeds the upward replenishment from the saturated zone. Hence, the VZC to ET occurs in a time step of $\Delta t = t_\mathrm{i} - t_{i-1}$ of the water content profile from hydrostatic equilibrium. Mathematically (modified from Shah N., *et al.* [39]),

$$ET_{\mathrm{vzc}} - P + q_{\mathrm{bot}} = \mathrm{VZC} \tag{18}$$

$$\mathrm{VZC} = [(\mathrm{TSM_{eq}} - \mathrm{TSM_{model}})|t_{\mathrm{i}-1} - (\mathrm{TSM_{eq}} - \mathrm{TSM_{model}})|t_\mathrm{i}]/(t_\mathrm{i} - t_{\mathrm{i}-1}) \tag{19}$$

where ET_{vzc} is the contribution of soil water to *ET*, *P* is precipitation, q_{bot} is the calculated bottom flux, and VZC is the contribution of soil water. $\mathrm{TSM_{eq}}$ is the soil water content in the column corresponding to DWT under hydrostatic equilibrium conditions, and $\mathrm{TSM_{model}}$ is the soil water content computed from simulated by Hydrus-1D for the corresponding time.

The calculated soil water contributions are shown in Figure 13. It can be seen that during the small rainfall period, the soil is dry and the *ET* is supplied by the groundwater. In contrast, after the rainfall, the soil becomes wet and the *ET* mainly comes from the soil water. This result confirmed the previous research results using the soil water balance approach at the same site [23]. However, during the simulation with heat flow, the cumulative amount of soil water and groundwater contribution to *ET* are 32.0 cm and 11.8 cm, which account for 73.1% and 26.9%, respectively, of the total ET (43.8 cm). In contrast, the cumulative soil water and groundwater contributions to ET without heat flow are 26.6 cm and 18.2 cm, which account for 59.4% and 40.6%, respectively, of the total ET (44.8 cm) without heat flow. The simulated results indicate that the contribution of groundwater to ET is overestimated without considering heat flow.

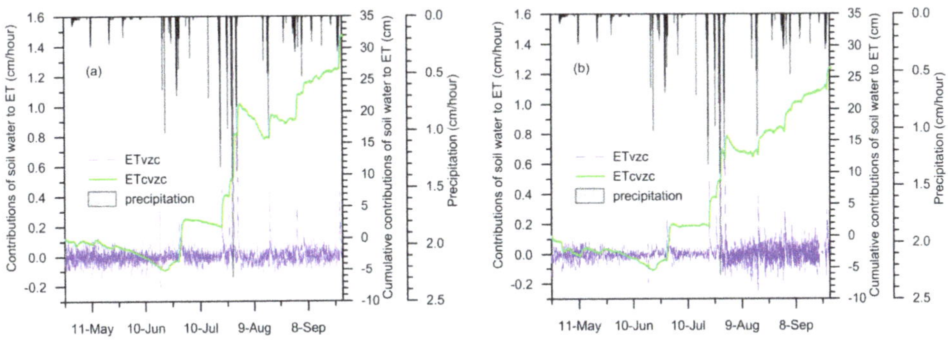

Figure 13. The contribution of soil water to ET during the simulation period (ET_{vzc} is the soil water contribution, and ET_{cvzc} is the cumulative contribution): (**a**) soil water movement with heat flow; and (**b**) soil water movement without heat flow.

3.8. Water Movement during Climate and LAI Change

Table 6 shows the calculated flux of liquid water and water vapor during periods of climate and LAI change compared to the results of the calibrated model. It can be seen that the maximum cumulative upward liquid water flux of LAI increased by 0.5, indicating that more water will be consumed if the vegetation becomes denser. However, the minimum cumulative downward liquid water was observed when the rainfall decreased by half. This result indicated that water use by the investigated bush will increase and that the percolation amount decreased as rainfall decreased. The cumulative water vapor flux did not change much in comparison with the liquid water flux with decreases in rainfall, increases in temperature, or increases in LAI. This result is consistent with the results presented in Figure 12, suggesting that the amount of water vapor was only slightly affected by a decrease in rainfall and by increases in temperature and LAI. Therefore, the vapor from groundwater will not allow a resilience of the investigated bush ecosystem in case of rainfall decrease, temperature increase and LAI increase.

Table 6. The cumulative flux of liquid water and water vapor during climate and *LAI* changes.

Water Flux	Calibrated Model	Rainfall*0.5	Temperaure + 5°C	*LAI**1.5
Cumulative upward liquid water flux (cm)	27.24	29.631	27.245	38.305
Cumulative downward liquid water flux (cm)	−28.322	−11.203	−28.312	−24.096
Cumulative upward vapor water flux (cm)	0.026	0.026	0.026	0.025
Cumulative downward vapor water flux (cm)	−0.071	−0.073	−0.071	−0.075

4. Discussion

In arid and semi-arid areas, thermally driven soil water movement is an important component of the soil-plant-atmosphere interaction. The temperature gradient changes due to diurnal temperature variations, resulting in the evaporation and condensation of water vapor in the soil. Zeng, *et al.* [41] found that the movement of water vapor can be described by three stages in shallow soil layers over the course of a day. Furthermore, the evaporation and condensation of water vapor are determined by the pressure head and temperature gradient. The results of water vapor movement simulation in this paper show similar characteristics as previously reported [41] during summer, when the air temperature was higher than the temperature of the groundwater (Figure 10b). However, the zero flux interface is formed in deeper soil in the autumn (Figure 11b). This phenomenon indicates that water vapor condensation is more frequently caused by both diurnal and seasonal air temperature changes under shallow groundwater table conditions in semi-arid desert regions. During the summer, both the air temperature and the groundwater temperature increase, so the alternation and condensation of vapor occur in shallow soil layers determined by diurnal soil temperature fluctuations. In contrast, the air temperature decreased and the groundwater temperature increased due to the soil temperature penetration time lag, resulting in the alternation of evaporation and condensation in shallow and deep soil layers.

Anthony, *et al.* [42] examined the magnitude of the water vapor flux reported in experimental studies by various authors and found that the maximum magnitude of moisture change due to vapor flow ranges from 7.2×10^{-3} to 2.5×10^{-2} cm/h. The simulated results in this study are in agreement with these values (Figures 10b and 11b). Regarding the water vapor contribution to total water flux, Parlange, *et al.* [43] found that water vapor contributed between 10% and 30% of the total flux based on one week of field observations in bare sandy loam soil. Deb, *et al.* [44] found that the total upward water vapor flux in the unsaturated soil layer of furrow-irrigated sandy loam soil was approximately 10.4% of the upward total water flux over a 50-day period. Compared to those results, the water vapor contribution to the total water flux in our simulation was much smaller: the upward and downward contributions were 0.11% and 0.28%, respectively. These smaller percentages are caused by the transpiration of the investigated bush from liquid water in semi-arid areas with shallow groundwater table conditions. In other words, the liquid water was dominant due to the large amount of transpiration, even though the rainfall decreased by half (Table 6). This phenomenon agrees with the results reported by Garcia, *et al.* [15]. Although the amount of water vapor flux is small, the mean pressure over the root zone is very low during dry periods with heat flow (Figure 7). Water vapor could provide a small but noteworthy source of water for plants during the driest period of the year [45]. Thus, simulations without heat flow did not capture the dynamics of pressure head changes that affect the rate and direction of water flow [15].

In semi-arid and arid regions, rainfall variations can lead to changes in water use strategy by plant species [11]. This study confirmed that an increase in rainfall could lead to an increased use of rain water and decreased groundwater use (Figure 13) during the growing season by the *Salix* bush. Nevertheless, quantifying groundwater use is another issue. The contribution of groundwater to vegetation water use varies by crop [46], grass [47], and trees [39] according to the Richards equation. However, these studies did not consider the effect of thermal driving flow. Based on our simulation results, the groundwater contribution to ET is different depending on heat flow condition (Figure 10).

This study is limited to thermally driven water movement on water use of *S. psammophila*. However, water use of ecological system includes competition and coordination mechanism. To evaluate vegetation competition for water and light limitation, Brolsma, *et al.* [48,49] created a coupled bio-physical-vegetation growth and variably saturated three-dimensional hydrological model to simulate both short-term and long-term vegetation dynamics along a hill slope in temperate forests. It found that once a tree is established under slightly unfavorable soil moisture conditions it cannot be outcompeted by smaller trees with better soil moisture uptake capabilities, both in dry as in wet conditions. To investigate water flow within and around a root network, Schneider, *et al.* [50] proposed a standalone root water uptake model to investigate the role of root architecture on the spatial distribution of root water uptake. Model simulations show a redistribution of water uptake from more densely to less densely rooted layers with time. Water use of the investigated bush vegetation system may involve both the competition and coordination mechanism referring to the bush ecological consistent as described in section 2.1. Moreover, Specific ecological features could also impact water-use strategies as point by Bertrand, *et al.* [51]. Therefore, these vegetation water use strategies should be taken into account in further modeling efforts.

5. Conclusions

A Hydrus-1D model was calibrated and validated using hourly field measurements of soil water content; the temperature over a period of 154 days; and liquid water, water vapor and heat transport. The soil water content and temperatures computed by the model fit well with the empirical values at all depths. The model was used to investigate the effects of heat flow on the soil water flux and the groundwater contribution to evapotranspiration by the *Salix* bush. The main conclusions of the study can be summarized as follows:

1. The mean pressure head over the root zone when heat flow is included is smaller than when heat flow is not included on dry days. This means that simulated heat transport varied with temperature and pressure head gradients, suggesting a mechanism for moisture redistribution within the root zone.
2. The zero flux interface of thermally driving water vapor flux varies daily and is affected seasonally by temperature gradients. During the summer, water vapor is condensed in shallow soil layers. However, the zero flux interfaces are formed in both shallow and deeper soil layers in the autumn. This will help us understand the pattern of water vapor movement over space and time.
3. In semi-arid areas, the water use of the *Salix* bush depends on rainfall infiltration. During the driest period, more groundwater is used for transpiration.
4. The percentages of groundwater contribution to *ET* were 26.9% and 40.6% with heat and without heat flow over the course of the 154-day simulation, respectively. Therefore, the groundwater contribution will be overestimated when thermally driven water vapor flow is not taken into account.

Acknowledgments: This research was supported by the National Natural Science Foundation of China (41102160, 51209064), the Ministry of Land and Resources of the People's Republic of China (201311076), and the Natural Science Foundation of Shaanxi Province (2014JQ5187, 2015KJXX-69). The authors are grateful to the reviewers; the comments and suggestions of the reviewers have contributed significantly to the improvement of the manuscript.

Author Contributions: Yangxiao Zhou, Jochen Wenninger and Jinting Huang conceived and designed the experiments; Jinting Huang performed the experiments; Yangxiao Zhou and Jinting Huang analyzed the data; Yangxiao Zhou, Rongze Hou and Jinting Huang wrote the paper.

Conflicts of Interest: The authors declare no conflict of interest.

References

1. Banimahd, S.A.; Zand-Parsa, S. Simulation of evaporation, coupled liquid water, water vapor and heat transport through the soil medium. *Agric. Water Manag.* **2013**, *130*, 168–177.
2. Li, X.; Zhang, Z.; Huang, L.; Wang, X. Review of the ecohydrological processes and feedback mechanisms controlling sand-binding vegetation systems in sandy desert regions of China. *Chin. Sci. Bull.* **2013**, *58*, 1483–1496.
3. Zhang, L.X.; Song, Y.Q. Efficiency of the three-north forest shelterbelt program. *Acta Scicentiarum Nat. Univ. Pekinesis* **2003**, *39*, 594–600. (In Chinese)

4. Wang, B.W.; Yang, Q.K.; Liu, Z.H. Effect of conversion of farmland to forest or grass land on soil erosion intensity changes in Yanhe River Basin, Loess Plateau of China. *Front. For. China* **2009**, *4*, 68–74.
5. Zhou, Y.; Wenninger, Z.; Yang, L.; Yin, L.; Huang, J.; Hou, L.; Wang, X.; Zhang, D.; Uhlenbrook, S. Groundwater-surface water interactions, vegetation dependencies and implications for water resources management in the semi-arid Hailiutu River catchment, China—A synthesis. *Hydrol. Earth Syst. Sci.* **2013**, *17*, 2453–2477.
6. Cao, S.X.; Chen, L.; Shankman, D.; Wang, C.M.; Wang, X.B.; Zhang, H. Excessive reliance on afforestation in China's arid and semi-arid regions: Lessons in ecological restoration. *Earth Sci. Rev.* **2011**, *104*, 240–245.
7. Li, X.R.; Tian, F.; Jia, R.L.; Zhang, Z.S.; Liu, L.C. Do biological soil crusts determine vegetation changes in sandy deserts? Implications for managing artificial vegetation. *Hydrol. Process.* **2010**, *24*, 3621–3630.
8. Schwinning, S.; Sala, O.E. Hierarchy of responses to resource pulses in arid and semi-arid ecosystems. *Oecologia* **2004**, *141*, 211–220.
9. Zhao, W.; Liu, B. The response of sap flow in shrubs to rainfall pulses in the desert region of China. *Agric. For. Meteorol.* **2010**, *150*, 1297–1306.
10. Williams, D.G.; Ehleringer, J.R. Intra- and interspecific variation for summer precipitation use in pinyon-juniper woodlands. *Ecol. Monogr.* **2000**, *70*, 517–537.
11. Kray, J.A.; Cooper, D.J.; Sanderson, J.S. Groundwater use by native plants in response to changes in precipitation in an intermountain basin. *J. Arid Environ.* **2012**, *83*, 25–34.
12. Stromberg, J.C.; Tiller, R.; Richter, B. Effects of groundwater decline on riparian vegetation of semiarid regions: The San Pedro, Arizona. *Ecol. Appl.* **1996**, *6*, 113–131.
13. Grantz, K.; Rajagopalan, B.; Clark, M.; Zagona, E. Seasonal shifts in the North American Monsoon. *J. Clim.* **2007**, *20*, 1923–1935.
14. Chimner, R.A.; Cooper, D.J. Using stable oxygen isotopes to quantify the water source used for transpiration by native shrubs in the San Luis Valley, Colorado USA. *Plant Soil* **2004**, *260*, 225–236.
15. Garcia, C.A.; Andraski, B.J.; Stonestrom, D.A.; Cooper, C.A.; Simunek, J.; Wheatcraft, S.W. Interacting vegetative andthermal contributions to water movement in desert soil. *Vadose Zone J.* **2011**, *10*, 552–564.
16. Philip, J.R.; de Vries, D.A. Moisture movement in porous materials under temperature gradients. *Transactions* **1957**, *38*, 222–231.
17. Saito, H.; Simunek, J.; Mohanty, B.P. Numerical analysis of coupled water, vapor, and heat transport in the vadose zone. *Vadose Zone J.* **2006**, *5*, 784–800.
18. Zeng, Y.J.; Su, Z.B.; Wan, L.; Wen, J. A simulation analysis of the advective effect on evaporation using a two-phase heat and mass flow model. *Water Resour. Res.* **2011**, *47*, 582–582.
19. Saito, H.; Šimůnek, J. Effect of meteorological models on the solutions of the surface energy balance and soil temperature variations in bare soils. *J. Hydrol.* **2009**, *373*, 545–561.
20. Yin, J.; Young, M.H.; Yu, Z. Effects of paleoclimate and time-varying canopy structures on paleowater fluxes. *J. Geophys. Res. Atmos.* **2008**, *113*, 304–312.

21. Šimůnek, J.; Šejna, M.; Saito, H.; Sakai, M.; van Genuchten, M.T. *HYDRUS software*; Version 4.15; The HYDRUS-1D software package for simulating the movement of water, heat, and multiple solutes in variably saturated media; Department of Environment Science, University of California, Riverside: Riverside, CA, USA, 2012.
22. Yang, Z.; Zhou, Y.X.; Wenninger, J.; Uhlenbrook, S. The causes of flow regime shifts in the semi-arid Hailiutu River, Northwest China. *Hydrol. Earth Syst. Sci.* **2012**, *16*, 87–103.
23. Huang, J.T.; Zhou, Y.X.; Yin, L.H.; Wenninger, J.; Zhang, J.; Hou, G.C.; Zhang, E.Y.; Uhlenbrook, S. Climatic controls on sap flow dynamics and used water sources of *Salix psammophila* in a semi-arid environment in northwest China. *Environ. Earth Sci.* **2015**, *73*, 289–301.
24. gvSIG. The Instruction of gvSIG. Available online: www.gvsig.org (accessed on 9 December 2015).
25. Sutanto, J.S.; Wenninger, J.; Coenders-Gerrits, A.M.J.; Uhlenbrook, S. Partitioning of evaporation into transpiration, soil evaporation and interception: A comparison between isotope measurements and a hydrus-1d model. *Hydrol. Earth Syst. Sci.* **2012**, *16*, 2605–2616.
26. Feddes, R.A.; Kowalik, P.J.; Zarandy, H. *Simulation of Field Water Use and Crop Yield*; John Wiley & Sons: New York, NY, USA, 1978.
27. Van Genuchten, M.T. A closed-form equation for predicting the hydraulic conductivity of unsaturated soils. *Soil Sci. Soc. Am. J.* **1980**, *44*, 892–898.
28. De Marsily, G. *Quantitative Hydrogeology*; Academic Press: London, UK, 1986.
29. Chung, S.O.; Horton, R. Soil heat and water flow with a partial surface mulch. *Water Resour. Res.* **1987**, *23*, 2175–2186.
30. Van Genuchten, M.T. *A Numerical Model for Water and Solute Movement in and below the Root Zone*; USA Salinity laboratory, United states Department of Agriculture (USDA), Agriculture Research Service (ARS): Riverside, CA, USA, 1987.
31. Liu, S.; Graham, W.D.; Jacobs, J.M. Daily potential evapotranspiration and diurnal climate forcings: Influence on the numerical modeling of soil water dynamics and evapotranspiration. *J. Hydrol.* **2005**, *309*, 39–52.
32. Willmott, C.J. On the validation of models. *Phys. Geogr.* **1981**, *2*, 184–194.
33. Willmott, C.J. Some comments on the evaluation of model performance. *Bull. Am. Meteorol. Soc.* **1982**, *63*, 1309–1313.
34. Wang, Y.W. Evolution of Surface Water and Vegetation in Inner Mongaolia Energy Base of Erodos Plateau. Master's Thesis, China University of Geoscience, Beijing, China, 25 May 2009. (In Chinese).
35. Oosterbaan, R.J.; Nijland, H.J. Determing the Saturated Hydraulic Conductivity. In *Drainage Principles and Applications*; International Institute for Land Reclamation and Improvement: Wageningen, The Netherlands, 1994.
36. Schaap, M.G.; Leij, F.J.; van Genuchten, M.T. ROSETTA: A computer program for estimating soil hydraulic properties with hierarchical pedotransfer functions. *J. Hydrol.* **2001**, *251*, 163–176.

37. Martinez, J.J.; Skaggs, T.H.; van Genuchten, M.T.; Candela, L. A root zone modeling approach to estimating groundwater recharge from irrigated areas. *J. Hydrol.* **2009**, *367*, 138–149.
38. Skaggs, T.H.; Shouse, P.J.; Poss, J.A. Irrigating forage crops with saline waters: 2. Modeling root uptake and drainage. *Vadose Zone J.* **2006**, *5*, 824–837.
39. Zhu, Y.; Ren, L.; Skaggs, T.H.; Lu, H.; Yu, Z.; Wu, Y.; Fang, X. Simulation of *Populuseuphratica* root uptake of groundwater in an arid woodland of the Ejina Basin, China. *Hydrol. Process.* **2009**, *23*, 2460–2469.
40. Shah, N.; Nachabe, M.; Ross, M. Extinction depth and evapotranspiration from groundwater under selected land covers. *Ground Water* **2007**, *45*, 329–338.
41. Zeng, Y.; Wan, L.; Su, Z.; Saito, H.; Huang, K.; Wang, X. Diurnal soil water dynamics in the shallow vadose zone (field site of China university of geosciences, China). *Environ. Geol.* **2009**, *58*, 11–23.
42. Anthony, T.C.; Marc, B.P. On water vapor transport in field soils. *Water Resour. Res.* **1998**, *34*, 731–739.
43. Parlange, M.B.; Cahill, A.T.; Nielsen, D.R.; Hopmans, J.W.; Wendroth, O. Review of heat and water movement in field soils. *Soil Tillage Res.* **1998**, *47*, 5–10.
44. Deb, S.K.; Shukla, M.K.; Mexal, J.G.; Sharma, P. Coupled liquid water, water vapor, and heat transport simulations in an unsaturated zone of a sandy loam field. *Soil Sci.* **2011**, *176*, 387–398.
45. Stewart, J.B. Evaporation from the wet canopy of a pine forest. *Water Resour. Res.* **1977**, *13*, 915–921.
46. Karimov, A.K.; Šimůnek, J.; Hanjra, M.A.; Avliyakulov, M.; Forkutsa, I. Effects of the shallow water table on water use of winter wheat and ecosystem health: Implications for unlocking the potential of groundwater in the Fergana valley (central Asia). *Agric. Water Manag.* **2014**, *131*, 57–69.
47. Soylu, M.E.; Istanbulluoglu, E.; Lenters, J.D.; Wang, T. Quantifying the impact of groundwater depth on evapotranspiration in a semi-arid grassland region. *Hydrol. Earth Syst. Sci.* **2011**, *15*, 787–806.
48. Brolsma, R.J.; Karssenberg, D.; Bierkens, M.F.P. Vegetation competition model for water and light limitation. I: Model description, one-dimensional competition and the influence of groundwater. *Ecol. Model.* **2010**, *221*, 1348–1363.
49. Brolsma, R.J.; Karssenberg, D.; Bierkens, M.F.P. Vegetation competition model for water and light limitation. II: Spatial dynamics of groundwater and vegetation. *Ecol. Model.* **2010**, *221*, 1364–1377.
50. Schneider, C.L.; Attinger, S.; Delfs, J.O.; Hildebrandt, A. Implementing small scale processes at the soil-plant interface-the role of root architectures for calculating root water uptake profiles. *Hydrol. Earth Syst. Sci. Discuss.* **2009**, *6*, 4233–4264.
51. Bertrand, G.; Masini, J.; Goldscheider, N.; Meeks, J.; Lavastre, V.; Celle-Jeanton, H.; Gobat, J.; Hunkeler, D. Determination of spatiotemporal variability of tree water uptake using stable isotopes ($\delta^{18}O$, $\delta^{2}H$) in an alluvial system supplied by a high altitude watershed, Pfyn forest, Switzerland. *Ecohydrology* **2014**, *7*, 319–333.

Method of Relating Grain Size Distribution to Hydraulic Conductivity in Dune Sands to Assist in Assessing Managed Aquifer Recharge Projects: Wadi Khulays Dune Field, Western Saudi Arabia

Oliver M. Lopez, Khan Z. Jadoon and Thomas M. Missimer

Abstract: Planning for use of a dune field aquifer for managed aquifer recharge (MAR) requires that hydraulic properties need to be estimated over a large geographic area. Saturated hydraulic conductivity of dune sands is commonly estimated from grain size distribution data by employing some type of empirical equation. Over 50 samples from the Wadi Khulays dune field in Western Saudi Arabia were collected and the grain size distribution, porosity, and hydraulic conductivity were measured. An evaluation of 20 existing empirical equations showed a generally high degree of error in the predicted compared to the measured hydraulic conductivity values of these samples. Statistical analyses comparing estimated *versus* measured hydraulic conductivity demonstrated that there is a significant relationship between hydraulic conductivity and mud percentage (and skewness). The modified Beyer equation, which showed a generally low prediction error, was modified by adding a second term fitting parameter related to the mud concentration based on 25 of the 50 samples analyzed. An inverse optimization process was conducted to quantify the fitting parameter and a new empirical equation was developed. This equation was tested against the remaining 25 samples analyzed and produced an estimated saturated hydraulic conductivity with the lowest error of any empirical equation. This methodology can be used for large dune field hydraulic conductivity estimation and reduce planning costs for MAR systems.

Reprinted from *Water*. Cite as: Lopez, O.M.; Jadoon, K.Z.; Missimer, T.M. Method of Relating Grain Size Distribution to Hydraulic Conductivity in Dune Sands to Assist in Assessing Managed Aquifer Recharge Projects: Wadi Khulays Dune Field, Western Saudi Arabia. *Water* **2015**, *7*, 6411–6426.

1. Introduction

Within the context of water management, dune aquifers have been used to store, treat and recover water for more than 60 years [1–7]. In arid lands, rainfall events are sparse and commonly produce flash flood events that are not conductive to recharge of most aquifers (e.g., wadi aquifers) [8]. Dune aquifers are known to be effective at allowing natural recharge [9–13]. Increasingly, dune aquifers are being considered for use in storing captured storm water or treated wastewater for future use [14]. Therefore, there is a need to understand the hydraulic properties of dune aquifers to assess water storage and recovery potential using the concept of aquifer storage and recovery (ASR).

Dunes are an important part of the landscape in arid regions, especially within the desert areas of Australia, the Sahara, and Arabia, where dune field cover constitutes 31%, 28%, and 26%, respectively, of the total land area [15]. Investigations of dune fields have been conducted to assess the physics of wind-transported sediment mechanisms [16], the sediment grain size properties [17–19], and the movement of the dunes because of geohazard potential [20]. Little consideration has been given to assessment of the sediment hydraulic properties in terms of hydraulic conductivity (saturated), porosity, and specific yield [21–23].

Empirical equations are commonly used to estimate saturated hydraulic conductivity from grain size data in various sediment types to populate grids in models or to set limits for the maximum rate of recharge for unsaturated zone flow [24–26]. There are over 30 published empirical equations commonly used to perform this task with each one having a different set of assumptions for its use. Unfortunately, many of these equations were developed using very limited databases with the comparative number of measured grain size distributions analyses and hydraulic conductivity measurements being limited to a population as small as 20. Considerable improvements to the accuracy of these empirical equations for analysis of specific depositional environments have been made recently [26]. Herein, we analyze the use of the inversion mathematical method to calibrate a new equation for relating the grain size distribution to the hydraulic conductivity of interior dune sediment at a specific location and compare the developed equation to the improved method of Rosas *et al.* [26].

In this research, the grain size distribution, porosity, and hydraulic conductivity variations in an interior dune field are measured in an effort to (1) provide a characterization of these parameters along the cross-section of a dune, and (2) develop an improved empirical relationship between grain size distribution and hydraulic conductivity for this dune field, using a mathematical inversion method, (3) compare the developed equation to other empirical equations used to estimate interior dune field hydraulic conductivity, and (4) investigate the impact of the mud percentage (silt and clay) fraction on the hydraulic conductivity. Furthermore, a calibration procedure was proposed to estimate the fitting parameter of the developed empirical relationship by considering twenty five samples collected in the Khulays dune field, then applying the equation to prediction of another set of 25 samples for assessment of predictive error. This methodology could be applied to various other depositional environments to facilitate estimates of hydraulic conductivity without having to measure it in a large numbers of samples in the laboratory. This research applies to prediction of saturated hydraulic conductivity. It is clearly understood that the packing of the dune sediment has an impact on the saturated hydraulic conductivity and that the empirical equations do not take this into consideration, so the calculated values can be assessed to be similar to dune sands located within 10 to 20 m of the surface, which are not well compacted.

Four research hypotheses where developed for testing and are: (1) most empirical equations currently used to estimate saturated hydraulic conductivity from grain size distribution of dune sands have large errors; (2) a significant statistical correlation exists between the percentage of mud (silt and clay) in dune sand and the hydraulic conductivity; (3) a significant statistical relationship exists between saturated hydraulic conductivity and the dispersion (and other statistical moments) in interior dune sands, and (4) inversion modeling can be applied to calibrate a field specific empirical equation relating saturated hydraulic conductivity and grain size distribution in interior dune sands.

2. Background and Methods

2.1. Site Description

The Wadi Khulays aquifer system, located about 50 km northeast of Jeddah, Saudi Arabia (Figure 1), consists of several valleys that drain from the western slope of the Arabian Shield into the Wadi Khulays plain near the Red Sea coast [27]. A dune field was formed by the prevailing northwesterly wind transporting alluvial deposits from the northern sheet wash and Wadi plains of Wadi Khulays to the south side, centered at about 22°6′ N, 39°14′ E [28,29]. Dune heights range from 3.5 to 8 m, depending on the location within the field. The dune sediment is predominantly quartz sand with some other siliciclastic minerals. Dune sediments lie on desert pavement that dips slightly westward and contains a significant percentage of mud and has a low hydraulic conductivity. Moisture was found at the base of the dune sediments. No significant vegetation occurs within the dune field.

2.2. Field Study and Laboratory Methods

The sediments were evaluated using standard laboratory methods for the determination of hydraulic conductivity, grain size distribution and porosity. First, samples were collected from the upper 5 to 10 cm across several dunes at five locations beginning at the toe and ending downwind of the avalanche face within the inter-dune area (Figure 2A). The dunes sampled occurred in two 800 m long transects (green markers on Figure 2B,C). A total of 50 samples were collected for full analysis. An additional 24 samples were collected from dune crests covering a surface area of about 37,500 m^2 and were analyzed only for grain size distribution (Figure 2B, blue markers).

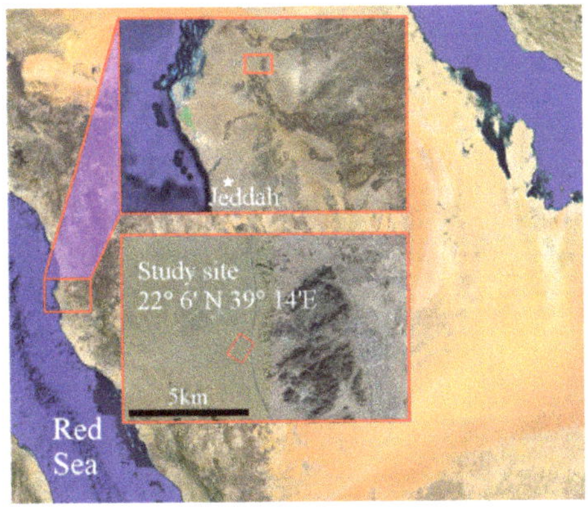

Figure 1. Map of Saudi Arabia showing the location of the dune field site.

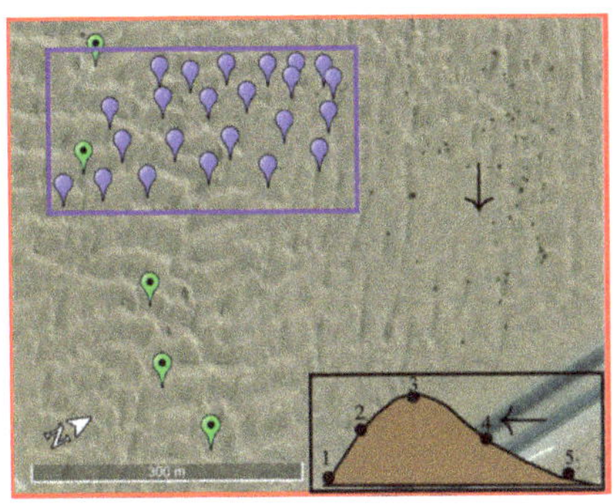

Figure 2. Wadi Khulays dune field sample locations (black arrows show the predominant wind direction). Green markers show the location of dunes selected for full sampling (analysis of grain size, porosity and hydraulic conductivity across the dune). Blue markers show a second sampling for grain size analysis only from dune crests. The blue region corresponds to the analysis shown in Figure 8. Inset: Diagram showing the sample location across individual dunes.

The grain size distribution was determined using the standard sieving technique as described by Tanner and Balsillie [29] and the American Society of Testing and Materials [30]. About 70 g of sample were analyzed using 34 standard sieves with size increments corresponding to 0.25-phi units (phi is equal to −log to the base 2 in mm). The amount of mud (silt and clay) in a sample was determined as the weight of the sample left in the pan after passing through all of the sieves with the smallest being 0.0063 mm. The sieving process involved the continuous rotation and tapping of the sieves by the use of a Ro-Tap machine for 30 min as recommended by Tanner and Balsillie [29]. The weight retained within each sieve was measured to an accuracy of 0.01g and the statistical moments were calculated following the approach described in Tanner and Balsillie [29], using custom software developed by Rosas *et al.* [26].

The hydraulic conductivity was measured using a standard constant head permeameter based on the methodology described by Wenzel [31] and following the American Society for Testing and Materials standard D2434-68 [32]. A constant head was applied to an approximate 10 cm column filled with saturated sediment sample and water flow was from bottom to top through the cylinder. The hydraulic conductivity was obtained by applying Darcy's Law of flow through porous media and was reported in meters per day.

In order to obtain an estimate of total porosity of each sample, a 250 mL (cc) graduated cylinder was first filled partially with a known volume of water. Sediment was carefully added to the cylinder and allowed to settle and compact. Additional water and sediment were added to the column. Care was taken to not allow air entrapment within the saturated column. The column was also compacted slightly by tapping the side of the cylinder with a rubber mallet to approximate natural system packing conditions near surface. The volume of sediment was determined along with the volume of water added. When the water level in the cylinder rose above the surface of the sediment, a correction was made to the water volume added (subtraction). Then, the estimated porosity was determined by dividing the volume of water by the volume of sediment.

2.3. Mathematical Analysis

Among the 21 empirical equations that estimate the hydraulic conductivity from the grain size distribution, most of them use a relationship in the form shown in Equation (1), where n is the porosity and β is a coefficient that may depend on the shape of the grain, type of depositional environment or other factors, d_{10} is the effective grain diameter (at which 10% of the sample is finer), g is the gravitational acceleration and ν is the kinematic viscosity of water.

$$K = \beta \frac{g}{\nu} \xi(n)\, d_{10}^2 \tag{1}$$

Rosas *et al.* [26] compared the effectiveness of these equations for different depositional environments, and provided depositional environment-dependent β coefficients, as well as offset values (bias) for beach sands, offshore marine sands, and dune sands (coastal and interior), as well as for river sediments. In this study, the four equations that showed the best fit for interior dune sands, as recommended in Rosas *et al.* [26], were selected as target equations to be modified by the inclusion of other grain size distribution parameters, based on the significance of their correlation with the hydraulic conductivity values. The inclusion was done by adding a second term to the target equations, which contains the mud percentage and a fitting parameter. The inclusion of a second term in the Beyer Equation [33] is shown below:

$$K\left[\frac{m}{day}\right] = P_1\frac{g}{v}log\left[\frac{500}{C}\right] + P_2exp\left(-mud\right) \quad (2)$$

where C is the coefficient of uniformity defined as $C = d_{60}/d_{10}$ (d_{60} is the diameter at which 60% of the sample is finer), P_1 and P_2 are the fitting parameters to be optimized. The inverse problem was formulated in the sense of root mean square error and the objectivity function was accordingly defined as:

$$\min \varnothing\left(\mathbf{b}\right) = \sqrt{\frac{\sum_i(K\left(i\right) - K^*\left(i,\mathbf{b}\right)^2}{n}} \quad (3)$$

where $K\left(i\right)$ and $K^*\left(i,\mathbf{b}\right)$ are the vectors containing the measured and modeled hydraulic conductivity, respectively, $\mathbf{b} = [P_1,P_2]$ is the parameter vector to be estimated, and, n is the number of samples. An inversion the first twenty-five measurements was used to estimate the fitting parameters P_1 and P_2. To properly deal with the nonlinearity of the objective function during minimization and, in particular, avoid being trapped in local minima, we adopted a global optimization approach. Hence, we used the shuffled complex evolution (SCE-UA) algorithm for the global optimization. This global optimization algorithm has been widely used in hydrologic modeling [34], and proved to be consistent and efficient for finding the global optimum parameter values of integrated hydrogeophysical and hydrologic models [35,36].

3. Results

3.1. *Grain Size, Porosity, and Hydraulic Conductivity Analyses*

Measured grain size statistical parameters, d_{10}, mud percentage, measured porosity and measured hydraulic conductivity of the 50 samples collected across the individual windward and leeward dunes are summarized in Figure 3. The grain size statistical properties d_{10}, mud (%), standard deviation, skewness, and kurtosis, and the measured hydraulic conductivity vary in a regular pattern across the dune profile, while mean grain diameter does not vary in the same pattern (Figure 3). The dunes crests have significantly higher hydraulic conductivities and d_{10} values, corresponding to the lowest percentage of mud content. Porosity of the dune sands was found to be relatively uniform with a mean of 0.39 and a standard deviation of 0.02. The red-cross marks on the figure show the location of outlier samples.

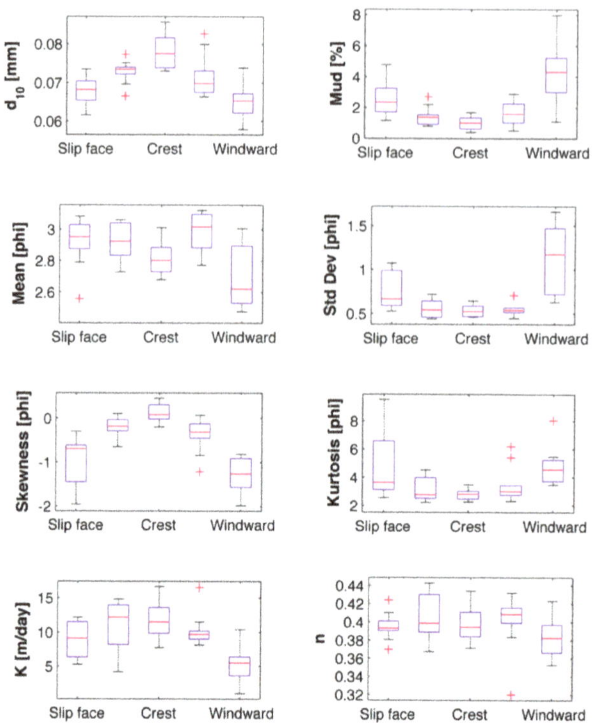

Figure 3. Summary of grain size and hydraulic properties along the cross-section of the dunes. Red crosses are outlier samples.

3.2. Empirical Hydraulic Conductivity Estimates from Grain Size Distribution

An analysis of 21 empirical methods used to estimate hydraulic conductivity from grain size analysis shows that there is a large error between measured and estimated values using most of the methods (Figure 4A) [37–51]. Each of the 21 methods contains somewhat different assumptions and specific equations should be applied to specific recommended sediment types, but many of these methods have not been used properly in numerous investigations, particularly by practitioners in the engineering field. It is evident in viewing this graphic that the estimate accuracy is based on the assumptions of each equation and perhaps on the properties of the sediments at the specific location being investigated. Equations for the listed methods and assumptions for use are contained in Rosas *et al.* [26]. The Beyer, Harleman, Hazen (modified) and Kozeny-Carman methods showed the best fit for hydraulic conductivity of interior dune sands in this analysis. Coefficients used by these methods were modified for dune sands in Rosas *et al.* [26] and showed a significant improvement (Figure 4B). As shown, there is still room for additional improvement of these modified equations.

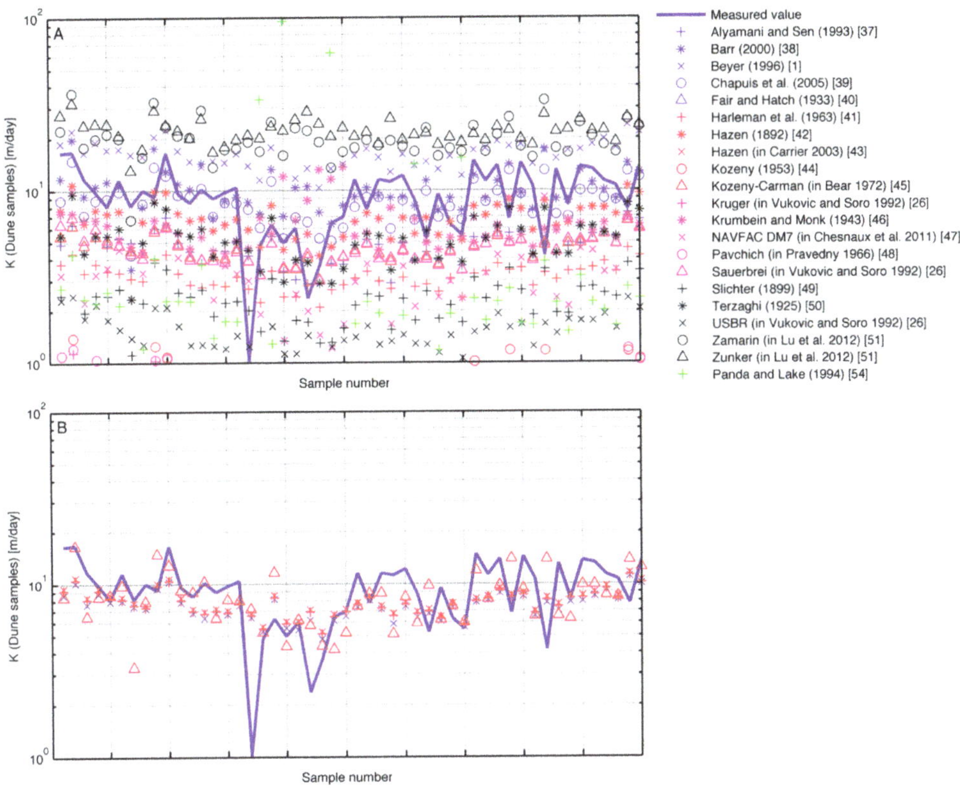

Figure 4. Estimated hydraulic conductivity using 21 empirical methods for all Wadi Khulays dune sample compared to the measured values (blue line) (**A**). (**B**) Comparison of the estimated hydraulic conductivities to the measured values using the modified methods for dune interior dune sands using the modified equations in Rosas *et al.* [26].

3.3. Statistical Analyses between Sediment Grain Size Properties and Hydraulic Conductivity

Linear relationships between the statistical moments and other properties of the grain size distribution *vs.* measured are shown as plots in Figure 5. The correlation coefficients R^2 and p are shown in Table 1. The p-values are reported as $-\log(p)$ in order to compare the order of magnitude of the correlations, with a significant correlation being considered to have a value of $-\log(p)$ of 2 or more ($p < 0.01$). Even though the R^2 values were very low for all the parameters considered ($R^2 < 0.4$), significant p-values were found for the correlation between the hydraulic conductivity and d_{10}^2, the standard deviation, skewness and kurtosis of the distribution in phi units ($-\log_2$ mm; [52]), as well as for the mud percentage of the sample (fraction <0.0625 mm).

Table 1. Correlation coefficients between hydraulic conductivity and grain size distribution parameters.

Correlated Parameters	R^2	$-\log(p\text{-Value})$
K, d_{10}^2	0.34	4.9
K, mean	0.04	0.8
K, standard deviation	0.30	4.4
K, skewness	0.39	6.0
K, kurtosis	0.22	3.2
K, mud	0.37	5.6
Mud, mean	0.16	2.4
Mud, standard deviation	0.72	14.2
Mud, skewness	0.45	7.1
Mud, kurtosis	0.11	1.7
K, $F = d_{10}^2 \exp(-Mud)$	0.56	4.8

The mud percentage in the samples was positively correlated with the standard deviation (R^2 = 0.72) and negatively correlated with the skewness of the distribution (R^2 = 0.45) (Figure 5).

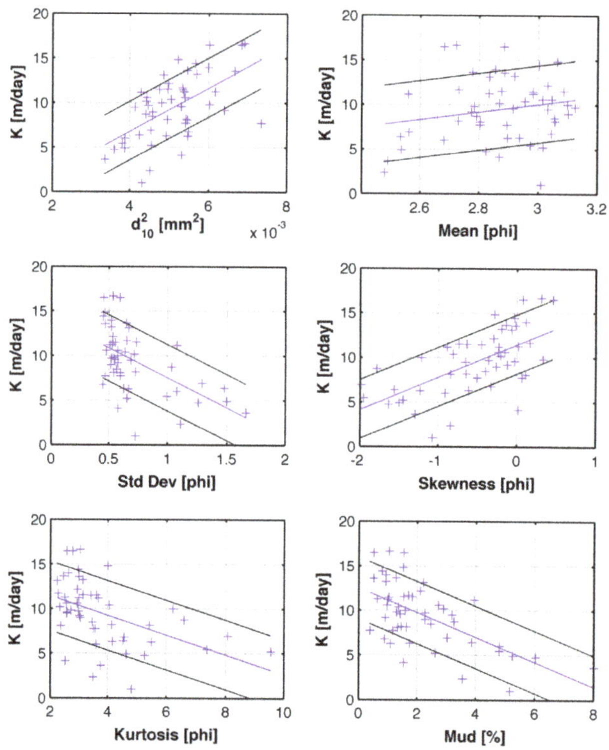

Figure 5. Relationship between mud percentage and other grain size distribution statistical parameters.

3.4. Optimization of the Target Equations by Inclusion of a Second Parameter

The percentage of mud was used for development of the fitting parameters (P_1 and P_2) for Equation (2). Figure 6A,B shows the convergence of the P_1 and P_2 parameters, respectively, to the global optimum values. For the inversion, the optimization parameter space was set to $-1 \times 10^{-2} \leqslant P_1 \leqslant 1 \times 10^{-2}$ and $-10 \leqslant P_2 \leqslant 10$. In the first 180 iterations the optimization algorithm searched different parameters set in the provided parameter space and later converged to the optimum values. The stopping criterion for the inversion was specified with respect to the convergence criterion in which an optimum was assumed to be reached when the objective function did not improve by more than 0.01% in 10 successive evolution loops. The estimated parameters P_1 and P_2 were 4.0×10^{-4} and -0.595.

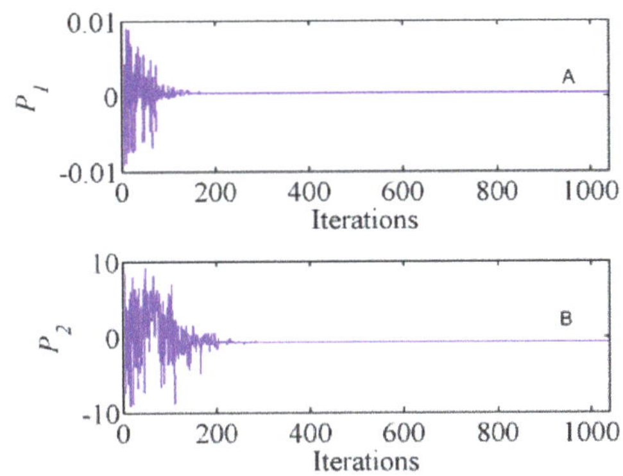

Figure 6. Convergence in the optimization of P_1 and P_2.

Figure 7 shows the response surfaces of the objective functions for the P_1 and P_2 parameter plane. The range of each parameter has been divided into 500 discrete values resulting in 250,000 objective function calculations for the plot. The white star marker corresponds to the solution found by the SCE optimization algorithm. Response surface analysis of the objective function is important as it provides valuable insights into the uniqueness of the inverse solution, the sensitivity of the model to the different parameters, and parameter correlations. The topography of the objective function shows that the parameters P_1 and P_2 are not correlated and a U-valley can be observed in the direction of the P_2 parameter. Furthermore, in the direction of parameter P_1, the V-shape valley of decreasing and increasing trend of the objective function shows that the objective function is more sensitive to the parameter P_1 as compared to the P_2.

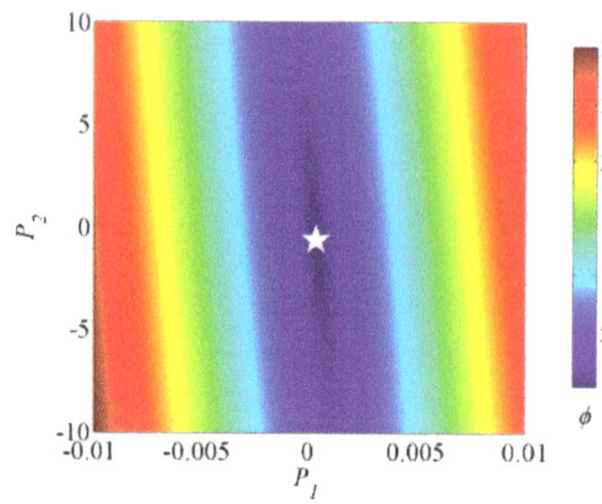

Figure 7. Response surfaces of the objective functions for P_1 and P_2 parameter plane.

Table 2 shows a comparison of the R-square and RMSE values obtained for the four best fitted empirical equations for interior dune sands: Beyer [33], Harleman *et al.* [41], Hazen [42] and Kozeny-Carmen [45], and the new equation by including mud (%) and two fitting parameters. The fitting parameters were estimated by inversion using first 25 measurements and for the rest of the measurements model hydraulic conductivity was estimated by using the estimated fitting parameters in the new equation. Values of R^2 and RMSE indicate that the new equation estimates modeled hydraulic conductivity close to the measured hydraulic conductivity. The RMSE improvement is calculated with respect to Equation (2).

Table 2. Performance of empirical equations.

Equation	R^2	RMSE (m/day)	RMSE Improvement (%)
Beyer	0.60	3.04	25%
Harleman	0.59	3.01	23%
Hazen	0.59	3.03	24%
Kozeny-Carmen	0.36	3.38	38%
Equation (2)	0.61	2.44	

4. Discussion

The patterns along the cross-section of the dunes were similar for the d_{10} value, the mud percentage and the hydraulic conductivities; dune crests showed higher hydraulic conductivities and d_{10} values, as well the lowest mud content, which were progressively lower (hydraulic conductivity and d_{10}) and higher (mud percentage) towards the extremities of the dune.

Since the dunes are constantly being turned over by physical processes (wind action and gravity), the patterns reveal the vertical distribution of the properties inside the dune. Sediments within the interdune areas were poorly sorted as indicated by a higher standard deviation. Coarser gravelly grains tend to accumulate in these areas due to gravity, and these larger grains offer a protection for the finer particles against wind erosion. Most of the samples were negatively skewed except for a number of samples collected from the crests.

Most of the equations found in the literature for the determination of hydraulic conductivity from grain size distribution use the square of the d_{10} value and a function of the porosity [26]. The porosity of the dune sands from the Wadi Khulays dune field was determined to be relatively uniform and therefore the effect of porosity on the hydraulic conductivity could not be evaluated.

The skewness of the grain distribution, as well as the mud content showed a statistically significant correlation with the hydraulic conductivity. None of the equations used in the Rosas *et al.* [26] study make use of the skewness in the distribution. Panda and Lake [53] modified the Kozeny-Carman equation to include a corrective factor based on the skewness and coefficient of variation ($Cdp = \sigma/\overline{dp}$ where $\overline{dp}$ is the mean grain diameter) of the particle size distribution (psd). The equation was found to underestimate the hydraulic conductivity of the dune sands and is problematic for a number of samples with high standard deviation in the psd (Figure 4). Krumbein and Monk [46] use the standard deviation of the distribution in phi units, however, the estimated hydraulic conductivities by this method were shown to be ineffective in determining the values for dune sands (Figure 4; [26]). Moreover, evaluating the statistical moments of the grain size distribution is not a straightforward task. Because the mud percentage is significantly correlated with both the hydraulic conductivity values and the skewness of the distribution, it was selected as a proxy parameter that can be used to improve the effectiveness in determining the hydraulic conductivity from grain size distribution.

The ability to accurately predict the hydraulic conductivity of several samples from grain size distribution should significantly reduce the effort in the characterization of large areas, such as dune fields. In this study, the inclusion of the mud term significantly reduced the RMSE of the predicted hydraulic conductivity values. This allowed for the calculation of the spatial distribution of the hydraulic properties over a large area of the dune field by measuring only the grain size distribution parameters and applying Equation (2) (Figure 8). Since the porosity of the dune sands was relatively uniform, the function of the porosity was assumed to be constant. This approach may be used in large dune areas to determine the appropriate locations for water infiltration or can be correlated with the infiltration rate in dune recharge studies.

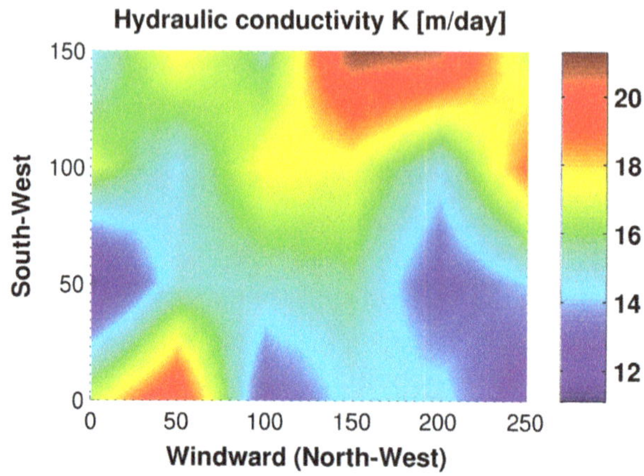

Figure 8. Spatial distribution of the hydraulic conductivity along the dune crests in a selected site within the dune field.

Since the samples used to develop the equation were relativity small in volume compared to the full dimensions of the dunes, the issue of scale needs to be addressed in that small samples commonly tend to underestimate the bulk hydraulic conductivity of an aquifer [54]. Larger scale measurement methods used to assess saturated aquifer systems, such as standard aquifer performance tests or tracer tests, almost always yield higher average hydraulic conductivity values based on the occurrence of bedding or diagenetic enhancement of hydraulic conductivity within an aquifer. While dune aquifers are not homogeneous and isotopic, they approach this condition better than almost any other type of aquifer, especially when located near or at land surface. Therefore, use of the new equation to determine the average hydraulic conductivity of a dune system will yield a reasonable estimate, perhaps on the low end of the actual average value. Within other types of aquifer systems, the error of using grain size distribution to estimate hydraulic conductivity becomes greater, especially as the heterogeneity of the sediment increases, as found by Rosas *et al.* [26].

5. Conclusions

Evaluating the hydraulic properties of dune sands is an important task in understanding recharge and in designing artificial storage and recovery systems in dune aquifer systems. The vertical distribution of these properties inside most dunes, except where there are internal mud layers, can be obtained by measuring the grain size distribution along the cross-section of the dune. Furthermore, there is interest in reducing the effort in estimating the hydraulic properties of dune sands over a large area. This can be achieved by the use of empirical relationships between grain size distribution parameters and the hydraulic conductivity.

Significant correlations between the hydraulic conductivity values of dune sands and their grain size distribution parameters were found, especially for the skewness of the distribution and the mud content, as well as with the d_{10} value, which is commonly used in empirical equations for determining the hydraulic conductivity. While the relationship of the hydraulic conductivity and the skewness is significant, its influence can be accounted for by the use of a proxy parameter, the mud content, which showed high correlations with both the skewness and hydraulic conductivity of the samples. Finally, the porosity of the dune sands is very uniform at a high porosity of 0.39, with a standard deviation of 0.02. Therefore, an accurate estimation of the hydraulic conductivity can be obtained by employing an empirical equation using the d_{10} value, the mud percentage of the sample and two optimized coefficients. This approach can be used to study the spatial distribution of dune sands in larger areas. Application of the new empirical equation developed from this research will improve estimates of hydraulic conductivity for interior dune fields, but may not be applied to coastal dune fields that tend to contain coarser sand, have a higher d_{10} value, and contain little or no mud.

The data and analysis confirms the first research hypothesis that large errors occur between estimated and measured saturated hydraulic conductivity using the published empirical equations. The second hypothesis concerning the statically significant relationship between the measured and estimated hydraulic conductivity and mud percentage is confirmed. The hypothesis concerning a significant statistical relationship between the hydraulic conductivity and the dispersion is not supported, but a statistically significant relationship with skewness was confirmed. A confirmation of the last hypothesis shows that the use of inversion modeling using the mud percentage with one of the existing empirical equations leads to improved estimation of hydraulic conductivity. The technique did make a slight reduction in error of prediction compared to the modified equations developed by Rosas *et al.* [26].

Detailed design of aquifer recharge systems in dune field settings does require that a number of field and laboratory tests should be performed to assess the saturated and unsaturated hydraulic conductivity (vertical and horizontal) using more sophisticated methods, such as double-ring infiltrometer tests and measurements made using undisturbed cores (if they can be collected) and aquifer performance tests where the sand is saturated. However, the relationship between grain size distribution and hydraulic conductivity can be used as a rapid scanning tool, to add data between detailed tests (populating a model grid), and for estimation in locations where large amounts of grain size data have been published (data mining and reanalysis). The estimation of saturated hydraulic conductivity from grain size analysis is therefore useful particularly in the planning of large-scale managed aquifer recharge projects, but cannot replace more detailed analytical methods used in design.

Acknowledgments: Funding for this research was provided by the Water Desalination and Reuse Center, King Abdullah University of Science and Technology and from discretionary faculty funding provided by the university.

Author Contributions: Oliver Lopez collected and analyzed the samples and worked on the manuscript. Khan Jadoon performed the mathematical analysis of the samples using inverse modeling. Thomas Missimer coordinated the research and wrote most of the text.

Conflicts of Interest: The authors declare no conflicts of interest.

References

1. Roosma, E.; Stakelbeek, A. Deep-well infiltration in north-Holland dune area. *Groundwater* **1990**, *28*, 778–782.
2. Martin, R.R.; Gerges, N.Z. Replenishment of an urban dune system by stormwater recharge. In Water Down Under 94: Groundwater Papers, Proceedings of the 25th Congress of the International Association of Hydrogeologists, Adelaide, Australia, 21 November 1994; Schonfeldt, C., Ed.; Institution of Engineers: Barton, Australia, 1994; pp. 21–25.
3. Van Breukelen, B.M.; Appleo, C.A.J.; Olsthhoorn, T.N. Hydrogeochemical transport modeling of 24 years of Rhine water infiltration in the dunes of the Amsterdam water supply. *J. Hydrol.* **1998**, *209*, 281–296.
4. Van Houtte, E.; Verbauwhede, J. Reuse of wastewater effluent for indirect production of drinking water in Flanders, Belgium. In Proceedings of the AWWA Membrane Technology Conference, Atlanta, GA, USA, 2–5 March 2003.
5. Dillion, P. Aquifer storage and recovery with stormwater and reclaimed water in Australia. In Proceedings of the UNESCO-VIETNAM Workshop on Augmenting Groundwater Resources by Artificial Recharge in South East Asia, Ho Chin Minh City, Vietnam, 15–17 December 2005; pp. 286–312.
6. Thoa, N.T.K.; Arduino, G.; Giang, N.V.; Van, P.T.K.; Vuong, B.T. The model for management of aquifer recharge in sand dune area of Bac Binh, Binh Thuan. *J. Earth Sci.* **2006**, *28*, 389–410.
7. Maliva, R.G.; Missimer, T.M. *Aquifer Storage and Recovery and Managed Aquifer Recharge: Planning, Hydrogeology, Design, and Operation*; Schlumberger Corporation: Houston, TX, USA, 2010.
8. Missimer, T.M.; Drewes, J.; Amy, G.; Maliva, R.G.; Keller, S. Restoration of wadi aquifers by artificial recharge with treated wastewater. *Groundwater* **2012**, *50*, 514–527.
9. Dincer, T.; Al-Muhgrin, A.; Zimmerman, U. Study of infiltration and recharge through sand dunes in arid zones with special reference to stable isotopes and thermonuclear tritium. *J. Hydrol.* **1974**, *23*, 79–109.

10. Allison, G.B.; Stone, W.J.; Hughes, M.W. Recharge in karst and dune elements of a semi-arid landscape as indicated by natural isotopes and chloride. *J. Hydrol.* **1985**, *76*, 1–25.
11. Winter, T.C. Effect of groundwater recharge on configuration of the water table beneath sand dunes and on seepage in lakes in the Sandhills of Nebraska. *J. Hydrol.* **1986**, *86*, 221–237.
12. Al-Hassoun, S.A.; Al-Turbak, A.S. Infiltration in two sand dune areas in Saudi Arabia. In Proceedings of the National Conference on Hydraulic Engineering and International Symposium on Engineering Hydrology, San Francisco, CA, USA, 25–30 July 1993; pp. 42–48.
13. Al-Hassoun, S.A. Simulation of moisture and recharge in sand dune using "SWIM" model. In Proceedings of the 20th IASTED International Conference on Modelling and Simulation (MS2009), Banff, AB, Canada, 6–8 July 2009; pp. 13–16.
14. Maliva, R.G.; Missimer, T.M. *Arid Lands Water Evaluation and Management*; Springer: Heidelberg, Germany, 2012.
15. Breed, C.S.; Fryberger, S.G.; Andrews, S.; McCauley, C.; Lennartz, F.; Gebel, D.; Horstman, K. Regional studies of sand seas using Lansat (ERTS) imagery. *A Study of Global Sand Sea*; U.S. Geological Survey Professional Paper 1052; McKee, E.D., Ed.; United States Government Printing Office: Washington, DC, USA, 1979; pp. 305–398.
16. Bagnold, R.A. *The Physics of Wind Blown Sand and Desert Dunes*; Dover Publications, Inc.: New York, NY, USA, 1954.
17. Ahlbrandt, T.S. Textural parameters of eolian deposits. *A Study of Global Sand Seas*; United States Geological Survey Professional Paper 1120-A; McKee, E.D., Ed.; United States Government Printing Office: Washington, DC, USA, 1979; pp. 21–51.
18. Lancaster, N. Grain size characteristics of linear dunes in the southwestern Kalahari. *J. Sediment. Pet.* **1986**, *56*, 395–400.
19. Lancaster, N. *Geomorphology of Desert Dunes*; Routledge: London, UK, 1995.
20. Al-Harthi, A.A. Geohazard assessment of sand dunes between Jeddah and Al-Lith, western Saudi Arabia. *Environ. Geol.* **2002**, *42*, 360–369.
21. Haruyuki, F.; Mitsuhiro, I.; Tahei, Y.; Takashi, T. Determining unsaturated hydraulic conductivity of dune sand a low pressure head from a steady-state evaporation experiment. *Sand Dune Res.* **1999**, *46*, 15–26.
22. Van den Berg, E.H.; de Vries, J.J. Influence of grain fabric and lamination on the anisotropy of hydraulic conductivity in unconsolidated dune sands. *J. Hydrol.* **2003**, *283*, 244–266.
23. Wang, T.; Zlotnik, V.A.; Wedin, D.; Wally, K.D. Spatial trends in saturated hydraulic conductivity of vegetated dunes in the Nebraska Sand Hills: Effects of depth and topography. *J. Hydrol.* **2008**, *349*, 88–97.
24. Kasenow, M. *Determination of Hydraulic Conductivity from Grain Size Analysis*; Water Resources Publications LLC: Littleton, CO, USA, 2002.
25. Vukovic, M.; Soro, A. Determination of hydraulic conductivity of porous media from grain-size composition. In *Determination of Hydraulic Conductivity from Grain Size Analysis*; Kasenow, M., Ed.; Water Resources Publications: Littleton, CO, USA, 1992.

26. Rosas, J.; Lopez, O.; Missimer, T.M.; Dehwah, A.H.A.; Sesler, K.; Coulibaly, K. Hydraulic conductivity and grain size distribution: Comparison of methods for different sedimentary depositional environments. *Groundwater* **2014**, *52*, 399–413.
27. Sagga, A.M.S. Barchan dunes of Wadi Khulays, western region of Saudi Arabia: Geomorphology and sedimentology relationships. *J. King Abdulaziz Univ. Earth Sci.* **1998**, *10*, 105–114.
28. Roobol, M.J.; Kadi, K.K. *Cenozoic Faulting and Traces of Tertiary Gabbro Dikes in the Rabigh Area*; Saudi Geological Survey Technical Report SGS-TR-2008-6 (geological map); Saudi Geological Survey: Jeddah, Saudi Arabia, 2007.
29. Tanner, W.F.; Balsillie, J.H. *Environmental Clastic Granulometry*; Florida Geological Survey Special Publication 40; Florida Geological Survey: Tallahassee, FL, USA, 1995.
30. American Society for Testing and Materials (ASTM). *Standard Test Method for Particle Size Analysis of Soils*; ASTM Standard D422-63; ASTM: West Conshohocken, PA, USA, 1995.
31. Wenzel, L.K. *Methods for Determining the Permeability of Water-Bearing Materials with Special Reference to Discharging-Well Methods*; U.S. Geological Survey Water-Supply Paper 887; United States Government Printing Office: Washington, DC, USA, 1942.
32. American Society for Testing and Materials (ASTM). *Standard Test Method for Permeability of Granular Soils*; ASTM Standard D2434-68; ASTM: West Conshohocken, PA, USA, 2006.
33. Beyer, W. Zur Bestimmung der Wasserdurchlassigkeit von Kieson und Sanduen aus derKornverteilung. *Wasserwirtsch.-Wassertech.* **1964**, *14*, 165–169. (In German)
34. Lambot, S.; Slob, E.C.; Rhebergen, J.; Lopera, O.; Jadoon, K.Z.; Vereecken, H. Remote estimation of the hydraulic properties of sand using full-waveform integrated hydrophysical inversion of time-lapse, off-ground GPR data. *Vad. Zone J.* **2009**, *8*, 743–754.
35. Jadoon, K.Z.; Lambot, S.; Slob, E.C.; Vereecken, H. Analysis of horn antenna transfer functions and phase-center position for modeling off-ground GPR. *IEEE Trans. Geosci. Remote Sens.* **2011**, *49*, 1649–1662.
36. Jadoon, K.Z.; Weihermuller, L.; Scharnagl, B.; Kowalsky, M.B.; Bechtold, M.; Hubbar, S.S.; Vereecken, H.; Lambot, S. Estimation of soil hydraulic conductivity parameters in the filed by integrated hydrogeophysical inversion of time-lapse ground-penetrating radar data. *Vad. Zone* **2012**, *11*.
37. Alyamani, M.S.; Sen, Z. Determination of hydraulic conductivity from complete grain-size distribution. *Groundwater* **1993**, *31*, 551–555.
38. Barr, D.W. Coefficient of permeability determined by measurable parameters. *Groundwater* **2001**, *39*, 356–361.
39. Chapuis, R.P.; Dallaire, V.; Marcotte, D.; Chouteau, M.; Acevedo, N.; Gagnon, F. Evaluating the hydraulic conductivity at three different scales within an unconfined sand aquifer at Lachenaie, Quebec. *Can. Geotech. J.* **2005**, *42*, 1212–1220.
40. Fair, G.M.; Hatch, L.P. Fundamental factors governing the stream-line flow of water through sand. *J. Am. Water Work. Assoc.* **1933**, *25*, 1551–1565.
41. Harleman, D.R.F.; Melhorn, P.F.; Rumer, R.R. Dispersion-permeability correlation in porous media. *J. Hydraul. Div. Am. Soc. Civ. Eng.* **1963**, *89*, 67–85.

42. Hazen, A. *Some Physical Properties of Sands and Gravels, with Special Reference to Their Use in Filtration*; 24th Annual Report; Massachusetts State Board of Health: Boston, MA, USA, 1892; pp. 539–556.
43. Carrier, W.D. Goodbye Hazen; Hello, Kozeny-Carman. *J. Geotech. Eng.* **2003**, *129*, 1054–1056.
44. Kozeny, J. Das wasser in boden, grundwasserbewegung. In *Hydraulik*; Springer: Vienna, Austria, 1953; pp. 280–445. (In German)
45. Bear, J. *Dynamics of Fluids in Porous Media*; Dover Publications, Inc.: New York, NY, USA, 1972.
46. Krumbein, W.C.; Monk, G.D. Permeability as a function of the size parameters of unconsolidated sands. *Trans. Am. Inst. Min. Metall. Pet. Eng.* **1943**, *151*, 153–163.
47. Chesnaux, R.; Baudement, C.; Hay, M. Assessing and comparing the hydraulic properties of granular aquifers on three different scales. In Proceedings of Geohydro 2011, Quebec City, QC, Canada, 28–31 August 2011; p. 9.
48. Pravedny, G.H. *Design and Selection of Grain Size Composition of Filter Beds for the Transition Zones of Large Dams*; Energiia: Moscow, Russia, 1966.
49. Slichter, C.S. *Theoretical Investigation of the Motion of Ground Waters*; U.S. Geological Survey 19th Annual Report, Part 2; United States Government Printing Office: Washington, DC, USA, 1899; p. 322.
50. Terzaghi, C. Principles of soil mechanics. *Eng. News-Rec.* **1925**, *95*, 832.
51. Lu, C.; Chen, X.; Cheng, C.; Ou, G.; Shu, L. Horizontal hydraulic conductivity of shallow streambed sediments and comparison with grain-size analysis results. *Hydrol. Process.* **2012**, *26*, 454–466.
52. Krumbein, W.C. Size frequency distribution of sediments. *J. Sed. Petrol.* **1934**, *4*, 65–77.
53. Panda, M.N.; Lake, L.W. Estimation of single-phase permeability from parameters of particle-size distributions. *Am. Assoc. Pet. Geol. Bull.* **1994**, *78*, 1028–1039.
54. Schulze-Makuch, D.; Carlson, D.A.; Cherkauer, D.S.; Malik, P. Scale dependency of hydraulic conductivity in heterogeneous media. *Groundwater* **1999**, *37*, 904–919.

Experimental Measurement of Diffusive Extinction Depth and Soil Moisture Gradients in a Dune Sand Aquifer in Western Saudi Arabia: Assessment of Evaporation Loss for Design of an MAR System

Iqra Mughal, Khan Z. Jadoon, P. Martin Mai, Samir Al-Mashharawi and Thomas M. Missimer

Abstract: A component of designing a managed aquifer recharge system in a dune aquifer is the control of diffusive evaporative loss of water which is governed by the physical properties of the sediments and the position of the water table. A critical water table position is the "extinction depth", below which no further loss of water occurs via diffusion. Field experiments were conducted to measure the extinction depth of sediments taken from a typical dune field in the region. The soil grain size characteristics, laboratory porosity, and saturated hydraulic conductivity were measured. The sand is classified as well-sorted, very fine sand with a mean grain diameter of 0.15 mm. Soil moisture gradients and diffusion loss rates were measured using sensors in a non-weighing lysimeter that was placed below land surface. The sand was saturated carefully with water from the bottom to the top and was exposed to the natural climate for a period of about two months. The moisture gradient showed a gradual decline during measurement until extinction depth was reached at about 100 cm below surface after 56 days. Diurnal temperature changes were observed in the upper 75 cm of the column and were negligible at greater depth.

Reprinted from *Water*. Cite as: Mughal, I.; Jadoon, K.Z.; Mai, P.M.; Al-Mashharawi, S.; Missimer, T.M. Experimental Measurement of Diffusive Extinction Depth and Soil Moisture Gradients in a Dune Sand Aquifer in Western Saudi Arabia: Assessment of Evaporation Loss for Design of an MAR System. *Water* **2015**, *7*, 6967–6982.

1. Introduction

Western Saudi Arabia is a hyper-arid area that receives only about 50 mm of rainfall each year in the coastal area and slightly greater accumulations in the interior higher altitude areas [1]. When rainfall does occur, it is an intense event that causes flash floods through ephemeral channels, known as wadis [2,3]. Flood water passing through the wadi channels allows some recharge to the underlying alluvial aquifer, but most of the water is lost to discharge to tidal water or evaporates with only 3%

to 10% being captured as recharge [4–6]. Recent research on future climatic changes in this region suggests that a greater degree of variability will occur with longer drought periods and more intense rainfall events with corresponding floods [7,8]. Therefore, it is necessary to assess a water management system that will capture freshwater from flood events and place the captured water into storage for use during drought periods.

One strategy being implemented is the construction of large dams within the ephemeral channel system to capture and store stormwater, such as the recently constructed dam in Wadi Khulays (Figure 1). While this strategy is effective at containing the captured water, the climate of the area is very hot and dry, resulting in a high degree of evaporative loss from the reservoir, on the order of 50% or greater [9,10]. To mitigate this loss of water to free surface evaporation, two managed aquifer recharge (MAR) schemes have been suggested, one using the storage within the wadi alluvial aquifer in downstream parts of the wadi system [10] and a second using a down-gradient dune field (Figure 1) [11].

(a) (b)

Figure 1. Wadi Khulays Dam and reservoir (**a**) and the downstream dune field (**b**), Lopez *et al.* [11].

Dune aquifers have been used in many regions for the storage of various types of treated freshwater [12,13] to impaired quality water [11,14]. There are several important issues in the use of a dune aquifer for stormwater storage in western Saudi Arabia. The aquifer must have adequate storage capacity (porosity or specific yield) to receive the stormwater, satisfactory hydraulic conductivity to allow infiltration of the water and recovery through wells or underdrains, a degree of stability to maintain the subsurface reservoir, a low hydraulic gradient to keep the water in storage, and a known extinction depth to allow a design that minimizes evaporative loss. A method of determining dune aquifer hydraulic properties and a rapid screening process in this region is presented in this special issue by Lopez *et al.* [15]. The primary purpose of our research is to experimentally measure the evaporative extinction depth of dune sands in this region of Saudi Arabia.

Previous research has estimated extinction depth in a variety of soil types [16–19]. The grain size characteristics of the soil or sediment, position of the water table, and climate are known to affect the extinction depth as demonstrated in actual measurements [20–22]. Commonly, estimates of extinction depth research have been conducted considering total evapotranspiration without a separation of diffusive evaporation loss and plant impacts via transpiration [16–19]. Furthermore, Soylu *et al.* [23] used different numerical models to investigate the sensitivity of land surface evapotranspiration to water table depth and soil texture. They concluded that further field-based studies and modeling are essential to improve the predictability of shallow water table and land surface interactions in numerical models, particularly as it relates to the parameterization of soil hydraulic properties. An excellent review of the evaporation from bare sandy soil in arid environments can be found in Wang [24].

In this research, we measured the diffusive evaporation loss under a condition of bare sand as would occur within an MAR site in a dune field without any influence of vegetation. The climatic conditions and soil moisture profiles were recorded. We then applied the hydrological simulator HYDRUS-1D to model the soil evaporation loss. The measured and modeled extinction depth in bare dune sand was compared to measurements made in other locations.

2. Background and Methods

2.1. Collection of Sand Samples from Dune Fields

A field experiment was carried out to measure the diffusive extinction depth of dune sand and to monitor the soil moisture and temperature dynamics during summer for 59 days (18 July 2012–14 September 2012). The sand used in the experiment was excavated from a desert dune in the southwestern region of Saudi Arabia (22°08′7.50″N, 39°16′17.35″E). The samples removed from the dune were rather homogeneous but mixed and not in a natural condition. However, dune sands are typically moving and unstable to a degree, therefore making the samples representative of the dune characteristics. The minimum and maximum temperatures in the field were 24 and 43 °C, respectively, recorded by the meteorological station installed nearby the experimental site. No rainfall events occurred during the period of the field experiment.

2.2. Grain Size Distribution Analyses

The grain size distribution of the dune sand was measured using the standard sieving technique as described by Tanner and Balsillie [25] and the National Soil Survey Laboratory Methods Manual [26]. A total of 34 sieves were used in the analysis. Five samples were randomly selected from the full volume of collected sediment for analysis.

2.3. Laboratory Porosity and Hydraulic Conductivity Measurements

Porosity measurements were made in the laboratory using the water saturation method. A known volume of water was placed into a graduated cylinder. A dry sediment sample was then added to the water within the cylinder. The sediment addition was carefully controlled to avoid creation of air bubbles within the column and was terminated before the volume of sediment exceeded the top of the water in the graduated cylinder. The column was tapped carefully with a rubber mallet until the sediment volume ceased to reduce in height. This approximated the natural packing of the sediment within the dune. The volume of water was determined be subtracting the excess water found in the cylinder from the initial volume of water added. The porosity was determined by dividing the water volume by the sediment volume. This analysis was conducted on three samples collected from the excavated sand and 50 samples on which grain size distribution and hydraulic conductivity were measured from the same dune field by Lopez *et al.* [15]. Additional porosity measurement were made during moisture calibration.

The hydraulic conductivity was measured in the laboratory with a standing head (constant head) permeameter, using standard laboratory methods [27]. Measurements were conducted on the same five samples on which porosity and grain size distribution analyses were made.

2.4. Calibration of Moisture Sensors

For monitoring the vertical profile of soil moisture content and temperature, 5TM soil moisture sensors were used, which are manufactured by Decagon Devices, Inc. (Pullman, WA, USA). These sensors measure the soil moisture content by measuring the permittivity of the soil, which is a strong function of water content, by using the Topps equation [28] given as below:

$$\theta_w\,(m^3/m^3) = 4.3 \times 10^{-6}{\varepsilon_r}^3 - 5.5 \times 10^{-4}{\varepsilon_r}^2 + 2.92 \times 10^{-2}\varepsilon_r - 5.3 \times 10^{-2} \quad (1)$$

where θ_w is soil moisture content and ε_r is permittivity of soil. The natural variation in the characteristics of most soils causes the generic calibration for ECH2O soil moisture sensors to have an error of approximately $\pm$ 3%–4%. The error can be reduced to $\pm$ 1%–2% for all soils by using a soil specific calibration. Thus, the sensors were calibrated to obtain the best possible accuracy in volumetric water content measurements before using them in the experiment.

The procedure used to calibrate the sensors is: (1) a 1 kg of dune sand was collected from the excavated sand and was dried and cleaned to remove rocks and clods; (2) a container with a diameter of 19.2 cm and a 18.1 cm height was filled and packed with the dried sand; (3) the sensors were inserted vertically into the sand within the container; (4) each sensor was inserted three to four times into the packed sand to verify the readings and an average of all moisture content readings was recorded in case minor variability was observed; (5) a volume of the undisturbed soil was carefully removed using a metal ring with a diameter of 79 mm and height of 55 mm and was transferred to a foil pan which was weighed before and after transfer of the soil; (6) the dried sand was saturated in 10 episodes by adding 400 mL of water to each and then the sand was compacted with sensor readings taken in the same manner as the dried sand; (7) the sand samples were removed from the container after each episode and transferred to 10 separate foil pans, each weighed before and after filling; (8) a total of 11 pans, one dry and the others containing wet sand, were placed in an oven at a temperature of 110 °C and left to dry for 24 h; (9) after 24 h, the pans were removed from the oven and were weighed to determine the water content (θ_w) of the sand for each sample using Equation (2) for each sample; (10) after calculating the:

$$(\theta_w) = \frac{(weight\ of\ pan + sand + water) - (weight\ of\ pan + dry\ sand)}{(weight\ of\ pan + dry\ sand) - (weight\ of\ pan)} \tag{2}$$

moisture content of the samples, a scatter plot was generated with the sensor data (permittivity (ε_r)) on the x-axis and the water content (moisture, (permittivity (ε_r)) on the y-axis, and (11) regression analysis was applied to a find best-fitting cubic polynomial between the parameters resulting in the best-fit curve shown in Figure 2.

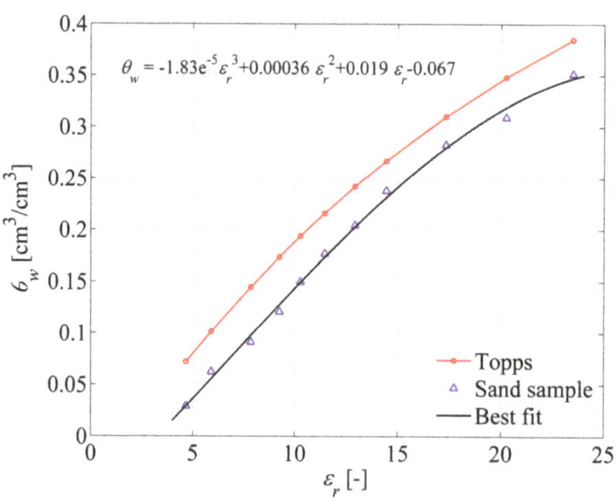

Figure 2. Plot showing the relationship between permittivity (ε_r) and soil moisture content (θ_w).

The red line in the plot represents the Topps equation [26] which the sensors use to calculate the soil moisture content and the black line is the best-fit of a cubic equation through the measurement points of the sand samples. The calibration function constructed by this plot was later applied on the raw sensor data that was collected during the experiment. A similar calibration procedure has been used successfully for geophysical measurements in the field to relate relative permittivity to soil water content [29,30].

2.5. Experimental Design and Instrumentation

The experimental setup consisted of a plastic barrel with diameter 150 cm and height of 150 cm that was positioned under the ground level. Figure 3 shows the experimental arrangement and instrumentation in the field.

PVC pipes with a 5 cm diameter were perforated by drilling small holes in them and the pipes were then wrapped with a thin cotton fabric to restrict entry of small sand particles into the pipes and to avoid clogging. Two perforated vertical pipes were connected to each other by horizontal perforated pipes inside the barrel. They were fixed inside the barrel to allow saturation of the sand from bottom upwards to avoid trapping air bubbles and to accurately monitor the water level fluctuation.

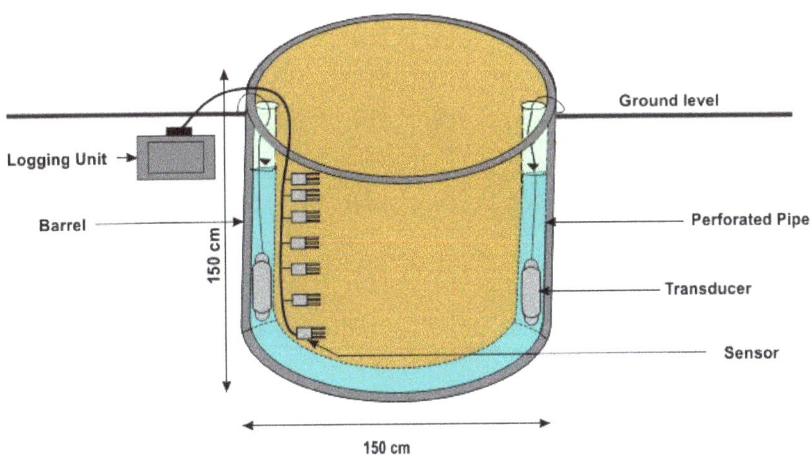

Figure 3. Schematic diagram of the setup to monitor water level, temperature and soil moisture gradients.

During the course of adding sand in the barrel, the already calibrated 5TM sensors were installed at different depths. Fourteen sensors were placed in the barrel at depths ranging from 3 to 100 cm. They were set to record data automatically, at a time step of five minutes, using three EM50 digital/analogue data loggers manufactured by Decagon Devices, Inc. Each logging unit has six ports, five to connect sensors and one port to transfer the data from the sensors to the computer.

A submersible pressure transducer was suspended inside the perforated pipes to monitor the varying water level during natural atmospheric conditions (evaporation, precipitation). The transducers used in the experiment were U20 titanium water level data logger manufactured by HOBO. The transducers were also set to record readings with a regular time step of five minutes.

The filled and buried barrel setup is shown in Figure 4 with the sand being dry in Figure 4a and saturated in Figure 4b. When the installation was completed, water was poured into the barrel through the vertical perforated pipes (bottom to top) to avoid the entrapment of air during saturation. The sand inside the barrel was wetted in several episodes until full saturation was reached and total volume of 0.928 m^3 water was injected in the barrel. Soil moisture content and temperature were recorded throughout the filling process. Upon being satisfied that the sand was fully saturated with no significant trapped air in the barrel, the system was left to record evaporation loss from the sand.

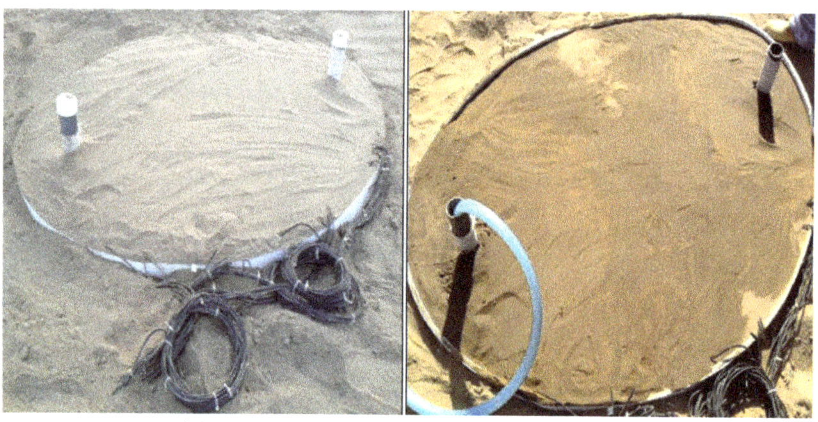

(**a**) (**b**)

Figure 4. The sand barrel before (**a**) and after (**b**) saturation.

2.6. Hydrodynamic Modeling

A variety of different hydrodynamic modeling tools are readily available to simulate water fluxes in the vadose zone, as compared in Scanlon *et al.* [31]. Hydrodynamic models mainly vary with regard to inclusion of vapor and heat fluxes as well as the internal coupling approaches of these fluxes. In arid and semi-arid regions, the vapor movement is often an important part of the total water flux when the soil moisture becomes relatively low in the topsoil. In this study, we choose to apply the hydrological simulator HYDRUS-1D because of its robust numerical scheme and short calculation times [31]. The hydrological simulator HYDRUS 1-D simulates non-isothermal water and vapor flow as described by Saito *et al.* [32]:

$$\frac{\partial\theta(h)}{\partial t} = \frac{\partial}{\partial z}\left[(K_{lh} + K_{vh})\frac{\partial h}{\partial z} + K_{lh} + (K_{LT} + K_{vT})\frac{\partial T}{\partial z}\right] - S\,(h) \quad (3)$$

In Equation (3), h is the time and depth dependent pressure head (cm), t is the time (s), θ is the total volumetric water content, being the sum ($\theta = \theta_v + \theta_l$) of the volumetric water vapor content, θ_v, and the volumetric liquid water content, θ_l (both expressed as an equivalent water content), $\theta(h)$ is the water content as a function of pressure head ($cm^3 \cdot cm^{-3}$), T is temperature (K), K_{lh} is isothermal hydraulic conductivity for the liquid phase ($cm \cdot s^{-1}$), K_{LT} is thermal hydraulic conductivity for the liquid phase ($cm \cdot K^{-1} \cdot s^{-1}$), K_{vh} is isothermal vapor hydraulic conductivity ($cm \cdot s^{-1}$), z is the depth positive downward in cm, and S is a sink term usually used to account for root water uptake (s^{-1}).

The classical Mualem-van Genuchten (MVG) model was used to describe the water retention and isothermal unsaturated hydraulic conductivity functions of the subsurface system [33,34]. The water retention is given by:

$$\theta_l\,(h) = \theta_r + (\theta_s + \theta_r)\,[1 + |\alpha h|^n]^{-m} \quad (4)$$

where θ_r and θ_s are the residual and saturated water contents ($cm^3 \cdot cm^{-3}$), respectively, α (cm^{-1}) and n (–) are shape parameters which are inversely related to the air entry value and the width of the pore size distribution, respectively, and m is defined as $m = 1-1/n$ with $n > 1$ [28]. The isothermal unsaturated hydraulic conductivity is given by:

$$K_{lh}(h) = K_s\left(\frac{\theta - \theta_r}{\theta_s - \theta_r}\right)^{\lambda}\left[1 - \left(1 - \left(\frac{\theta - \theta_r}{\theta_s - \theta_r}\right)^{\frac{1}{m}}\right)^{m}\right]^2 \quad (5)$$

where K_s is the saturated hydraulic conductivity ($cm \cdot h^{-1}$), λ (–) represents pore tortuosity, and r determines the shape of the hydraulic conductivity function. θ_r is the residual water content. It is used as an empirical parameter which is either given the value of zero or the value which best fits the water retention data or it can be derived from soil texture using the pedo-transfer function. θ_s can be obtained by fully saturating the sand sample and then drying the sample in an oven for 24 h at 110 °C. The value of λ generally used is 0.5 or it can be estimated from bulk density and hydraulic conductivity with which it is highly correlated [35]. Values of α and n were obtained from the Rosetta database and are 0.145 and 2.18, respectively [Rosetta].

The thermal hydraulic conductivity function, K_{LT}, in Equation (3) is defined as follows [36]:

$$K_{LT} = K_{Lh}\left(hG_{wT}\frac{1}{\gamma_0}\frac{d\gamma}{dT}\right) \tag{6}$$

where G_{wT} is the grain factor (unitless), which quantifies the temperature dependence of the soil water retention curve [37], γ is the surface tension of soil water ($J \cdot m^{-2}$), and γ_0 is the surface tension at 25 °C (= 71.89 $g \cdot s^{-2}$). The temperature dependence of γ is given by:

$$\gamma = 75.6 - 0.1425T - 2.38 \times 10^{-4}T^2 \tag{7}$$

where γ is in $g \cdot s^{-2}$ and T in °C. The isothermal, K_{vh} and thermal, K_{vT}, vapor hydraulic conductivities are described by Nassar and Horton [38].

Equation (3) is solved numerically using the hydrological simulator HYDRUS-1D [32]. The depth of the domain was fixed to z = 150 cm, the size of the barrel, and the vertical profile was discretized into 150 equally sized elements (dz = 1 cm). The time step for the top boundary condition (evaporation) was fixed to one hour, whereas the lower boundary condition was fixed to zero water flux. The top boundary condition for temperature was assumed to be equal to the air temperature and lower boundary temperature was fixed at 35 °C.

3. Results

3.1. Physical Properties of the Media

The grain size data from the five randomly selected samples showed rather uniform properties. The sediment is very well sorted and lies within the very fine sand classification of the Wentworth scale [39].

Laboratory porosity measurements also showed some degree of variation with the full range of the samples being 0.33–0.41 (Table 1). The average value was about 0.38. This is comparative to the large number of measurements made on samples from the same dune field by Lopez *et al.* [15]. They found a mean value of 0.39 ± 0.02 for 50 samples. However, a large number of additional samples were measured during moisture calibration with the average being 0.35. This value was consistent with the volume of water placed into the cylinder.

Table 1. Soil porosity and hydraulic conductivity data.

Parameter	Sample 1	Sample 2	Sample 3	Sample 4	Sample 5
Porosity	0.41	-	0.33	-	0.40
Hydraulic Conductivity (m/day)	6.48	3.30 (?)	6.75	7.57	7.54

The measured hydraulic conductivity values ranged from 6.48 to 7.54 m/day. A small value of 3.30 m/d was measured which is considered to be an anomaly. The range in hydraulic conductivity occurring across the dunes at the site where the sample was collected ranges from 5 to 11 m/day [15].

3.2. Water Level Profile

The fluctuation of the water level inside the barrel was monitored by the transducers installed in the perforated pipes as shown in Figure 5.

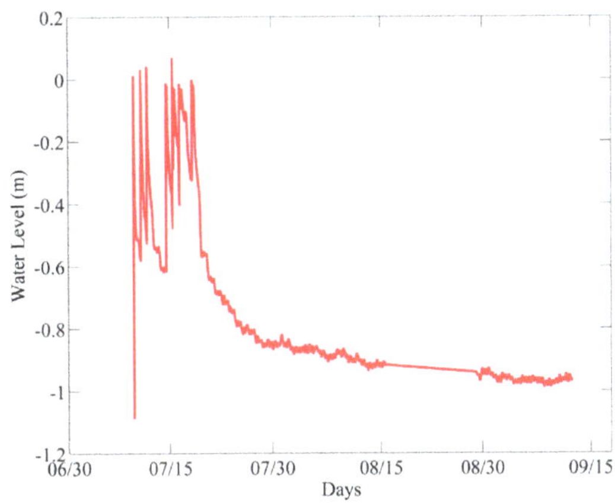

Figure 5. Water level fluctuations monitored with a time step of 15 min.

In response to the first wetting episode, a rapid increase in the water level was recorded by the transducers. The abrupt rise in water level was followed by a decline as the water added through the perforated pipe filled the pipes quickly and then took time for imbibition in sand and to become fully stabilized. The seven prominent peaks ranging from 0 to −0.61 m (Figure 5), observed in the initial days of the experiment, were caused by the continuous wetting-infiltration cycles in the sand. After several wetting-infiltration episodes, the range of water level fluctuation shows a decrease, as the sand was gradually being saturated. After saturation, the sand was exposed to natural conditions (evaporation, infiltration) for a few days. This helped to stabilize the water level in the pipe and the barrel due to the slow imbibition of water in the sand. Consequently, a gradual decline can be noticed which lasted for almost 10 days (from 20 to 30 July) as evident in Figure 5. After declining for 35 more days, the gradient eventually stabilized at approximately 0.97 m and then remained at this depth for the duration of the experiment. Thus, this depth can be established as the extinction depth of dune sand. The data gap between August 15 and 20 can be interpolated using the average gradient during this period based on the trend.

3.3. Soil Moisture Content Profile

Figure 6a–c shows the changes in soil moisture content recorded by 5TM sensors installed at various depths in the barrel. Figure 6a,b depict the instantaneous rise in soil moisture content due to the first sand wetting session. The initial high range in fluctuations of the soil moisture content were caused by the simultaneous wetting–infiltration cycles of the sand. The soil moisture content shows a negative gradient for 20 days, when the barrel was left under natural conditions (evaporation). The gradient then shows a gentle decline through 30 more days. Figure 6c depicts the soil moisture variation recorded by the deepest sensors installed in the barrel which document a decline of soil moisture content first at a depth of 40 cm, followed by the declines at 45 and 50 cm. The sensors installed deeper than 45 cm show slight declines and steadiness of the soil moisture content trend compared to the sensors that were installed at shallow depths (Figure 6c). The symmetric diurnal fluctuations are evident in the shallow profiles, which can be seen more prominently in the upper 25 cm.

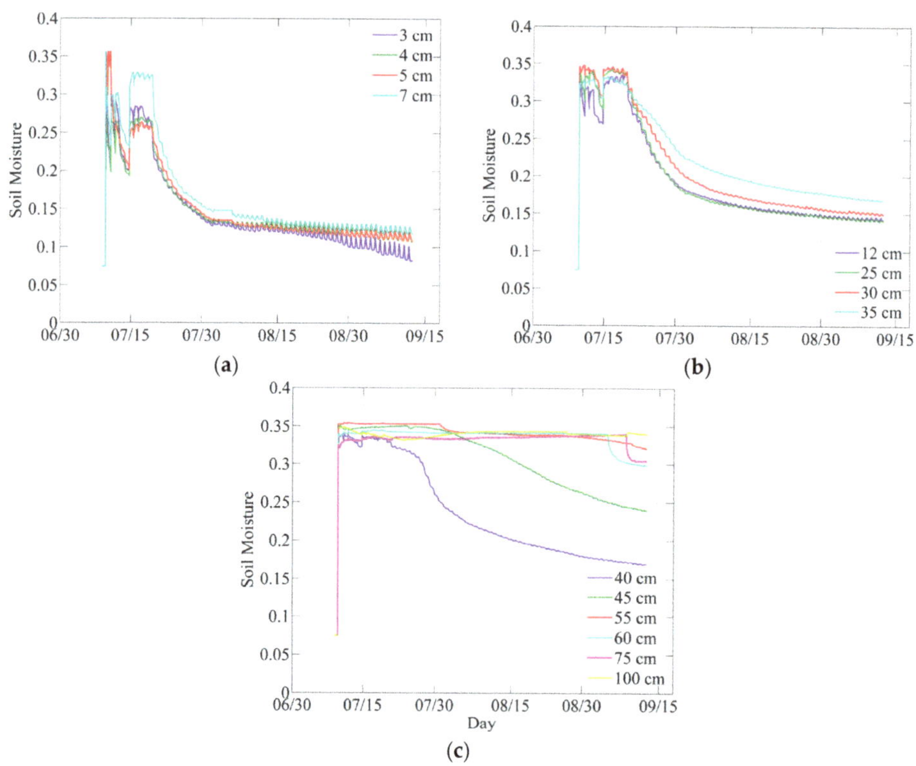

Figure 6. Soil moisture content variations recorded by sensors (**a**) from 3 to 7 cm of depth; (**b**) 12 to 35 cm of depth; and (**c**) from 40 to 100 cm of depth.

3.4. Temperature Profile

The temperature profiles were recorded using the Decagon sensors installed at depths ranging from 3 to 100 cm in the barrel (Figure 7). At the start of the experiment, the sand temperature was high as the sand was dry and hot. When the sand was wetted, a decline in the temperature was observed. An average diurnal temperature fluctuation of 10–20 °C occurred in the upper 10 cm (Figure 7a). The diurnal temperature fluctuations declined in magnitude with depth in the soil profile and between 12 and 35 cm depth the diurnal variation is the range of 3–7 °C (Figure 7b). A temperature of over 35 °C persisted throughout the profile of the dune sand event to a depth of 100 cm (Figure 7c).

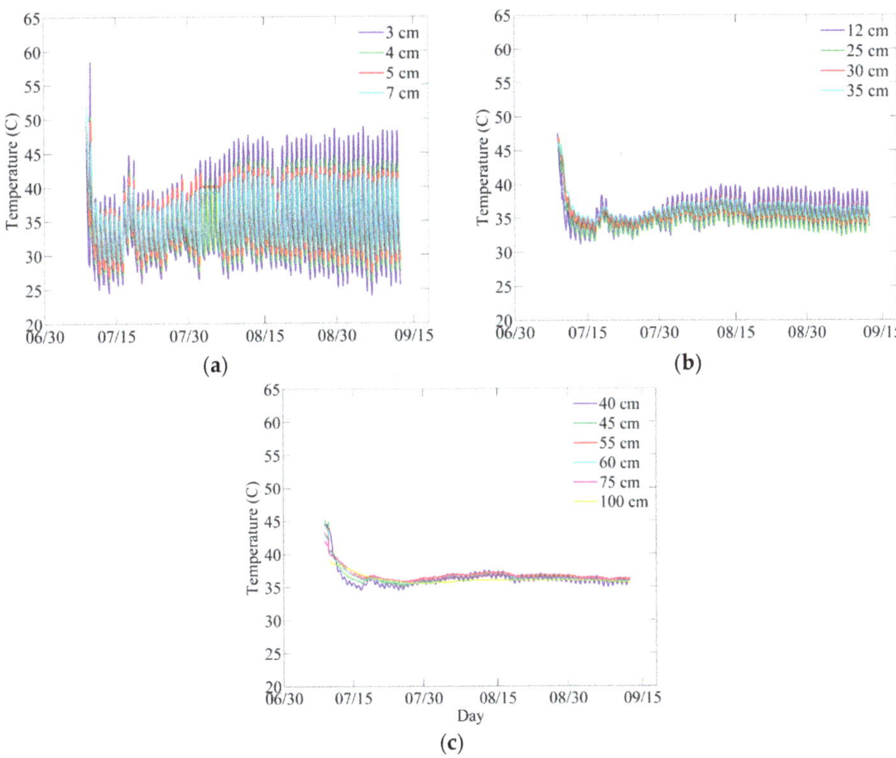

Figure 7. Temperature fluctuations recorded by the sensors (**a**) from depths of 3–7 cm; (**b**) from depths of 12–35 cm; and (**c**) from depths of 40–100 cm.

3.5. Hydrodynamic Modeling Results

Like any hydrological simulator, HYDRUS-1D requires top and initial boundary conditions. In this study, natural conditions (precipitation and evaporation) were considered as the boundary condition at the top. Meteorological variables were recorded at a weather station located near the site. No precipitation occurred during the period of the field experiment. The air temperature recorded during the evaporation period is shown in Figure 8. The maximum and minimum air temperatures were 43 and 24 °C, respectively.

The FAO-56 Penman-Monteith equation was used to determine the potential evaporation following a similar previous study [16]. The hourly potential evaporation "E_p" was estimated using the FAO-56 method as described by Allen *et al.* [40] (Figure 9). The method used first calculated the potential evaporation from a grass ground-cover reference E_r using the modified Penman-Monteith equation [40]. Hourly average values of various meteorological variables, including air temperature, relative humidity, wind speed, incoming shortwave radiation, and barometric pressure, were input into the model. Then, the reference evapotranspiration is scaled with an empirical coefficient which is $E_p = 1.15E_r$ [34]. This coefficient defines the higher rate of diffusive evaporation potential of bare soil as compared to the reference grass cover [41].

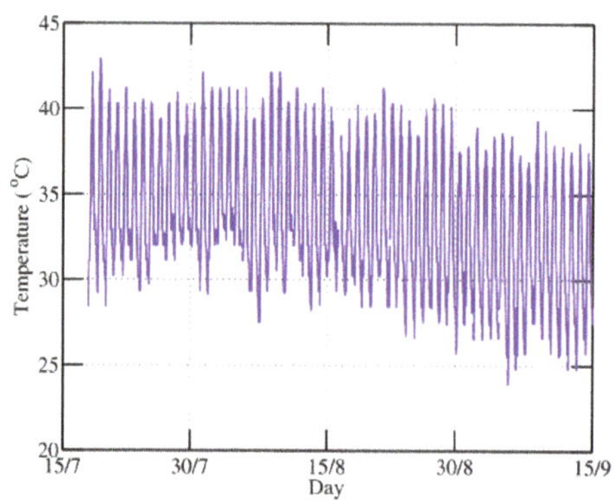

Figure 8. Air temperature recorded during experiment.

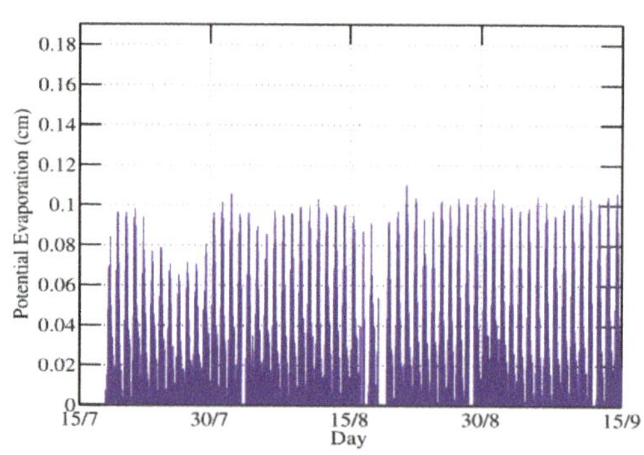

Figure 9. Estimated potential evaporation during the field experiment.

In practical applications, the depth-dependent initial conditions in the hydrodynamic model are poorly known, or not at all. The initial condition used in numerical simulation with respect to the soil water content is shown in Figure 10.

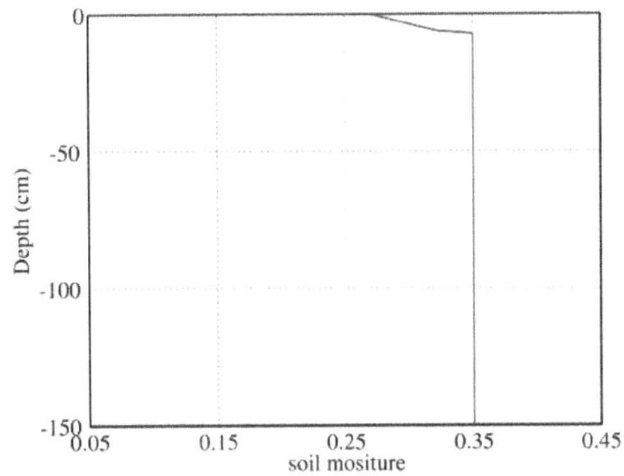

Figure 10. Soil moisture content recorded in the sand column at the beginning of the experiment.

It is derived from the depth-dependent water content measured by the sensors exactly after the last imbibition. The sensors showed almost full saturation below a depth of 7 cm, and the value of fully saturated sand was considered to be 0.35 ($cm^3 \cdot cm^{-3}$) obtained from the laboratory experiment. The sand in the top 7 cm was not completely saturated because water was added from the perforated pipes from the bottom upwards (to avoid air entrapment). Furthermore, the water added from the bottom via the perforated pipes needed more time to get imbibed in the sand to have a stable level near the surface.

3.6. Comparison of Sensor and MVG Model Results

The soil moisture content was calculated with a HYDRUS-1D simulation (Figure 11). The figure shows the soil water content profile follows the same trend as observed in the sensor data (Figure 6). The water content of the soil is lowest at 5 cm. The trends from depths 7–45 cm show increasing saturation and thus increasing moisture content until the 100 cm depth is reached where the soil moisture is highest at complete saturation (0.35).

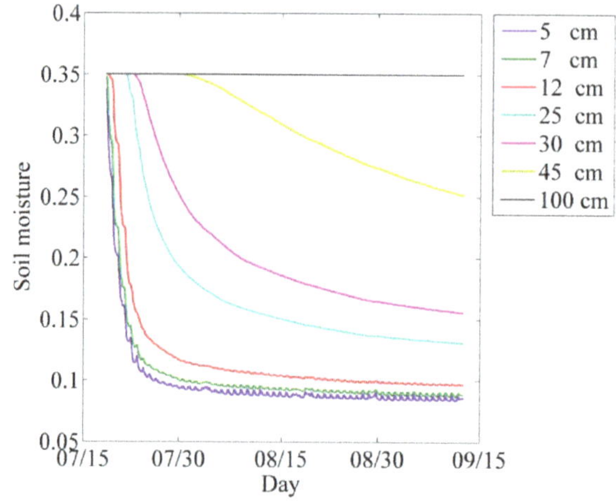

Figure 11. Soil moisture content profile obtained by hydrodynamic simulation.

The diurnal variation of soil water content can be observed in the top 5–12 cm depth. The soil moisture content data obtained from the sensors at seven depths were plotted against the water content obtained from the Mualem van-Genuchten (MVG) model by using HYDRUS 1-D at these depths (Figure 12).

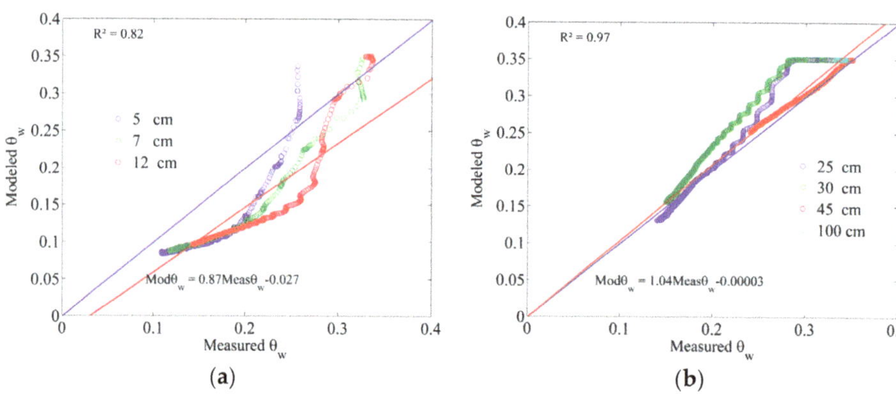

Figure 12. Soil water content θw measured by sensors *versus* soil water content obtained by hydrodynamic modeling at shallow (**a**) and greater (**b**) depths. Blue line represents 1:1 line and red is the best linear fit.

It is observed the water content obtained from the MVG model overestimated the sensor derived measurements in the top 12 cm depth (Figure 12a), whereas, at greater depths, the MVG estimated water content was overestimated (Figure 12b). The data points are somewhat scattered and yield an r^2 value of 0.84 and 0.71, for shallow and greater depths, respectively. Comparing model results with the measured data shows a model overestimation at depths of 25–45 cm and an underestimation from 5 to 12 cm. Detailed comparisons of the sensor and model estimations are shown in Figure 12a,b.

4. Discussion

The field experiment and HYDRUS 1-D modeling of bare dune sand under local climatic conditions showed the diffusive extinction depth to be about 1 m. The mean grain diameter of the sand used in the experiment was at about 0.15 mm or about 2.9 phi units. For comparison, the experimental work conducted by Hellwig [22] found that the evaporative extinction depth of bare soil with a mean grain diameter of 0.53 mm was about 0.6 m. The results of the experiments of Hellwig [20,22] on the effects of particle size, position of the water table, and climate are generally consistent with the results herein obtained. However, our experiments were conducted using a finer grain size of the tested soil and was at higher average surface temperatures and low humidity compared to the Hellwig experiments.

Modeling work using HYDRUS 1-D on estimating the extinction depth of soils with various compositions and land covers was conducted by Shah *et al.* [18]. They concluded that pure, bare sand with no significant content of fines produced an extinction depth of 0.5 m. However, they concluded that fine-textured soils would have a greater extinction depth compared to coarse-grained soils for similar land covers. Since the dune sand tested has some mud and a fine to very fine sand mean grain diameter, the results are expected to yield a deeper extinction depth. Therefore, the experimental work conducted on the bare dune sand is in general agreement with the research conducted by Shah *et al.* [18], yet specific to the climatic conditions found at the site.

It is very likely that the extinction depth found in this experimental method applies to dune sands with the mean grain diameter and measured porosity values in the ranges found in most desert areas. Compaction of fine-grain, well-rounded, predominantly quartz composition dune sands is not relevant above a vertical effective stress of less than 1 to 2 MPa which corresponds to a depth of about 100 m [42]. Further, the loss of porosity between land surface and 100 m in very fine dune sand, based on experimental results, is less than 1% at about 100 m [42]. This has no impact on the measured extinction depth in the field experiment.

5. Conclusions

Experimental studies were conducted to estimate the diffusive extinction depth of dune sand and to monitor the water level, soil moisture and temperature gradients. The sand used in the experiment was excavated from dunes located in the southwestern region of Saudi Arabia. Several samples of the sand were subjected to porosity, hydraulic conductivity and grain size distribution laboratory analyses. In a field experiment, the dune sand was placed in a barrel with a height of 150 cm and a diameter of 150 cm and was left exposed under natural conditions (precipitation, evaporation) after saturation. Transducers were used to record the water level fluctuations and 5TM sensors were used to capture the dynamics of soil moisture content and temperature profiles.

The water level gradient showed stabilization at a depth of about 1 m in the sand column which was established as the extinction depth of the dune sand under a bare condition. Moreover, it was found that soil specific calibration of the soil moisture content sensors is useful in obtaining accurate volumetric water content measurements.

It was found that the hydrodynamic simulator HYDRUS 1-D, when provided with accurate hydraulic and meteorological parameters and specific initial and top boundary conditions, provides a reasonably accurate representation of vadose zone processes, making this method highly applicable in arid and semi-arid hydrological processes especially for those related to evaporation and extinction depths. The results from sensor derived soil moisture content data and that obtained from HYDRUS 1-D using the single porosity model of Maulem–van Gunechten (MVG) was compared. It was found that both datasets are correlated as the soil moisture profile obtained from hydrodynamic simulation follows the same trend as the soil moisture content profile recorded by the sensors.

The experimental and modeling results found are in general agreement with previous studies conducted on measurement of the diffusive evaporative extinction depth. Experiments conducted on bare sandy soils by Hellwig [20–22] and Shah *et al.* [40] showed the extinction depth to be 0.6 and 0.5 m respectively for coarser sands compared to those studied. This is consistent with the concept that the extinction depth increases based on a reduction in mean grain diameter and addition of clay or fine-grained organic material. The finer grain sediment tends to increase the height of the capillary fringe and aids in diffusion.

Based on the measurement of the extinction depth for this sand, an MAR project designed to store water in the source dune field would require that the water table position be maintained below a depth of 1 m. Therefore, the loss of water during the storage cycle could be minimized and a more efficient system could be designed.

Acknowledgments: The funding for this research was provided by the Water Desalination and Reuse Center at the King Abdullah University of Science and Technology and discretionary faculty funding. The authors thank the local farmer Khadr Bin Sayed AlGamdi for providing access to his farm where the field experiments were conducted.

Author Contributions: Iqra Mughal conducted the background research, the field work, and wrote part of the text. Samir Al-Mashharawi worked with Iqra Mughal on the field work and helped collect and test the samples. Khan Z. Jadoon worked with Iqra Mughal on the HYDRUS1 modeling. P. Martin Mai and Thomas M. Missimer wrote and reviewed the manuscript with added references and background.

Conflicts of Interest: The authors declare no conflict of interest.

References

1. Subyani, A.M. Geostatistical study of annual and mean rainfall patterns in southwest Saudi Arabia. *Hydrol. Sci. J.* **2004**, *49*, 803–817.
2. Sen, Z. *Wadi Hydrology*; CRC Press, Taylor & Francis Group: Boca Raton, FL, USA, 2008.
3. Maliva, R.G.; Missimer, T.M. *Arid Lands Water Evaluation and Management*; Springer: Heidelberg, Germany, 2012.
4. Abdulrazzak, M.J.; Morel-Seytoux, H.J. Recharge from an ephemeral stream following wetting front arrival to water table. *Water Resour. Res.* **1983**, *19*, 194–200.
5. Sorman, A.U.; Abdulrazzak, M.J. Infiltration-recharge through wadi beds in arid regions. *Hydrol. Sci. J.* **1993**, *38*, 173–186.
6. Missimer, T.M.; Drewes, J.E.; Amy, G.; Maliva, R.G.; Keller, S. Restoration of wadi aquifers by artificial recharge with treated wastewater. *Groundwater* **2012**, *50*, 514–527.
7. AlSarmi, S.; Washington, R. Recent observed climate change over the Arabian Peninsula. *J. Geophys. Res.* **2011**, *166*, D11109.
8. Almazroui, M. Simulation of present and future climate of Saudi Arabia using a regional climate model (PRECIS). *Int. J. Climatol.* **2013**, *9*, 2247–2259.
9. Al-Muttair, F.F.; Sendil, U.; Al-Turbak, A.S. Management of recharge dams in Saudi Arabia. *J. Water Res. Plan. Manag.* **1994**, *120*, 749–763.
10. Missimer, T.M.; Guo, W.; Maliva, R.G.; Rosas, J.; Jadoon, K.Z. Enhancement of wadi recharge using dams coupled with aquifer storage and recovery wells. *Environ. Earth Sci.* **2014**, *73*, 1303–1315.
11. Lopez, O.; Stenchikov, G.; Missimer, T.M. Water management during climate change using aquifer storage and recovery in a dunefield in western Saudi Arabia. *Environ. Res. Lett.* **2014**, *9*.
12. Vandenbohede, A.; van Houtte, E.; Lebbe, L. Sustainable groundwater extraction in coastal areas: A Belgian example. *Environ. Geol.* **2009**, *57*, 735–747.
13. Thoa, N.T.K.; Arduino, G.; Giang, N.V.; van, P.T.K.; Vuong, B.T. The model for management of aquifer recharge in sand dune area of Bac Binh, Binh Thuan, Vietnam. *J. Earth Sci.* **2006**, *28*, 389–410.
14. Dillion, P. Aquifer storage and recovery with stormwater and reclaimed water in Australia. In Proceedings of the UNESCO-VIETNAM Artificial Recharge in South East Asia, Ho Chin Minh City, Virtnam, 15–17 December 2004.
15. Lopez, O.M.; Jadoon, K.Z.; Missimer, T.M. Method of relating grain size distribution to hydraulic conductivity in dune sands to assist in assessing managed aquifer recharge projects: Wadi Khulays dune field, western Saudi Arabia. *Water* **2015**, *7*, 6411–6426.
16. Droogers, P.; Allen, R.G. Estimating reference evapotranspiration under inaccurate data conditions. *Irrig. Drain. Syst.* **2002**, *16*, 33–45.
17. Inmark, S.; Irmak, A.; Allen, R.; Jones, J. Solar and net radiation-based equations to estimate reference evapotranspiration in humid climates. *J. Irrig. Drain. Eng.* **2003**, *129*, 336–347.

18. Shah, N.; Nachabe, M.; Ross, M. Extinction depth and evapotranspiration from ground water under selected land covers. *Ground Water* **2007**, *45*, 329–338.
19. Xiao, X.; Li, B.; Sauer, T.; Heitman, J.; Ren, T. Cumulative soil water evaporation as a function of depth and time. *Vadose Zone J.* **2000**, *10*, 1016–1022.
20. Hellwig, D.H.R. Evaporation of water from sand, 1: Experimental set-up and climate influence. *J. Hydrol.* **1973**, *18*, 93–108.
21. Hellwig, D.H.R. Evaporation of water from sand, 2: Diurnal variations. *J. Hydrol.* **1973**, *18*, 109–118.
22. Hellwig, D.H.R. Evaporation of water from sand, 4: The influence of the depth of the water table and the particle size distribution of the sand. *J. Hydrol.* **1973**, *18*, 109–118.
23. Soylu, M.E.; Istanbulluoglu, E.; Lenters, J.D.; Wang, T. Quantifying the impact of groundwater depth on evapotranspiration in a semi-arid grassland region. *Hydrol. Earth Syst. Sci.* **2011**, *15*, 787–806.
24. Wang, X. Vapor flow resistance of dry soil layer to soil water evaporation in arid environment: An overview. *Water* **2015**, *7*, 4552–4574.
25. Tanner, W.F.; Balsillie, J.H. *Environmental Granulometry*; Special Publication No. 40; Florida Geological Survey: Tallahassee, FL, USA, 1995.
26. United States Department of Agriculture-Natural Resources Conservation Service (USDA-NRCS). *Soil Survey Laboratory Methods Manual*; Report No. 42, Version 4.0; National Soil Center: Lincoln, NE, USA, 2004.
27. American Society for Testing and Materials (ASTM). *Standard Test Method for Permeability of Granular Soils, ASTM Standard D2434–68*; ASTM: West Conshocken, PA, USA, 2006.
28. Topp, S.; Grasmueck, M.; Vega, S.; Viggiano, D.A. Electromagnetic determination of soil water content: Measurement in coaxial transmission lines. *Water Res.* **1980**, *16*, 574–582.
29. Jadoon, K.Z.; Weihermüller, L.; Scharnagl, B.; Kowalsky, M.B.; Bechtold, M.; Hubbard, S.S.; Vereecken, H.; Lambot, S. Estimation of soil hydraulic parameters in the field by integrated hydrogeophysical inversion of time-lapse ground-penetrating radar data. *Vadose Zone J.* **2012**, *11*, 1–17.
30. Dimitrov, M.; Vanderborght, J.; Kostov, K.G.; Jadoon, K.Z.; Weihermuller, L.; Jackson, T.J.; Bindlish, R.; Pachepsky, Y.; Schwank, M.; Vereecken, H. Soil hydraulic parameters and surface soil moisture of a tilled bare soil plot inversely derived from L-band brightness temperatures. *Vadose Zone J.* **2014**, *13*, 1–18.
31. Scanlon, B.R.; Christman, M.; Reedy, R.C.; Porro, I.; Simunek, J.; Flerchinger, G.N. Intercode comparisons for simulating water balance of surficial sediments in semiarid regions. *Water Res.* **2002**, *38*, 1323–1339.
32. Saito, H.J.; Simunek, J.; Mohanty, B. Numerical analyses of coupled water, vapor and heat transport in the vadose zone. *Vadose Zone J.* **2006**, *5*, 784–800.
33. Mualem, Y. A new model for predicting the hydraulic conductivity of unsaturated porous media. *Water Res.* **1976**, *12*, 513–522.
34. Van Genuchten, M.T. A closed form equation for predicting the hydraulic conductivity of unsaturated soils. *Soil Sci. Soc. Am. J.* **1980**, *44*, 892–898.

35. Vereecken, H. Estimating the unsaturated hydraulic conductivity from theoretical models using simple soil properties. *Geodema* **1995**, *65*, 81–92.
36. Noborio, K.; McInnes, K.J.; Heilman, J.L. Two-dimensional model for water, heat, and solute transport in furrow-irrigated soil: II. Field evaluation. *Soil Sci. Soc. Am. J.* **1996**, *60*, 1010–1021.
37. Nimmo, J.R.; Miller, E.E. The temperature dependence of isothermal moisture *vs.* potential characteristics of soils. *Soil Sci. Soc. Am. J.* **1986**, *50*, 1105–1113.
38. Nassar, I.N.; Horton, R. Simultaneous transfer of heat, water, and solute in porous media: I. Theoretical development. *Soil Sci. Soc. Am. J.* **1992**, *56*, 1350–1356.
39. Wentworth, C.K. A scale of grade and class terms for clastic sediments. *J. Geol.* **1922**, *30*, 377–392.
40. Allen, R.G.; Pereira, L.S.; Smith, M. *Crop Evapotranspiration: Guidelines for Computing Crop Water Requirements*; Irrigation and Drainage Paper 56; Food and Agriculture Organization of the United Nations: Roma, Italy, 1998.
41. Scharnagl, B.; Vrugt, J.; Vereecken, H.; Herbst, M. Inverse modeling of *in situ* soil water dynamics: Investigation the effect of different prior distributions of the soil hydraulic parameters. *Hydrol. Earth Syst. Sci.* **2011**, *15*, 3043–3059.
42. Chuhan, F.A.; Kjeldstad, A.; Bjorlykke, K.; Hoeg, K. Experimental compression of loose sands: Relevance to porosity reduction during burial in sedimentary basins. *Can. Geotech. J.* **2003**, *40*, 997–1011.

Chapter 2:
Water Policy and Management in Drylands

Improving the Performance of Water Policies: Evidence from Drought in Spain

Mohamed Taher Kahil, Jose Albiac, Ariel Dinar, Elena Calvo, Encarna Esteban, Lorenzo Avella and Marta Garcia-Molla

Abstract: Water scarcity is a critical environmental issue worldwide, especially in arid and semiarid regions. In those regions, climate change projections suggest further reductions in freshwater supplies and increases of the recurrence, longevity and intensity of drought events. At present, one important question for policy debate is the identification of water policies that could address the mounting water scarcity problems. Suitable policies should improve economic efficiency, achieve environmental sustainability, and meet equity needs. This paper develops and applies an integrated hydro-economic model that links hydrological, economic and environmental elements to such issues. The model is used to conduct a direct comparison of water markets, water pricing and institutional cooperation, based on their economic, environmental and equity outcomes. The analysis is performed in the Jucar Basin of Spain, which is a good natural experiment for studying water scarcity and climate change policies. Results indicate that both institutional and water market policies are high performing instruments to limit the economic damage costs of droughts, achieving almost the same social benefits. However, the environmental effects of water markets are worrying. Another important finding is that water pricing is a poor policy option not only in terms of private and environmental benefits but also in terms of equity.

Reprinted from *Water*. Cite as: Kahil, M.T.; Albiac, J.; Dinar, A.; Calvo, E.; Esteban, E.; Avella, L.; Garcia-Molla, M. Improving the Performance of Water Policies: Evidence from Drought in Spain. *Water* **2016**, *8*, 34.

1. Introduction

Water scarcity and water quality degradation are becoming widespread problems in most regions around the world. The reasons are the large increase in global water extractions in the last century from 600 to 3900 km^3, driven by the intensive growth of population and income, coupled with a questionable performance of water governance and policies [1].

The scale of the global growing overexploitation indicates that water mismanagement is quite common, and that sustainable management of basins is a complex and difficult task. At first, water scarcity resulted from surface extractions, but recently it is worsening because of the unprecedented depletion of groundwater brought about by falling pumping costs. Between 1960 and 2000, groundwater extractions rose from 310 to 730 km^3 per year pushing depletion up to 150 km^3 [2]. This staggering annual depletion ranges from 50 km^3 in the Indus-Ganges-Brahmaputra region to 24 km^3 in the USA, and is 13 km^3 in the Tigris-Euphrates region, and 9 km^3 in Northern China (NASA GRACE data estimations).

Water scarcity is increased gradually by the decisions on water extractions in river basins linked to land use and economic activities. The problems arising from water scarcity could become critical during drought periods. Climate change is projected to aggravate the severity and recurrence of drought events, especially in arid and semiarid regions [3]. In those regions, the combined effects of human-induced permanent water scarcity and climate change-induced droughts portend unprecedented levels of water resources degradation.

The sustainable management of water is quite challenging because of the different types of goods and services provided by water. These goods and services can be classified as private goods, common pool resources, or public goods, depending on the degree of exclusion and rivalry in consumption among consumers. A good is non-excludable when individuals cannot be excluded from their use, and a good is non-rival when consumption by one individual does not reduce availability to others. Treated drinkable water in urban networks is close to a private good (rivalry & exclusion), water in surface watercourses and aquifers is close to a common pool resource (rivalry & non-exclusion), while water sustaining ecosystems comes close to a public good (non-rivalry & non-exclusion) [4]. The management of water is governed by public policies because pure competitive markets fail to account for the common pool and public good characteristics of water.

The contribution of this paper is to develop and apply an innovative approach to inform the ongoing policy discussion addressing water scarcity and droughts. A hydro-economic model of the Jucar Basin in Spain is used to conduct a direct comparison of policies based on their economic, environmental and equity effects. Three policy alternatives are considered: (1) an institutional approach based on stakeholders' cooperation; (2) a water market policy; and (3) a water pricing policy. The assessment of the three policies provides information to stakeholders and decision makers about the tradeoffs between the policies in the allocation of water among sectors and locations. The paper is organized as follows. The three types of water policies are reviewed in Section 2. Then, the Jucar River Basin and the modeling framework are described in Section 3. Section 4 presents the drought and policy scenarios and the simulation results. Finally, Section 5 concludes with the summary and policy implications.

2. Types of Policy Instruments

Three types of policy instruments could address the market externalities created by the common pool and public good characteristics of water. The first type is the

"Pigou solution", based on taxation of water extractions [5]. This is the water pricing approach that is being implemented in the European Water Framework Directive (WFD) [6]. The second type is the "Coase solution", which is based on privatizing the resource and trading [7]. This is the water market approach that has been implemented in Australia [8]. The third type is the common property governance [9], based on the evidence that coercive government rules can fail because they lack legitimacy and knowledge of local conditions. This is the institutional cooperative approach, where affected stakeholders design the rules and enforcement mechanisms for the sustainable management of common pool resources [10], although this approach has not received widespread attention in either research or policy circles.

Mainstream water policies in some countries are derived from the Dublin Statement on Water, which declares water an economic good [11], and are based on so-called economic instruments such as water markets or water pricing. Besides the European Union and Australia, both water pricing and water markets are being considered at present for solving the acute water scarcity problems in China [12].

These economic instruments can work well when water exhibits private good characteristics such as in urban networks, but work less well when water exhibits common pool resource or public good characteristics. There is a strong consensus among experts that water pricing could achieve sizable gains in efficiency and welfare in urban and industrial water networks [13], although implementation could face technical and political difficulties. Irrigation water from surface watercourses and aquifers exhibits common pool resource characteristics, and the use of economic instruments requires transforming the resource into a private good. This transformation is quite difficult, especially in arid and semiarid regions under strong water scarcity pressures, and would require the support of stakeholders.

Water pricing in irrigation, to achieve water conservation, has been the subject of debate since the 1990s. A string of the literature finds that irrigation water pricing has limited effects on water conservation [14,15], and some authors indicate that water markets seem far more effective than water pricing for allocating irrigation water [16]. Several studies in Spain support those previous findings, but also find that water pricing policy involves disproportionate costs to farmers [17,18]. In contrast, Tsur *et al.* [19] indicate that water pricing could achieve an efficient allocation of irrigation water without damaging farmers' benefits, if the pricing policy guarantees that all or part of the revenue collected by water agencies remains in the area and is reinvested in improving water use efficiency.

In recent decades, the water market approach has been gaining ground in some parts of the world to allocate water to irrigation such as in Australia and Chile. Previous studies in the literature consider that water trading is a flexible and efficient way to address water allocation problems [20–22]. These studies indicate that water markets may increase water use efficiency, avoid the development of new costly water resources, and achieve significant welfare gains by reallocating water from crops with low to high marginal value of water. Numerous pre-requisites are needed for the design of well-functioning water markets such as the definition of water rights, the creation of legal and institutional frameworks for trade, and investments in infrastructure to facilitate water transfer [23].

The Murray-Darling Basin in Australia is the main agricultural area in the country. It is at present the most active water market in the world, and during the drought of 2002–2012, this market generated benefits in the range of several hundred million to 1 billion US dollars per year [24,25]. A challenge to water markets is the third party effects such as environmental impacts, which would reduce the benefits of trading. Water markets reduce streamflows because previously unused water allocations are traded, and also because gains in irrigation efficiency at a parcel level reduce drainage and return flows to the environment downstream. This reduction in basin return flows has been demonstrated in different settings [26–28]. Another worrying effect is the large surge in groundwater extractions, as shown in the last drought in the Murray-Darling Basin (Blewett [29] indicates that extractions between 2002 and 2007 were seven times above the allowed limits placed on groundwater users). The choice in Australia has been to mostly ignore the third party impacts of water markets [24].

Medellín *et al.* [30] estimate very large potential gains from water trading under droughts or climate change in California. These gains in the Central Valley of California are estimated at 1.4 billion US dollars. However, implementing these potential gains from trading is quite a challenge as the failure of the Water Bank experience in the 2009 drought shows. Water transfers were blocked by the water exporting regions and environmentalist NGOs.

Culp *et al.* [31] indicate that the highly complex institutional setting in western USA, including a set of restrictive laws and regulations on water rights, imposes significant obstacles to water trading. Thus, the transaction costs of trading are extremely high, limiting the achievement of the full potential of water markets. The issue of transaction costs has been analyzed by Regnacq *et al.* [32] for the case of California water markets. Their empirical results show the importance of the transaction costs linked to distance and institutional impediments in the decision to trade. Although part of these costs represents a legitimate means to avoid third party impacts (especially the natural environment), the rest of the costs could be reduced to increase trade. The attainment of the water markets solution seems to require well-functioning institutions, involving stakeholders' cooperation and more transparent administrative mechanisms.

In Spain, the approach to water management is based on institutional arrangements and relies on the river basin authorities. The basin authorities are responsible for water management, water allocation, control and enforcement, planning and waterworks. The special feature of this institutional arrangement is the key role played by stakeholders in managing the basin authority.

Stakeholders are part of the basin authorities, taking decisions in the basin governing bodies and in local watershed boards, and they are involved at all levels of decision making: planning, financing, waterworks, measures design, enforcement, and water management. The management of water is decentralized, with the basin authorities in charge of water allocation, and water user associations in charge of secondary infrastructure and water usage. The main advantage of this institutional setting is that stakeholders cooperate in the design and enforcement of decisions, rules and regulations, and therefore the implementation and enforcement processes are conducted smoothly [33].

Therefore, water allocation relies on the cooperation of stakeholders in basin authorities. Although water management in Spain is far from perfect, there have been recent mounting signs of successful experiences in the case of the La Mancha aquifers [34], where aquifer extractions have been curbed through stakeholders' cooperation.

Irrigated agriculture is the largest user of water in most arid and semiarid regions, and plays an important role in sustaining rural livelihoods and ecosystems. Adjustments to the shortfall of water supply in basins fall mainly on irrigation activities, which often trigger considerable economic and environmental impacts, and social conflicts. One important question for future policy debates is the identification of potential water management policies in irrigation. Suitable policies should improve economic efficiency, achieve environmental sustainability, and address equity when faced by growing scarcity, droughts and climate change.

Previous studies in the literature analyzed the advantages and limitations of the different approaches to allocate water in irrigation. We find that there is a gap in previous literature regarding the comparison between policy instruments in order to address the market externalities of water resources, and also to determine the relative efficacy of these different policy approaches. Filling this gap may improve the performance of water policies in many basins. The contribution of this paper is to apply state-of-art methodology in the direct comparison of three important water policy instruments: water markets, water pricing, and the status-quo institutional cooperation. The analysis is performed by formulating and applying an integrated hydro-economic model that links hydrological, economic, and environmental elements to assess the performance outcomes of these policies.

3. Materials and Methods

3.1. The Jucar River Basin

The Jucar River Basin (JRB) is located in the regions of Valencia and Castilla-La Mancha in Eastern Spain. It extends over 22,300 km^2 and covers the area drained by the Jucar River and its tributaries, mainly the Magro and the Cabriel Rivers (Figure 1). The Basin has an irregular Mediterranean hydrology, characterized by recurrent drought spells and normal years with dry summers.

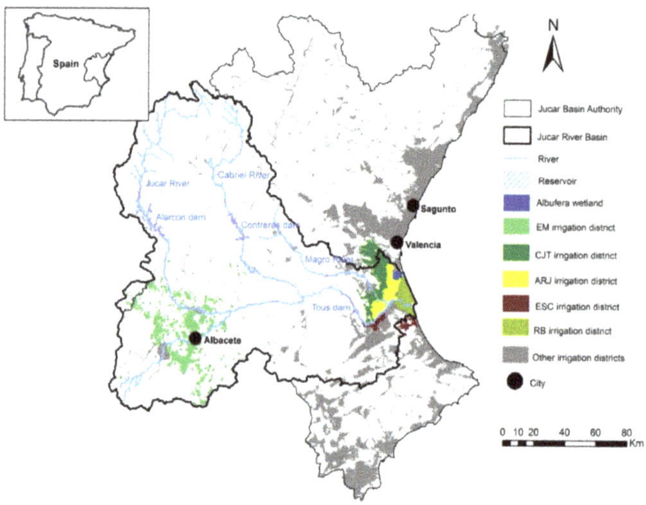

Figure 1. Map of the Jucar River Basin.

The JRB renewable water resources are nearly 1700 Mm^3/year but water extractions are very close to renewable resources, 1680 Mm^3, and the Basin is almost a closed water system. The main water use is irrigated agriculture with 1400 Mm^3, followed by urban and industrial uses of 270 Mm^3, which supply households, industries, and services of more than one million inhabitants (Table 1). There are also non-consumptive uses for hydropower, aquaculture and recreation.

Table 1. Water use by sector and origin in the JRB in a normal flow year (Mm^3). Source: CHJ [35].

Origin	Agriculture	Urban	Industrial	Total
Surface water	761	118	24	903
Groundwater	633	104	25	762
Reuse	11	0	1	12
Total	1405	222	50	1677

The irrigated area extends over 190,000 ha, and the main crops grown are rice, wheat, barley, garlic, lettuce, grapes, and citrus. There are three major irrigation areas located in the upper Jucar, the lower Jucar, and the bordering area of the Turia Basin. The Eastern La Mancha irrigation area (EM) is located in the upper Jucar, covering 100,000 ha. The irrigation districts of Acequia Real del Jucar (ARJ), Escalona y Carcagente (ESC), and Ribera Baja (RB) are in the lower Jucar, with an area of 35,000 ha. The irrigation district of Canal Jucar-Turia (CJT) is located in the bordering Turia Basin with an area of 22,000 ha (Table 2).

Table 2. The main water users in the JRB. Source. CHJ [35].

Water Users	Water Use (Mm^3)		
	Surface Water	Groundwater	Total
City of Albacete	17	0	17
EM aquifer irrigation district	13	386	399
Nuclear central of Cofrentes	14	0	14
City of Valencia	95	0	95
City of Sagunto	8	0	8
CJT irrigation district	70	91	161
ARJ irrigation district	213	0	213
ESC irrigation district	38	0	38
RB irrigation district	254	0	254
Other uses	193	285	478
Total JRB	915	762	1677

The expansion of water extractions in the Basin and the severe drought spells in recent decades have triggered considerable negative environmental and economic impacts. The growth of water extractions in recent decades has been driven especially by subsurface irrigation from the EM aquifer. The aquifer depletion, combined with other important water extractions in the Basin, and the recurrent drought spells have caused the water flows in the Jucar River to diminish. Environmental flows are dwindling in many parts of the Basin, resulting in serious damages to water-dependent ecosystems. There have been negative impacts on the downstream water users. For instance, the water available to the ARJ district has fallen from 700 to 200 Mm^3 in the last 40 years. Consequently, the dwindling return flows from the irrigation districts in the lower Jucar have caused serious environmental problems to the Albufera wetland, which is mostly fed by these return flows [36].

The Albufera wetland is the main aquatic ecosystem in the JRB. It is a fresh-water lagoon included in the RAMSAR list, and was declared a special protected area for birds (The RAMSAR convention is an international treaty for the conservation and sustainable utilization of wetlands). The Albufera wetland receives water from the return flows of the irrigation districts in the lower Jucar, mainly from the ARJ and the RB irrigation districts, and other flows originate from discharges of untreated and treated urban and industrial wastewaters. There is an important water quality problem driven by deficiencies in the sewage disposal and treatment systems in the adjacent municipalities, and by the reduced flows originating from the Jucar River that are used to improve the quality of wastewater discharges [37].

The increased frequency and intensity of drought spells during recent decades has been addressed by the Jucar Basin authority with investments in several long-term adaptation measures, such as construction of storage and regulation facilities, improvement of water efficiency through investment in irrigation systems, and installation of metering devices and special groundwater monitoring programs to control groundwater extractions.

3.2. The Modeling Framework

The comparison of policies is based on the hydro-economic model developed in Kahil *et al.* [38]. The model includes three components: (1) a reduced form hydrological sub-model; (2) a regional economic sub-model consisting of irrigation districts and urban centers; and (3) an environmental benefit sub-model. The reduced form hydrological sub-model is used to link the different components of the River Basin and to simulate the spatial hydrological impacts of droughts. The mathematical formulation of the reduced form hydrological sub-model is as follows:

$$Wout_d = Win_d - Wloss_d - Div_d^{IR} - Div_d^{URB} \tag{1}$$

$$Win_{d+1} = Wout_d + r_d^{IR} \cdot \left(Div_d^{IR}\right) + r_d^{URB} \cdot \left(Div_d^{URB}\right) + RO_{d+1} \tag{2}$$

$$Wout_d \geqslant E_d^{min} \tag{3}$$

where Equations (1)–(3) are the mass balance, the flow continuity, and the minimum-environmental flow constraints, respectively. These constraints determine the water available in the different river reaches that can be used after considering the environmental restrictions. $Wout_d$ is the water outflow from a river reach d; Win_d the water inflow to d; $Wloss_d$ the loss of water in d; Div_d^{IR} the water diversion to irrigation districts located in d; Div_d^{URB} the water diversion to urban and industrial activities located in d; Win_{d+1} the water inflow to the next river reach $d+1$; $\left[r_d^{IR} \cdot \left(Div_d^{IR}\right)\right]$ the return flows from irrigation districts; $\left[r_d^{URB} \cdot \left(Div_d^{URB}\right)\right]$ the return flows from urban and industrial activities; RO_{d+1} the runoff entering river reach $d+1$ from tributaries; and E_d^{min} the minimum environmental flow established for each river reach.

The regional economic sub-model accounts for the decision processes made by irrigation water users in the five major irrigation districts (EM, CJT, ARJ, ESC, and RB) and by urban users in the three main cities (Valencia, Albacete, and Sagunto). A farm-level programming component has been developed for each irrigation district, which maximizes farmers' private benefits from irrigation activities by choosing a crop mix subject to various technical and resource constraints. A Leontief production function technology is assumed with fixed input and output prices, in which farmers are price takers. The optimization problem is given by the following formulation:

$$Max\ B_k^{IR} = \sum_{ij} C'_{ijk} \cdot X_{ijk} \quad (4)$$

subject to

$$\sum_{i} X_{ijk} \leqslant Tland_{kj} \quad (5)$$

$$\sum_{ij} W_{ijk} \cdot X_{ijk} \leqslant Twater_k \quad (6)$$

$$\sum_{ij} L_{ijk} \cdot X_{ijk} \leqslant Tlabor_k \quad (7)$$

$$X_{ijk} \geqslant 0 \quad (9)$$

where B_k^{IR} is farmers' net benefits in irrigation district k. C'_{ijk} is a vector of coefficients of net income per hectare of crop i using irrigation technology j. The net income of each crop is equal to revenue minus direct and indirect costs, and amortizations. The decision variable in the optimization problem is X_{ijk}, corresponding to the area of crop i using irrigation technology j. Crops are aggregated into three representative crop groups: cereals, vegetables, and fruit trees. Irrigation technologies are flood, sprinkler, and drip.

Constraint Equation (5) represents the available area for irrigation equipped with technology j in irrigation district k, $Tland_{kj}$. The water constraint Equation (6) represents irrigation water availability in irrigation district k, $Twater_k$, which depends on surface and subsurface water extractions for that district. Parameter W_{ijk} is gross water requirements per hectare of each crop i using irrigation technology j. The labor constraint Equation (7) represents labor availability in irrigation district k, $Tlabor_k$. Parameter L_{ijk} is labor requirements per hectare of crop i using irrigation technology j.

For urban water uses, an economic surplus optimization scheme has been developed for each city in the Basin. The optimization problem maximizes social surplus given by the consumer and producer surplus from water use in each city, subject to several physical and institutional constraints. The optimization problem is:

$$Max\ B_u^{URB} = \left(a_{du} \cdot Q_{du} - \frac{1}{2} \cdot b_{du} \cdot Q_{du}^2 - a_{su} \cdot Q_{su} - \frac{1}{2} \cdot b_{su} \cdot Q_{su}^2 \right) \quad (9)$$

subject to

$$Q_{du} - Q_{su} \leqslant 0 \quad (10)$$

$$Q_{du}, Q_{su} \geqslant 0 \tag{11}$$

where B_u^{URB} is the consumer and producer surplus of city u. Variables Q_{du} and Q_{su} are water demand and supply by/to the city u, respectively. Parameters a_{du} and b_{du} are the intercept and slope of the inverse demand function, while parameters a_{su} and b_{su} are the intercept and slope of the water supply function. Equation (10) states that supply must be greater than or equal to demand. The quantity supplied, Q_{su}, is the connecting variable between urban use optimization components and the reduced form hydrological sub-model.

The environmental benefits sub-model accounts for the environmental benefits generated by the main aquatic ecosystem in the JRB, the Albufera wetland. The sub-model considers only water inflows to the Albufera wetland originating from irrigation return flows of the downstream ARJ and RB irrigation districts. Inflows and benefits of the Albufera wetland are given by the following expressions:

$$E_{Albufera} = \alpha \cdot r_{ARJ}^{IR} \cdot \left(D_{ARJ}^{IR}\right) + \beta \cdot r_{RB}^{IR} \cdot \left(D_{RB}^{IR}\right) \tag{12}$$

$$B_{Albufera} = \begin{cases} \rho_1 \cdot E_{Albufera} \; if \;\; 0 \leqslant E_{Albufera} \leqslant E_1 \\ \delta_2 + \rho_2 \cdot E_{Albufera} \; if \;\; E_1 < E_{Albufera} \leqslant E_2 \\ \delta_3 + \rho_3 \cdot E_{Albufera} \; if \; E_{Albufera} > E_2 \end{cases} \tag{13}$$

where Equation (12) determines the quantity of water flowing to the Albufera wetland, $E_{Albufera}$. Parameters α and β represent the shares of return flows that feed the wetland from the ARJ and RB irrigation districts, respectively. The products $[r_{ARJ}^{IR} \cdot (D_{ARJ}^{IR})]$ and $[r_{RB}^{IR} \cdot (D_{RB}^{IR})]$ are return flows from the ARJ and RB irrigation districts, respectively.

Equation (13) represents economic environmental benefits, $B_{Albufera}$, from the ecosystem services that the Albufera wetland provides to society. The environmental benefit function is assumed to be a piecewise linear function of water inflows, $E_{Albufera}$, to the wetland. This function expresses shifts in the ecosystem status when critical thresholds of water inflows E_1 and E_2 are reached, following the approach of Scheffer *et al.* [39]. The reason is that ecosystems do not always respond smoothly to changes in environmental conditions, and they may switch abruptly to a contrasting alternative state for certain critical levels. Time series data of various hydrological and chemical indicators have been collected to characterize the ecosystem health status of the wetland [35], along with economic valuation studies of the Albufera and other wetlands [40–42]. The specification and estimation of the environmental benefit function are described in Kahil *et al.* [38].

Detailed information on the technical coefficients and parameters of the hydro-economic model has been collected from field surveys, expert consultation, statistics, and reviewing the literature [35,43–47]. This information covers water inflows to the Basin, water diversion to users, urban water prices and costs, efficiency of primary and secondary conveyance channels, crop yields and prices, subsidies, production costs, amortizations, crop water requirements, crop labor requirements, land and labor availability, and groundwater extractions.

4. Comparison of Water Policies

Results from running the hydro-economic model are used to analyze the economic and environmental effects of the three alternative water policies designed to cope with scarcity and drought: the current institutional arrangement of the basin authority, water markets, and water pricing. Water markets and water pricing are implemented differently. First, the model is run to maximize the private benefits of irrigation and urban use. This solution entails the optimal water allocations and optimal shadow prices (Shadow prices reflect the economic value of water to users and their willingness to pay for it. In technical terms, the shadow price of water is the marginal value of water for the particular user, or the value the user obtains from applying an additional unit of water. In the water market policy, water is exchanged until the shadow prices are equalized among all users, following the equi-marginal rule). Two policies can be implemented to achieve this optimal solution. One is water markets, where trading among users leads to the optimal water allocations (which generate the corresponding shadow prices). The other policy is water pricing, where water taxes are used to align current water prices with the optimal shadow prices (which generate the corresponding water allocations). The resulting benefits for farmers are quite different under water markets and water pricing, since farmers make revenue when selling water with markets, but with water pricing, they lose revenue because of the water taxes.

The model provides results on the private benefits of users, environmental benefits, water use and return flows, and inflows to the Albufera wetland. Social benefits are assumed to be the sum of the private benefits from irrigation and urban use, and the environmental benefits (Table 3).

Two drought scenarios are considered, mild drought and severe drought. The reduction of water inflows over normal levels is 22 percent for mild droughts, and 66 percent for severe droughts. More information on the characterization of drought scenarios can be found in Kahil *et al.* [38]. The model simulates the outcomes of the three alternative policies to deal with these two drought scenarios.

Table 3. Policies under drought: institutional cooperation, water markets, and water pricing.

Drought Scenario	Normal Year	Mild Drought			Severe Drought		
Type of Water Policy	Current Situation (Institutional Cooperation)	Institutional Cooperation	Water Markets	Water Pricing	Institutional Cooperation	Water Markets	Water Pricing
			Water Use (Mm^3) [a]				
Irrigation districts	1030	908	908	908	683	683	683
EM	399	359	363	363	304	316	316
CJT	155	132	150	150	107	146	146
ARJ	200	180	197	197	131	185	185
ESC	33	30	32	32	18	31	31
RB	243	207	166	166	123	4	4
Urban use	119	105	105	105	74	74	74
Traded water	-	-	40	-	-	120	-
Environmental flows (inflows to Albufera)	60	52	50	50	34	29	29
			Private and Environmental Benefits (million Euros) [b]				
Private benefits							
Irrigation districts	190	171	175	93	136	148	54
EM	80	72	72	37	61	62	31
CJT	45	40	42	33	36	39	17
ARJ	34	31	32	17	23	25	4
ESC	7	7	7	5	4	5	2
RB	24	21	22	1	12	17	0
Urban use	283	276	276	276	241	241	241
Total	473	447	451	369	377	389	295
Environmental benefits	75	37	32	32	22	19	19
Social benefits	548	484	483	401	399	408	314

Note: [a] Water allocations to irrigation, urban use and environment in million cubic meters. [b] Private benefits from irrigation and urban use, and environmental benefits in million Euros.

Institutional cooperation is the baseline policy, and represents the current water management to cope with scarcity and droughts. The basin authorities are the main administrative bodies responsible for water management, and they are organized around the governing boards, the stakeholder boards, and the management services. An important feature of basin authorities is the involvement of stakeholders, which has been a permanent characteristic since their creation in the 1920's. Stakeholders include water users, public administrations, farmers' unions and environmental groups. The stakeholders' representatives are present in all governing and participation bodies at the basin scale, and run the watershed boards at the local scale [33].

This approach entails flexible adaptive changes in water allocations based on the negotiation and cooperation of users, where water stakeholders are involved in the decision making process, including the environmental concerns. The water allocations that result from cooperation are observed in the data from both normal and drought periods.

The *water market* policy opens up water trading between economic agents in irrigation districts and urban centers. Economic theory predicts that water markets achieve welfare gains by reallocating water from low to high marginal values of water, and this efficient use of water maximizes the total private benefits summed over agents. The model is used to test the water market policy alternative, and empirically estimate the market potential welfare gains. Water trade becomes more pronounced as drought severity intensifies, reaching 120 Mm^3 under severe drought. The main effect is the improvement of irrigation efficiency, but also the subsequent fall in irrigation return flows, which further reduce the environmental flows in the basin.

The *water pricing* policy achieves also the efficient use of water by adjusting water prices to balance water demand with the available water supply during drought. This policy alternative is in line with the water pricing policy advocated by the European Water Framework Directive, reiterated in the recent Blueprint to Safeguard Europe's Water Resources [6]. Water prices in each irrigation district and urban center are set equal to the marginal value of water at the efficient level of water use, which is the market-clearing price. This water tax revenue is collected by the public and private water agencies responsible for water supply. All or part of this revenue may be employed outside the basin areas, representing a loss of benefit from both individual farmer and basin perspectives. One advantage of the *water pricing* policy is that it assures the financial viability of the water agencies, which could guarantee their operation without the need of public subsidies. As indicated above, the water taxes levied with water pricing involve significant revenue losses for farmers.

Social benefits under the *institutional* or baseline policy in normal flow conditions amount to 548 million Euros. Private benefits are 190 million Euros for irrigation and 283 million for urban demand, from using 1030 and 119 Mm^3 of water, respectively. Environmental benefits provided by the Albufera wetland are 75 million Euros, and the Albufera wetland receives 60 Mm^3 of return flows from the ARJ and RB irrigation districts, which support the ecological status of the wetland.

4.1. Mild Drought Scenario

Mild drought events reduce social benefits by 65 million Euros under the institutional and water market policies, but the social benefits are reduced by 150 million Euros under water pricing. The environmental losses are close to 40 million Euros under all policies, cutting environmental benefits by half. The difference among policies is the irrigation losses, which are below 20 million under institutional and water market policies, but escalate to 100 million under water pricing. Therefore the large benefit losses from the water pricing policy are driven by the large impact of pricing on irrigation profits.

The environment sustains significant benefit losses derived from the reduction of water inflows to the Albufera wetland. These water inflows under water markets and water pricing fall below the critical threshold E_1, creating a regime shift in the wetland. The institutional policy achieves higher environmental benefits because it allocates more water to the Albufera wetland, avoiding further desiccation and ecosystem degradation.

The effects on the urban sector are moderate both in terms of water allocations and private benefits. The reason is the priority rules under the institutional policy, and also the availability of additional water sources at higher costs from neighboring basins in the case of Valencia and Sagunto (Turia Basin), or groundwater in the case of Albacete.

Farmers face diminishing water use from drought and reduced crop acreage, mostly cereals because these are the less profitable crops. The allocation of irrigation water to the RB, ARJ and CJT districts changes between the institutional and water market policies. Water markets allocate 40 Mm^3 less water to RB, and this water is assigned to ARJ and CJT. These water exchanges are driven by the differences among water shadow prices in districts. As indicated above, the shadow prices of water are the marginal values of water in each location, and therefore water exchanges reallocate water from locations with low marginal values of water to locations with high marginal values where water is more profitable. However, the private benefits of all irrigation districts are almost the same under both the institutional and water market policies.

The opportunity costs of policies incurred by farmers are the benefit losses sustained under each policy. A steep increase in the opportunity costs of a particular policy would be met by opposition from farmers leading to policy failure, given that other feasible policies are less costly. The costs of the water pricing policy are very high for farmers compared to the institutional or water market policies, with irrigation benefits falling by half when water pricing is implemented instead of the other policies. The reason for these high costs is the large losses sustained by farmers from taxing water. Opposition to the water pricing policy would be strong in the RB, EM and ARJ districts, where the opportunity costs of implementing water pricing are especially damaging to farmers. This empirical finding shows that the institutional and water market policy options are much more feasible and equitable than water pricing, because water pricing involves disproportionate costs to farmers.

4.2. Severe Drought Scenario

The effects of severe drought are more pronounced than those of mild drought, although they show similar patterns. The fall in social benefits is almost 150 million Euros under the institutional and water market policies, but social benefits losses escalate to almost 250 million under water pricing. Environmental benefits sustain quite large losses, although the institutional policy allocates slightly more environmental flows to the Albufera wetland.

The irrigation benefits by district are almost the same under the institutional and water market policies, and the main difference is the change in water allocation to the RB, ARJ and CJT districts. Compared to the institutional policy, water markets are driven by the shadow prices of water, reallocating water from locations with low marginal value of water to locations with high marginal value. Water trading allocates more water to the ARJ, ESC, EM and CJT districts by reducing the allocation of the RB district by 120 Mm^3.

Choosing the water pricing policy under severe drought is quite detrimental to farmers because water taxes escalate, and they cannot generate revenue by selling water. The implementation of water pricing instead of the institutional or water market policies, makes farmers lose two thirds of their private profits. In districts such as RB and ARJ, the private benefits of farmers are almost entirely wiped out. The opportunity costs for farmers of the water pricing policy are disproportionate.

The total costs and their distribution among those who bear them from confronting a severe drought in the Jucar Basin by the irrigation, urban and environmental sectors depend on the policy selected by decision makers, and these costs are given by the benefit losses incurred by each sector. These costs are 42 million Euros for the urban sector (283–241) and 53 million for the environment (75–22) regardless of the policy chosen, but these costs triple from 50 million Euros (190–140) to almost 150 million (190–54) for the irrigation sector by selecting the water pricing policy instead of the other policies.

4.3. Additional Measures to Protect The Environment

Protecting environmental flows, especially during droughts, is a major challenge in almost all basins in arid and semiarid regions. In these basins, regulators face a challenge to enforce environmental flows not only because they have to control surface and subsurface extractions, but also because the irrigation returns component of environmental flows is even more difficult to regulate than water extractions. Examples of these management difficulties include basins where water management efforts are quite sophisticated, such as the Jucar basin in Spain, the Murray-Darling Basin in Australia, and the Central Valley in California (In the Jucar Basin, there was a desiccation of the Jucar mainstem during the last drought [34]. In the Murray-Darling Basin, groundwater depletion reached 104 km^3 during the last drought [29]. In the Central Valley of California, groundwater depletion has reached 80 km^3 during the current drought [48]).

Two additional measures are considered for the JRB to protect environmental flows, one associated with water markets and the other with the institutional policy. The first measure follows the example of the Murray-Darling Basin, where a very expensive program is being implemented to recover water for the environment using a public water buyback program [49]. Although expensive, this seems to be a workable policy to reap most of the private benefits of pure water markets while protecting ecosystems, and this could be called the environmental water market. The second measure is to improve the current institutional stakeholder cooperation in Jucar, by including environmental stakeholders as full participants. These augmented environmental flows are achieved by the negotiation among all economic and environmental stakeholders, which appears to be a sustainable institutional policy.

Both the environmental water market and the sustainable institutional policies achieve large gains in environmental benefits, above 200 Million Euros in mild and severe droughts, with social benefits in the Basin reaching around 730 million Euros under mild drought and 660 million Euros under severe drought [38,50].

5. Conclusions

The sustainable use of water resources requires a reliable understanding of the main processes and their linkages, an accurate assessment of impacts, and improving management by stakeholders and governance by policy makers to deal with water scarcity, droughts and climate change. Sound management and governance is quite a challenge because of the wide and complex range of goods and services provided by water, including private goods, common pool resources, and public goods.

This paper presents an empirical assessment of three water policy instruments to address water scarcity and droughts: water pricing, water markets, and common property governance. A direct comparison of the three policies is made by developing and applying an integrated hydro-economic model of the Jucar Basin in Spain, analyzing the economic and environmental effects of each policy.

Water pricing and water markets are economic instruments that work well when water is a private good, but less well when water is a common pool resource or public good. Studies in California and Australia demonstrate the large gains of water markets, both potential gains in California [22,30] and actual gains in Australia [24,25].

We present evidence from Spain, a community with an ancient tradition of cooperation among stakeholders in water user associations dating back centuries. Evidence from Spain regarding alternative proposed policy instruments is derived from the Jucar Basin, where water markets, water pricing, and institutional policies are simulated under drought.

The empirical results highlight that both institutional and water market policies are economically-efficient instruments to limit the economic damage costs of droughts, achieving similar social benefits in terms of private and environmental benefits. This finding is important because it shows that in the case of Jucar, the status quo institutional policy can attain almost the same private benefits as water markets.

The advantages of water markets compared to the institutional policy of stakeholders' cooperation are a slight reduction in land fallowing, a small improvement in irrigation efficiency, and a more even distribution of drought losses among irrigation districts, important for equity concerns. Water markets minimize private economic damages from drought but disregard the environmental benefits. Results show that water markets entail a reduction of water for environmental purposes, causing faster ecosystem regime shifts compared to the current institutional setting. The reason lies with the public good characteristic of environmental flows, which are external to markets, leading to excessive ecosystem degradation. This is important when planning for a future with climate change and emerging social demands for and economic benefits from aquatic ecosystem protection.

Water pricing is the policy advocated by the European WFD. This policy poses important implementation challenges in arid and semiarid regions such as Spain, where irrigation is the largest user of water, with strong impacts on the supply of a wide range of ecosystem services. The water pricing policy for managing drought is detrimental to farmers. Implementing water pricing instead of water markets or institutional policies, increases farmers' losses by 80 and 100 million Euros, a high percentage of their base incomes, under mild and severe drought, respectively.

These benefit losses are the opportunity costs of the water pricing policy to farmers, and the steep opportunity costs of water pricing would be economically and politically damaging. The main empirical finding on water pricing is that farmers lose from half to two thirds of their net benefits when the water pricing policy is implemented during drought, instead of the water market or institutional policies. Enforcing water pricing will become a difficult task facing tough political and technical hurdles.

The empirical results show that water market and institutional policies are much more economically attractive and equitable than water pricing, because water pricing involves disproportionate costs to farmers. There are also additional measures for these two policies that could enhance the protection of environmental flows. One measure is public water buyback programs for water markets, in order to reap the benefits of water markets while protecting ecosystems. The other measure is greening the cooperation in the institutional policy, by including the environment as a full stakeholder in the process of water allocation among sectors and spatial locations. However, protecting the environment with water pricing will require adding further "environmental" and "resource use" costs to water prices (in WFD terminology), resulting in highly disproportionate costs to farmers.

Water management in the JRB is based on the negotiation and cooperation of stakeholders, which seems to provide a worthwhile prospect for sustainable water management in irrigation. In fact, this approach achieves better environmental outcomes compared to other policy instruments, and almost the same outcomes in terms of farmers' private benefits and social benefits compared to the water market policy. However, the status quo institutional-based approach poses difficult implementation challenges in real-world situations. The reasons are that institutions may involve asymmetric negotiation power among the stakeholders, while the severe scarcity of water resources may considerably reduce incentives for cooperation.

The evidence from the JRB highlights that, despite these limitations, the status quo institutional-based approach of stakeholders' cooperation was able to reduce environmental and economic damages during the last drought period, and to surrogate social conflicts by cooperation. The JRB experience suggests that the implementation of the institutional approach in managing water resources requires sufficient institutional capacity to deal with power asymmetry and resource scarcity, as well as available social capital supporting cooperation, which is particularly necessary for the promotion of self-regulation initiatives.

Acknowledgments: This study was financed by project INIA RTA2010-00109-C04 and INIA RTA2014-00050-00-00 from the Spanish Ministry of Economy and Competitiveness partially financed by ERDF funds, and project BIL/13/MA/072 from MAPFRE Foundation. The Ministry supported also a Ph.D scholarship of the first author. Among individuals, special assistance has been provided by Manuel Pulido-Velazquez (UPV), Alfonso Calera and David Sanz (IDR), and Maria Calera (UCL).

Author Contributions: Mohamed Taher Kahil, Ariel Dinar and Jose Albiac designed the methodology. All authors carried out the model setup, calibration and validation, as well as the simulation of water policies. Mohamed Taher Kahil, Jose Albiac and Ariel Dinar wrote the first draft of the manuscript, with further contributions from the rest of the authors.

Conflicts of Interest: The authors declare no conflict of interest.

References

1. United Nations Development Programme (UNDP). *Human Development Report 2006: Beyond Scarcity: Power, Poverty and the Global Water Crisis*; UNDP: New York, NY, USA, 2006.
2. Konikow, L. Contribution of global groundwater depletion since 1900 to sea-level rise. *Geophys. Res. Lett.* **2011**, *38*.
3. Intergovernmental Panel on Climate Change (IPCC). *Climate Change 2014: Impacts, Adaptation, and Vulnerability*; Contribution of Working Groups II to the Fifth Assessment Report of the IPCC; IPCC: Geneva, Switzerland, 2014.
4. Booker, J.; Howitt, R.; Michelsen, A.; Young, R. Economics and the Modeling of Water Resources and Policies. *Nat. Resour. Model.* **2012**, *25*, 168–218.
5. Pigou, A. *The Economics of Welfare*; Macmillan: New York, NY, USA, 1920.
6. European Commission (EC). *Communication from the Commission to the European Parliament, the Council, the European Economic and Social Committee and the Committee of the Regions. A Blueprint to Safeguard Europe's Water Resources*; COM(2012) 673 Final; European Commission: Brussels, Belgium, 2012.
7. Coase, R. The problem of social cost. *J. Law Econ.* **1960**, *3*, 1–44.
8. National Water Commission (NWC). *Water Markets in Australia: A Short History*; NWC: Canberra, Australian, 2011.

9. Ostrom, E. *Governing the Commons: The Evolution of Institutions for Collective Action*; Cambridge University Press: Cambridge, UK, 1990.
10. Ostrom, E.; Burger, J.; Field, C.; Norgaard, R.; Policansky, D. Revisiting the Commons: Local Lessons, Global Challenges. *Science* **1999**, *284*, 278–282.
11. International Conference on Water and the Environment (ICWE). *The Dublin Statement and Report of the Conference*; World Meteorological Organization: Geneva, Switzerland, 1992.
12. Che, Y.; Shang, Z. Water Pricing in China: Impact of Socioeconomic Development. In *Water Pricing Experiences and Innovations*; Dinar, A., Pochat, V., Albiac, J., Eds.; Springer: Dordrecht, The Netherlands, 2015.
13. Hanemann, W. Determinants of Urban Water Use. In *Urban Water Demand Management and Planning*; Bauman, D., Boland, J., Hanemann, W., Eds.; McGraw-Hill: New York, NY, USA, 1998.
14. Moore, M. The bureau of reclamations new mandate for irrigation water conservation-purposes and policy alternatives. *Water Resour. Res.* **1991**, *27*, 145–155.
15. Scheierling, S.; Young, R.; Cardon, G. Determining the price responsiveness of demands for irrigation water deliveries *versus* consumptive use. *J. Agric. Resour. Econ.* **2004**, *29*, 328–345.
16. Cornish, G.; Bosworth, B.; Perry, C.; Burke, J. *Water Charging in Irrigated Agriculture. An Analysis of International Experience*; FAO Water Report No. 28: Rome, Italy, 2004.
17. Garrido, A.; Calatrava, J. Trends in water pricing and markets. In *Water Policy in Spain*; Garrido, A., Llamas, M., Eds.; CRC Press: Leiden, The Netherlands, 2009.
18. Calatrava, J.; Guillem, A.; Martinez-Granados, D. Análisis de alternativas para la eliminación de la sobreexplotación de acuíferos en el Valle del Guadalentín. *Econ. Agrar. Recur. Nat.* **2011**, *11*, 33–62.
19. Tsur, Y.; Dinar, A.; Doukkali, R.; Roe, T. Irrigation water pricing: Policy implications based on international comparison. *Environ. Dev. Econ.* **2004**, *9*, 735–755.
20. Easter, K.; Dinar, A.; Rosegrant, M. The Performance of Water Markets: Transaction Costs, Interjurisdictional Barriers and Institutional Options. In *Conflict and Cooperation on Trans-Boundary Water Resources*; Just, R., Netanyahu, S., Eds.; Kluwer: Norwell, MA, USA, 1998.
21. Connor, J.; Schwabe, K.; King, D.; Kaczan, D.; Kirby, M. Impacts of climate change on lower Murray irrigation. *Aust. J. Agric. Resour. Econ.* **2009**, *53*, 437–456.
22. Howitt, R.; Medellín, J.; MacEwan, D.; Lund, J. Calibrating disaggregate economic models of agricultural production and water management. *Environ. Model. Softw.* **2012**, *38*, 244–258.
23. Dinar, A.; Rosegrant, M.; Meinzen-Dick, R. *Water Allocation Mechanisms: Principles and Examples*; Policy Research Working Paper No. WPS 1779; World Bank: Washington, DC, USA, 1997.
24. Kirby, M.; Bark, R.; Connor, J.; Qureshi, E.; Keyworth, S. Sustainable irrigation: How did irrigated agriculture in Australia's Murray-Darling Basin adapt to the Millennium drought? *Agric. Water Manag.* **2014**, *145*, 154–162.

25. Connor, J.; Kaczan, D. Principles for economically efficient and environmentally sustainable water markets: The Australian experience. In *Drought in Arid and Semi-Arid Environments: A Multi-Disciplinary and Cross-Country Perspective*; Schwabe, K., Albiac, J., Connor, J., Hassan, R., Meza, L., Eds.; Springer: Dordrecht, The Netherlands, 2013.
26. Qureshi, M.; Schwabe, K.; Connor, J.; Kirby, M. Environmental water incentive policy and return flows. *Water Resour. Res.* **2010**, *46*.
27. Howe, C.; Schurmeier, D.; Shaw, W. Innovative Approaches to Water Allocation: The Potential for Water Markets. *Water Resour. Res.* **1986**, *22*, 439–449.
28. Kahil, M.T.; Connor, J.; Albiac, J. Efficient water management policies for irrigation adaptation to climate change in Southern Europe. *Ecol. Econ.* **2015**, *120*, 226–233.
29. Blewett, R. *Shaping a Nation. A Geology of Australia*; Geoscience Australia-ANU Press: Canberra, Australia, 2012.
30. Medellín, J.; Howitt, R.; Lund, J. Modeling Economic-Engineering Responses to Drougth: The California Case. In *Drought in Arid and Semi-Arid Environments: A Multi-Disciplinary and Cross-Country Perspective*; Schwabe, K., Albiac, J., Connor, J., Hassan, R., Meza, L., Eds.; Springer: Dordrecht, The Netherlands, 2013.
31. Culp, P.; Glennon, R.; Libecap, G. *Shopping for Water: How the Market can Mitigate Shortages in the American West. Discussion Paper 2014-05*; Stanford Woods Institute for the Environment: Washington, DC, USA, 2014.
32. Regnacq, C.; Dinar, A.; Hanak, E. The gravity of water: Water trade friction in California. In Proceedings of the Meetings of the Allied Social Science Association, San Francisco, CA, USA, 3–5 January 2016.
33. Albiac, J.; Esteban, E.; Tapia, J.; Rivas, E. Water scarcity and droughts in Spain: Impacts and policy measures. In *Drought in Arid and Semi-Arid Environments: A Multi-Disciplinary and Cross-Country Perspective*; Schwabe, K., Albiac, J., Connor, J., Hassan, R., Meza, L., Eds.; Springer: Dordrecht, The Netherlands, 2013.
34. Esteban, E.; Albiac, J. The problem of sustainable groundwater management: The case of La Mancha aquifers, Spain. *Hydrogeol. J.* **2012**, *20*, 851–863.
35. Confederación Hidrográfica del Júcar (CHJ). *Esquema Provisional de Temas Importantes*; Ministerio de Medioambiente: Valencia, Spain, 2009.
36. García-Molla, M.; Sanchis, C.; Ortega, M.; Avellá, L. Irrigation Associations Coping with Drought: The Case of Four Irrigation Districts in Eastern Spain. In *Drought in Arid and Semi-Arid Environments: A Multi-Disciplinary and Cross-Country Perspective*; Schwabe, K., Albiac, J., Connor, J., Hassan, R., Meza, L., Eds.; Springer: Dordrecht, The Netherlands, 2013.
37. Sanchis, C. La Albufera de Valencia: Cincuenta años de eutrofia. *Métode* **2011**, *70*, 32–41.
38. Kahil, M.T.; Dinar, A.; Albiac, J. Modeling water scarcity and droughts for policy adaptation to climate change in arid and semiarid regions. *J. Hydrol.* **2015**, *522*, 95–109.
39. Scheffer, M.; Carpenter, S.; Foley, J.; Walker, B. Catastrophic shifts in ecosystems. *Nature* **2001**, *413*, 591–596.
40. Del Saz, S.; Perez, L. El valor de uso recreativo del parque natural de L'Albufera a través del método indirecto del coste de viaje. *Estud. Econ. Apl.* **1999**, *11*, 41–62.

41. Woodward, R.; Wui, Y. The economic value of wetland services: A meta-analysis. *Ecol. Econ.* **2001**, *37*, 257–270.
42. Brander, L.; Florax, R.; Vermaat, J. The empirics of wetland valuation: A comprehensive summary and a meta-analysis of the literature. *Environ. Resour. Econ.* **2006**, *33*, 223–250.
43. Generalitat Valenciana (GV). *Base de Datos 1T de Superficies de Cultivos por Término Municipal Para la Generalitat Valenciana 2009*; Departamento de Agricultura, Generalitat Valenciana: Valencia, Spain, 2009.
44. Gobierno de Castilla La Mancha (GCLM). *Base de Datos 1T de Superficies de Cultivos Por Término Municipal Para Castilla La Mancha 2009*; Departamento de Agricultura, Gobierno de Castilla La Mancha: Albacete, Spain, 2009.
45. Instituto Nacional de Estadística (INE). *Censo Agrario 2009*; INE: Madrid, Spain, 2009.
46. Confederación Hidrográfica del Júcar (CHJ). *Sistema Júcar: Datos Del Año Hidrológico 2012–2013*; Ministerio de Medioambiente: Valencia, Spain, 2012.
47. Ministerio de Medio Ambiente y Medio Rural y Marino (MARM). *Análisis de la Economía de Los Sistemas de Producción: Resultados Técnico-Económicos de Explotaciones Agrícolas de Valencia y Castilla La Mancha en 2009. Subsecretaria de Medio Ambiente, Medio Rural y Marino*; MARM: Madrid, Spain, 2010.
48. University of California Center for Hydrologic Modeling (UCCHM). *UCCHM Water Advisory # 1*; University of California: Irvine, CA, USA, 2014.
49. Wheeler, S.; Loch, A.; Zuo, A.; Bjornlund, H. Reviewing the adoption and impact of water markets in the Murray-Darling Basin, Australia. *J. Hydrol.* **2014**, *518*, 28–41.
50. Kahil, M.T.; Dinar, A.; Albiac, J. Cooperative water management and ecosystem protection under scarcity and drought in arid and semiarid regions. *Water Resour. Econ.* **2015**.

Simulation of Groundwater-Surface Water Interactions under Different Land Use Scenarios in the Bulang Catchment, Northwest China

Zhi Yang, Yangxiao Zhou, Jochen Wenninger, Stefan Uhlenbrook and Li Wan

Abstract: Groundwater is the most important resource for local society and the ecosystem in the semi-arid Hailiutu River catchment. The catchment water balance was analyzed by considering vegetation types with the Normalized Difference Vegetation Index (NDVI), determining evapotranspiration rates by combining sap flow measurements and NDVI values, recorded precipitation, measured river discharge and groundwater levels from November 2010 to October 2011. A simple water balance computation, a steady state groundwater flow model, and a transient groundwater flow model were used to assess water balance changes under different land use scenarios. It was shown that 91% of the precipitation is consumed by the crops, bushes and trees; only 9% of the annual precipitation becomes net groundwater recharge which maintains a stable stream discharge in observed year. Four land use scenarios were formulated for assessing the impacts of land use changes on the catchment water balance, the river discharge, and groundwater storage in the Bulang catchment. The scenarios are: (1) the quasi natural state of the vegetation covered by desert grasses; (2) the current land use/vegetation types; (3) the change of crop types to dry resistant crops; and (4) the ideal land use covered by dry resistant crops and desert grasses, These four scenarios were simulated and compared with measured data from 2011, which was a dry year. Furthermore, the scenarios (2) and (4) were evaluated under normal and wet conditions for years in 2009 and 2014, respectively. The simulation results show that replacing current vegetation and crop types with dry resistant types can significantly increase net groundwater recharge which leads to the increase of groundwater storage and river discharges. The depleted groundwater storage during the dry year could be restored during the normal and wet years so that groundwater provides a reliable resource to sustain river discharge and the dependent vegetations in the area.

Reprinted from *Water*. Cite as: Yang, Z.; Zhou, Y.; Wenninger, J.; Uhlenbrook, S.; Wan, L. Simulation of Groundwater-Surface Water Interactions under Different Land Use Scenarios in the Bulang Catchment, Northwest China. *Water* **2015**, *7*, 5959–5985.

1. Introduction

Water is the most important limiting factor for agricultural production and ecosystem protection in semi-arid conditions [1]. To maintain a delicate balance between the protection of the ecosystem and the sustainable development of local society is critical task in semi-arid regions. Optimizing water use efficiency for agricultural crops [2] is a key approach to mitigate water shortages and to reduce environmental problems in arid and semi-arid regions. Along with the water shortage limitation for agricultural productivity, the desertification has been controlled by planting shrubs as ecosystem rehabilitation measures in recent decades in the semi-arid parts of northwest China [3]. However, the water balance in semi-arid catchments can be significantly influenced by vegetation type [4], irrigation schedules [5], and groundwater irrigation [5]. Brown *et al.* [6] determined that the changes in water yield at various time scales were a result of permanent changes in the vegetation cover by means of mean annual water balance model. Jothityangkoon *et al.* [7] concluded that spatial variability of soil depths appears to be the most important controlling parameter for runoff variability at all time and space scales, followed by the spatial variability of climate and vegetation cover in semi-arid catchments. The differences in water balance components between a number of temperate and semi-arid catchments in Australia can be attributed to the variability of soil profile characteristics like water storage capacity and permeability, vegetation coverage and water use efficiency, rainfall, and potential evaporation [8]. Scott *et al.* [9] indicated that grassland relies primarily on recent precipitation, while the tree/shrub obtained water from deeper parts of the soil profile in the semi-arid riparian floodplain of the San Pedro River in southeastern Arizona. Planting of shrub seedlings can significantly enhance topsoil development on the dune surface and stabilizing the sand dunes [10], but the water consumption of artificially introduced plants and the effects on the water balance have not been investigated.

Land use management and rehabilitation strategies would have significant impact on the catchment water balance and hence on water yield and groundwater recharge [11]. Consequently, understanding impacts of land use change on the catchment water balance dynamics is critical for sustainable water resources management. Stream flow [12–14], flow regime [15], and the river water quality [16,17] can be influenced by the change of land use/land cover in the catchment. Compared with the mechanism of land use effects on the surface water, the groundwater recharge [18,19], discharge [20,21], levels [22], hydrochemistry and contamination [23], and nitrate concentrations [24–27] can be indirectly affected by the land use changes through infiltration. Krause *et al.* [28] assessed the impacts of different strategies for managing wetland water resources and groundwater dynamics of landscapes based on the analysis of model simulation results of complex scenarios for land-use changes and changes of the density of the drainage-network, but regional groundwater modeling studies are often hampered by data scarcity in space and time especially in semi-arid regions [29]. Thus, the estimation of areal inputs such as precipitation and actual evapotranspiration (ET) is essential for groundwater model studies. Actual ET corresponds to the real water consumption and is usually estimated by considering weather parameters, crop factors, management and environmental conditions [30]. Despite of some uncertainties and inconsistencies in the results, remote sensing is a useful technique for the study of groundwater hydrology and has aided the successful location of important groundwater resources [31]. Remote sensing and Geographic Information Systems (GIS) have been used for the investigation of springs [32], determining the groundwater dependent ecosystems [33], determining the recharge potential zones [34], mapping groundwater recharge and discharge areas [35], detecting potential groundwater flow systems [36], and monitoring infiltration rates in semi arid soils [37]. Remote sensing data can be employed for estimating the ET by means of energy balance methods, statistical methods using the difference between surface and air temperature, surface energy balance models, and spatial variability methods at different scales [38]. Remote sensing and the Normalized Difference Vegetation Index (NDVI) have been widely employed for estimating groundwater evapotranspiration [39,40], investigating the relationship between vegetation growth and depth to groundwater table [41–44], and accessing groundwater recharge fluxes [45]. Many researchers have conducted direct measurements like the sap flow method which uses the stem heat balance technique [46], and scaled these values to the catchment level transpiration [47]. The remote sensing could offer the relevant spatial data and parameters at the appropriate scale for use in distributed hydrological models [48] and groundwater models [49,50]. Those studies trend to utilize the areal ET calculated from the theory of surface energy balance with remote sensing data at lager scale. However, few studies have been conducted on simulating the groundwater response to different land use scenarios by determining areal ET using field measurements and remote sensing data in semi-arid regions particularly in Asia.

Most recent studies focus on the simulation of land use or climate change impacts for improving the water use efficiency for agricultural production, but few have been conducted on artificially introduced plants for stabilizing sand dunes and the effects on the water resources in semi-arid regions. Since evapotranspiration from crops and shrubs dominate groundwater discharge in semi-arid catchments [51], a balance must be achieved between land use (prevention of desertification) and water resources conservation for agricultural and other purposes. Sap flow measurements can directly provide transpiration rates of the individual vegetation in carefully selected sites. However, the uncertainty of estimating areal ET rates varies in space and time through the necessary up-scaling for the areal computation. Remote sensing technologies have been recently employed for calculating the areal ET by means of energy balance at large scale, but few were conducted using combined field measurements and remote sensing data (NDVI) in semi-arid regions. Multiple procedures have been employed in this study in order to evaluate the influence of land use management on hydrological processes in the semiarid Bulang catchment. The methods used in this study consisted of identifying vegetation types with NDVI values, determining evapotranspiration rates with sap flow measurements in the field, and computing the areal ET rates by a combination of remote sensing data (NDVI) and field measurements. Impacts of different land use scenarios on water resources were simulated by a groundwater model with the aid of combined field measurements, remote sensing, and GIS techniques, which reveals the catchment water balance in the sandy region of the middle section of the Yellow River Basin. The results provide scientific information to support rational land use management and water resources conservation in semi-arid areas.

2. Materials and Methods

2.1. Study Area

The Bulang River is a tributary of the Hailiutu River, which is located in the middle reach of the Yellow River in northwest China (Figure 1). Located in a semi-arid region, the total area of the Bulang catchment is 91.7 km^2. The land surface is characterized by undulating sand dunes, scattered desert bushes (*Salix Psammophila*), cultivated croplands (*Zea mays*), and a perennial river in the southwestern downstream area. The surface elevation of the Bulang catchment ranges from 1300 m at the northeastern boundary to 1160 m above mean sea level at the catchment outlet in the southwest. The long-term annual average daily mean temperature is 8.1 °C and the monthly mean daily air temperature is below zero in the winter time from November until March. The mean annual precipitation measured at the nearby Wushenqi meteorological station for the period 1984 to 2011 was 340 mm/year. Precipitation mainly falls in June, July, August and September. The mean annual pan evaporation (recorded with an evaporation pan with a diameter of 20 cm) is 2184 mm/year (Wushenqi metrological station, 1985–2004). The geological formations in the Bulang catchment mainly consist of four strata (1) the Holocene Maowusu sand dunes with a thickness from 0 to 30 m; (2) the upper Pleistocene Shalawusu formation (semi consolidated sandstone) with a thickness of 5 to ~90 m; (3) the Cretaceous Luohe sandstone with a thickness of 180 to ~330 m; and (4) the underlying impermeable Jurassic sedimentary formation.

2.2. *Methods*

2.2.1. Water Balance

The catchment water balance can be calculated with the components of precipitation (P), evapotranspiration (ET), discharge of the Bulang stream (Q), deep groundwater circulation discharging to the main Hailiutu River (D), and the change of storage (dS/dt) in the catchment as:

$$P\text{-}ET\text{-}Q\text{-}D = \frac{dS}{dt} \quad (1)$$

Precipitation, discharge of the Bulang River, and the change of the groundwater levels that represent the storage change in the system can be directly measured in the field, but not the ET and deep circulation fluxes. In this study, the areal ET was estimated from the site measurements of sap flow of maize, salix bush, and willow tree which were up scaled by using the NDVI generated vegetation cover from remote sensing data.

2.2.2. Groundwater Discharge

The groundwater discharge as the baseflow was separated from daily average river discharge measurements according to hydrograph separation results using isotopic tracers [51]. Daily river discharge can be considered as groundwater discharge during days without rainfall, while 74.8% of the river discharge originated from the groundwater in the two days after the recorded heavy rainfall based on the trace method [51].

2.2.3. Groundwater Model

In correspondence with the geological structure, the upper Maowusu sand dunes together with Shalawush deposits were simulated with one model layer as the unconfined aquifer and the underlying Cretaceous Luohe sandstone was simulated with another model layer as the confined aquifer. The outlet of the river valley was simulated as a head-dependent flow boundary, considered to be the deep groundwater discharge to the Hailiutu River. All other boundaries were taken as no-flow boundaries in line with the catchment water divide. The Bulang stream was simulated as a drain since groundwater always discharges to the stream. Net groundwater recharge was calculated as precipitation (P) minus evapotranspiration (ET).

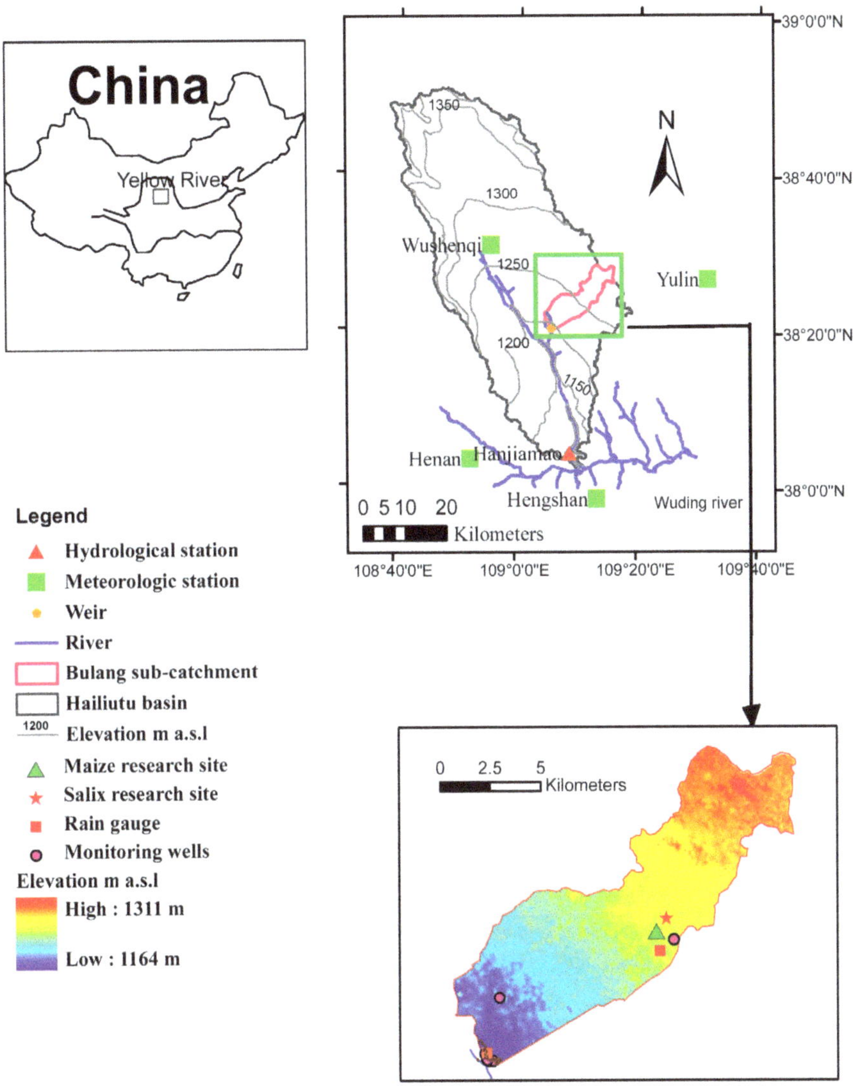

Figure 1. Location of the Bulang sub-catchment within the Hailiutu basin in Northwest China, the location of sap flow measurements for maize and salix, hydrological and meteorological stations, groundwater level monitoring wells, and the digital elevation model.

The popular numerical model code MODFLOW [52] was used for the flow simulation. The numerical model grid cell has a uniform size of 50 by 50 m, resulting in a model grid consisting of 310 rows and 350 columns. The top elevation of the model is the surface elevation which is taken from the Digital Elevation Model with a resolution of 30 by 30 m. The bottom elevation of the model is interpolated from the limited geological borehole data. The hydrogeological parameters such as hydraulic conductivity, specific yield, and specific storage were divided into three zones according to lithology. The General-Head Boundary (GHB) package of MODFLOW was used to simulate deep groundwater discharge to the Hailiutu River, where flow into or out of a GHB cell is calculated in proportion to the difference between the head in the model cell and the stage of the Hailiutu River. The MODFLOW Drain package was used to simulate groundwater discharge to the Bulang stream since the stream acts as a drain which always receives groundwater discharge. The Recharge package was used to simulate the net recharge which was calculated in an Excel sheet being precipitation minus evapotranspiration.

A steady state flow model was calibrated first with the annual average values of rainfall, ET rates, stream discharges, and the groundwater levels measured from November 2010 to October 2011. A transient flow model was then constructed and calibrated with the measured daily data from November 2010 to October 2011. The model calibration was performed with the optimization code PEST [53]. PEST found optimal values of hydraulic conductivities and storage parameters by minimizing the sum of squared differences between model-calculated and observed values of groundwater heads at the observation wells. The Drain conductance was further calibrated by comparing calculated and measured stream discharges.

2.2.4. Scenario Analysis

For semi-arid areas such as our study area, a delicate balance among vegetating sand dunes, agricultural production, and water resources conservation must be maintained [44]. Over-vegetating sand dunes with high water consumption species (such as poplar trees and salix bushes) would reduce net groundwater recharge and diminish stream flows. Desert grasses, such as *Artemisia Ordosica* and *Korshinsk Peashrub*, consume much less water and are better choice for vegetating sand dunes. These grasses are common in areas with a deep groundwater table. The dominant agricultural crop type in the area is maize. Maize needs to be irrigated six times [54] and consumes a lot of water. The local agricultural department has started a program to extend dry-resistant crops such as sorghum and millet. Considering the possible land use changes in the area, the following scenarios are proposed:

- Scenario 1 assumes a natural situation in which the catchment is covered by desert grasses. This scenario sets up a bench mark case to compare impacts of land use changes on groundwater and stream flow.
- Scenario 2 represents the current land use in 2011. Impacts of current land use changes on groundwater and stream flow can be analyzed by a comparison to scenario 1.

- Scenario 3 simulates effects of replacing maize crops with less water consumptive crops such as sorghum and millet in order to provide support to the policy of extending dry-resistant crops in the area.
- Scenario 4 proposes an ideal land use scenario where sand dunes are covered by desert grasses and dry-resistant crops are grown. This scenario gives direction to the future land use changes in the area.

Since there are inter-annual variations of precipitation and evapotranspiration accordingly, scenarios 2 and 4 were simulated under the dry, normal, and wet years. Analysis of long-term annual precipitation (Wushenqi meteorological station, 1959–2014) found that 2011, 2009, and 2014 represent dry (87.7%), normal (50.2%), and wet (27.8%) years, respectively. These two land use scenarios were further simulated with a multiple year transient groundwater model in order to assess impacts of inter-annual variations of climate.

2.2.5. Model Inputs for the Simulation of Land Use Scenarios

The calibrated transient groundwater flow model was used to simulate these four land use scenarios. Given inputs of 8-days net recharge for November 2010 to October 2011, the model simulates groundwater level changes and computes groundwater discharge to the Bulang River, deep groundwater circulation to the Hailiutu River and change of groundwater storage.

The net recharge was calculated to be the precipitation minus evapotranspiration. The evapotranspiration was estimated in the same procedure as before with equations of:

$$\mathrm{ET_i} = \frac{NDVI_i}{NDVI_{plant}}\mathrm{ET_{plant}}, \qquad 0 < \mathrm{NDVI_i} < 0.4 \tag{2}$$

$$\mathrm{ET_i} = \frac{NDVI_i}{NDVI_{crop}}\mathrm{ET_{crop}}, \qquad \mathrm{NDVI_i} \geqslant 0.4 \tag{3}$$

where $\mathrm{ET_i}$ is estimated model grid ET value. $\mathrm{ET_{plant}}$ and $\mathrm{ET_{crop}}$ were found from relevant studies [55] and are presented in Table 1.

In order to evaluate impacts of inter-annual variations of precipitation on the water balances, scenarios 2 and 4 were further simulated under a normal and wet years. The year of 2009 represents a normal year and 2014 represents a wet year. Daily precipitation of Wushenqi station was collected to compute net recharge. NDVI maps of 2009 and 2014 were processed to estimate ET values. These data are presented in Table 2.

Table 1. ET values for reference crops and plants for scenario simulations (mm/year).

Scenario	Rainfall	Crop ET	Plant ET	NDVI Map
1-Natural situation	214.8	*Artemisia Ordosica* (133.6)	*Artemisia Ordosica* (133.6)	August 2011
2-Current land use	214.8	Maize (503.1)	*Salix Psammophila* (245.1)	August 2011
3-Dry-resistant crop	214.8	Sorghum (402.5)	*Salix Psammophila* (245.1)	August 2011
4-Ideal land use	214.8	Sorghum (402.5)	*Artemisia Ordosica* (133.6)	August 2011

Table 2. Precipitation and ET values for reference crops and plants for scenarios 2 and 4 (mm/year).

Scenario	Rainfall	Crop ET	Plant ET	NDVI Map
2-Current land use	Dry (214.8) Normal (340.0) Wet (420.0)	Maize (503.1)	*Salix Psammophila* (245.1)	August 2011 August 2009 September 2014
4-Ideal land use	Dry (214.8) Normal (340.0) Wet (420.0)	Sorghum (402.5)	*Artemisia Ordosica* (133.6)	August 2011 August 2009 September 2014

2.3. *Field Measurements*

2.3.1. Precipitation and Evapotranspiration

In order to measure the hydrological variables, a set of instruments had been installed in the catchment. Precipitation was measured from November 2010 to October 2011 using two Hobo rain gauges (RG3-M Data Logging Rain Gauge, Onset Corporation, Bourne, MA, USA) at the outlet of the catchment and at Nanitan (Figures 1 and 2). Stem flow sensors (Flow 32, Dynamax, Houston, TX, USA) were used to measure sap flow in maize stems at the maize research site [54] and in salix stems at the salix research site [56] during the growing season in 2011. The NDVI map (resolution 30 m $\times$ 30 m) generated from remote sensing data have been utilized for classifying the land use and vegetation types in the area (Figure 2). NDVI values larger than 0.4 indicate crop land, between 0 and 0.4 are desert bushes, and smaller than 0 are considered as bare sand.

2.3.2. Discharge Gauge

One discharge gauging station was constructed at the outlet of the Bulang catchment (Figure 1). The Yujiawan gauging station consists of one permanent rectangular weir equipped with an e + WATER L water level logger (Type 11.41.54, Eijkelkamp Agrisearch Equipment, Giesbeek, The Netherlands) where water levels are recorded with a frequency of 30 min. Water depths are converted to discharges using a rating curve based on regular manual discharge measurements carried out with a current meter and the velocity-area method.

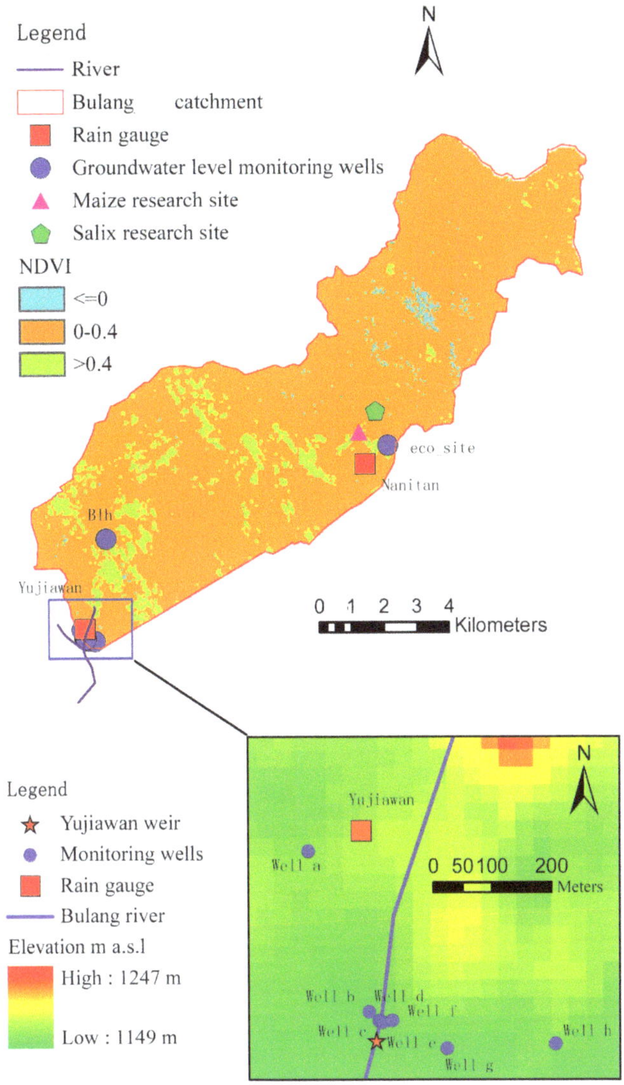

Figure 2. NDVI map of the Bulang catchment, interpretation of remote sensing data from TM on August 2011.

2.3.3. Observation Wells

Several groundwater level monitoring wells were installed in the Bulang catchment (Figures 1 and 2). Equipped with submersible pressure transducers (Type: MiniDiver, Eijkelkamp Agrisearch Equipment, Giesbeek, The Netherlands), the groundwater levels in these observation wells were recorded at 10-min intervals. Barometric compensation was carried out using air pressure measurements from a pressure transducer (Type: BaroDiver, Eijkelkamp Agrisearch Equipment, Giesbeek, The Netherlands) installed at the site. Groundwater levels were converted into the height of the water table above mean sea level with the calibration of the land elevation of wells, height of water column above the MiniDiver in the wells, and the depth of the MiniDiver in the boreholes.

3. Results

3.1. Estimation of Water Balance and Groundwater Discharge

3.1.1. Estimation of Catchment Water Balance

The total observed precipitation from November 2010 to October 2011 at Yujiawan and Nanitan station are 206.8 mm and 217.6 mm, respectively. The annual areal precipitation in the catchment was estimated to be 214.8 mm/year by means of area weighed average of two rainfall stations. Discharge at the Yujiawan weir varies from 0.01 to 0.23 m^3/s (Figure 3). Stream discharge is very stable during the winter months and varies in the summer months due to irrigation water use and sporadic rainfall. A heavy rain event occurred on 2 July and generated the highest discharge in the observation period.

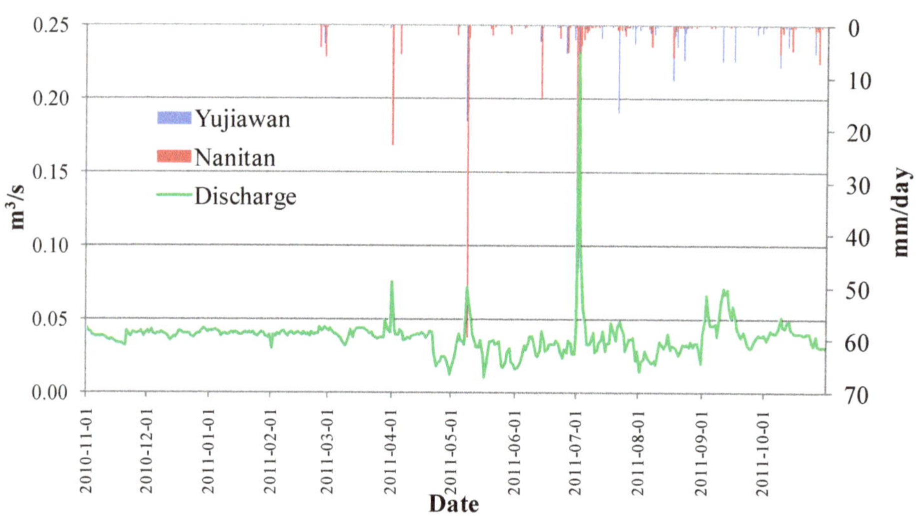

Figure 3. Rainfall at Yujiawan and Nanitan station, measured discharge at Yujiawan weir from November 2010 to October 2011.

ET rates measured at the maize and salix research site were 503.1 mm/year and 245.1 mm/year, respectively. The NDVI values at the sap flow measurement sites for maize and salix are 0.59 and 0.17, respectively. The NDVI was used to upscale evapotranspiration rate measured in the salix and maize research sites to the catchment using Equations (2) and (3).

The areal ET value of the catchment was calculated as the average ET of all cell ET values.

Figure 4a illustrates the up-scaling approach for areal ET with sap flow measurements and NDVI values. Figure 4b plots the estimated daily ET rates. The daily ET rates increases from mid April, peak in July and August, and decreases from September onwards. Daily ET rates were very low during rainy days. The total areal evapotranspiration rate in the Bulang catchment is estimated to be 186.2 mm/year.

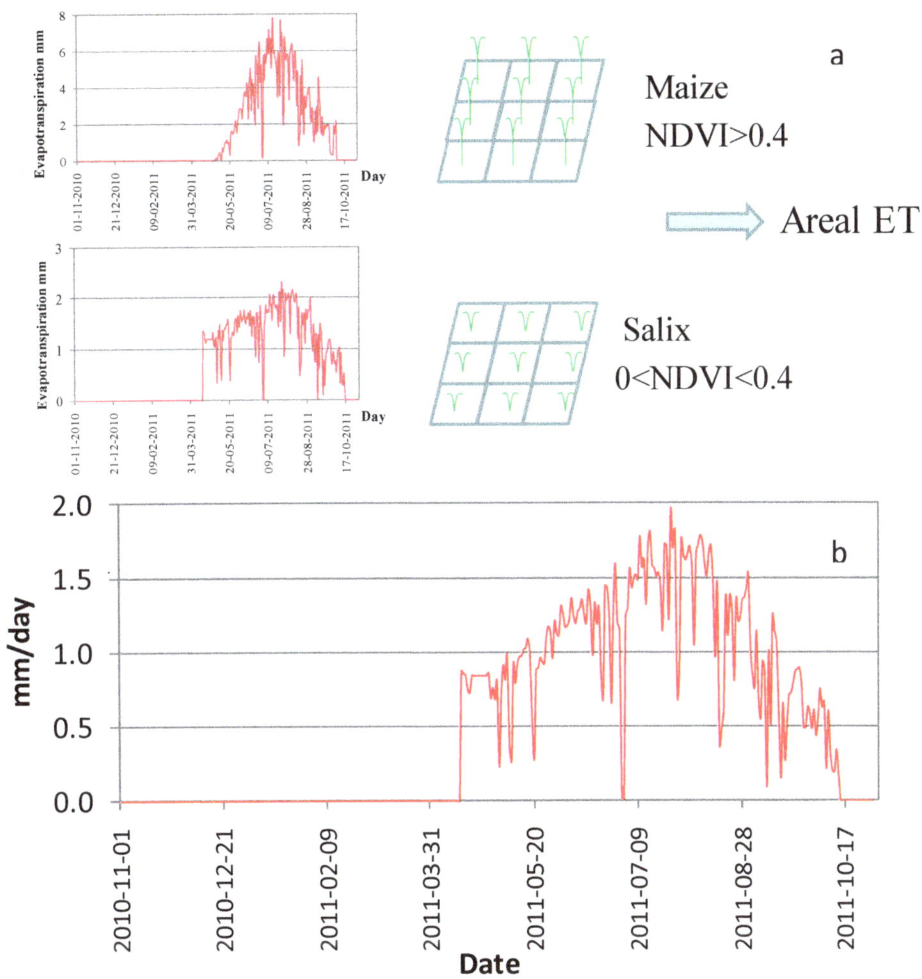

Figure 4. Up-scaling approach for areal evapotranspiration with sap flow measurements and NDVI values (**a**); and the estimated areal evapotranspiration (**b**).

The estimation of areal interception in the Bulang catchment followed the same approach as the estimation of the areal ET. The areal interception of 9.8 mm/year was obtained with the reference interception coefficient of 0.249 for salix [57] and 0.373 for maize [58].

Mu *et al.* [59] created a data set of the evapotranspiration rates in 8 days intervals from MOD16 Global Terrestrial Evaporation Data Set. Figure 5 compares the areal ET rates of 8 days estimated from this study and Mu's data set. The larges differences are found during the winter period. ET values estimated from the remote sensing methods show significant different distribution compared with the ET values estimated in this study based on sap flow measurements in the experimental period, where ET rates are zero in the winter period from December to March. There should be no ET in winter since the daily average air temperature is below zero and there are no crops and the trees do not transpire in this period.

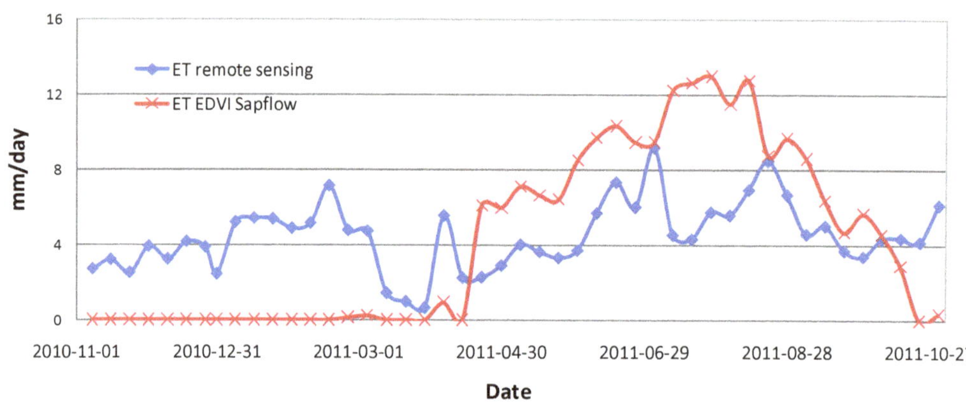

Figure 5. Comparison of evapotranspiration estimated by remote sensing and calculated by upscaling approach with NDVI and sap flow measurements from November 2010 to October 2011.

The change of the storage in the catchment is mainly based on the change of groundwater storage. Groundwater storage change can be estimated by the change of the groundwater levels multiplied by the porosity of the aquifer. Figure 6 shows that all groundwater levels at the end of the measuring period recovered back to the values at the beginning of the measuring period. Therefore, it can reasonably be assumed that there is no overall change of groundwater storage. The results of the water balance computation are shown in Figure 7 as daily water depth and are summarized in Table 3 as annual water depth in the catchment. The deep circulation in Table 3 refers to the regional groundwater flow to the Hailiutu River. The deep groundwater circulation was estimated in the water balance equation as the difference of total inflow minus total outflow. The net groundwater recharge equals to the precipitation minus total evapotranspiration.

Table 3. Annual water balance estimation in the Bulang catchment in mm/year.

Estimation method	Precipitation	ET	Net Recharge	Discharge	Deep Circulation	Change of Storage
Water Balance	214.8	196.0	18.8	12.6	6.2	0
Steady model	214.8	196.0	18.8	13.1	5.8	0

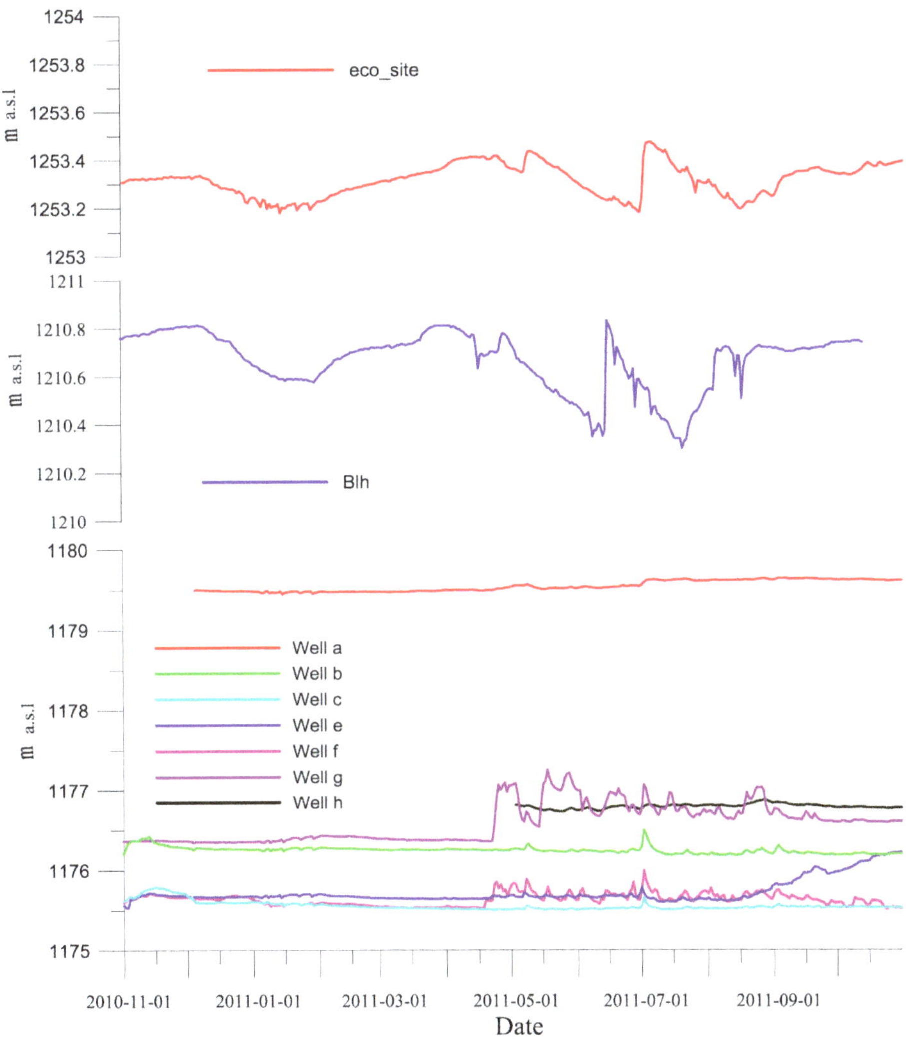

Figure 6. Observed groundwater levels in Bulang catchment.

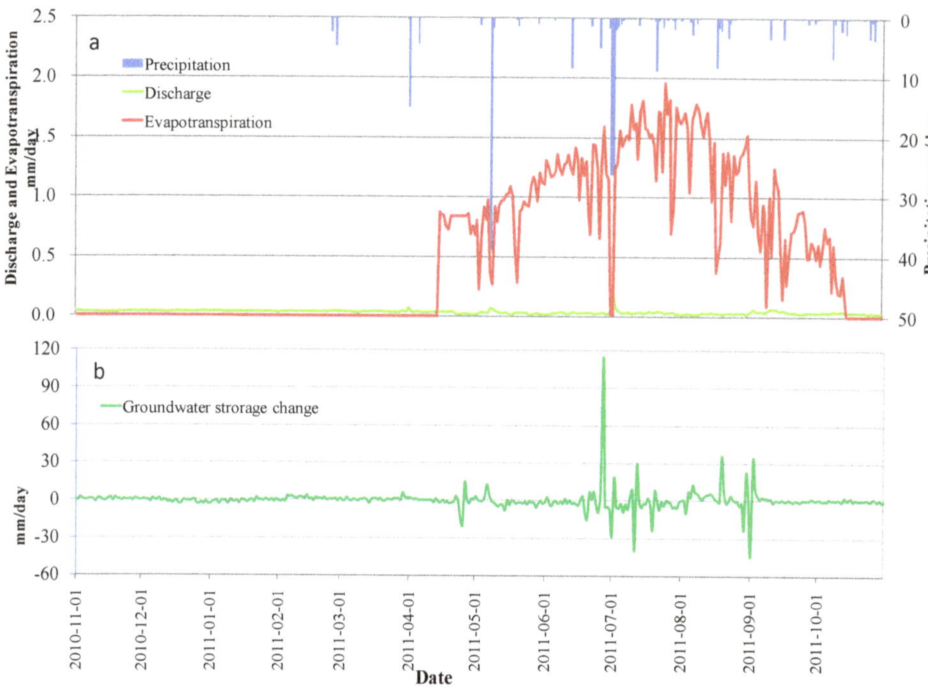

Figure 7. Average precipitation, river discharge, and estimated ET (**a**); groundwater storage change (**b**) in Bulang catchment from November 2010 to October 2011.

3.1.2. Estimation of Groundwater Discharge

Groundwater discharge as the base flow in the Bulang catchment can be derived from the river discharge measurements during the experimental period. A base flow separation was carried out based on the analysis of the isotopic tracers in the river, groundwater, and the heavy rainfall occurred in July 2011 [51]. Results of the hydrograph separations illustrate that the pre-event component accounts for 74.8% of the total discharge during rainfall events. Compared to other graphical and mathematical hydrograph separation methods, the event based isotopic separation method provides more accurate estimation of the groundwater discharge. It is estimated that 96.4% of total stream flow is composed of groundwater based on the event separation results from November 2010 to October 2011. Figure 8 illustrates the relations among the discharge, base flow, and the rainfall in the Bulang catchment.

3.2. Calibration of the Groundwater Model

3.2.1. Calibration of the Steady Groundwater Model

Since the change of groundwater storage in the observed period from November 2010 to October 2011 can be neglected, a steady state groundwater flow model was constructed and calibrated first with annual average values of the net recharge, groundwater discharge, and groundwater levels. The purposes of the steady state model were to calibrate the hydraulic conductivity and to create initial conditions for the transient model. Average values of groundwater levels measured in nine observation wells (Figure 2) were used to compare model-calculated groundwater levels for the model calibration. The optimization code, PEST was used to optimize values of hydraulic conductivity in three zones so that the sum of squared differences between calculated and measured groundwater levels are minimized.

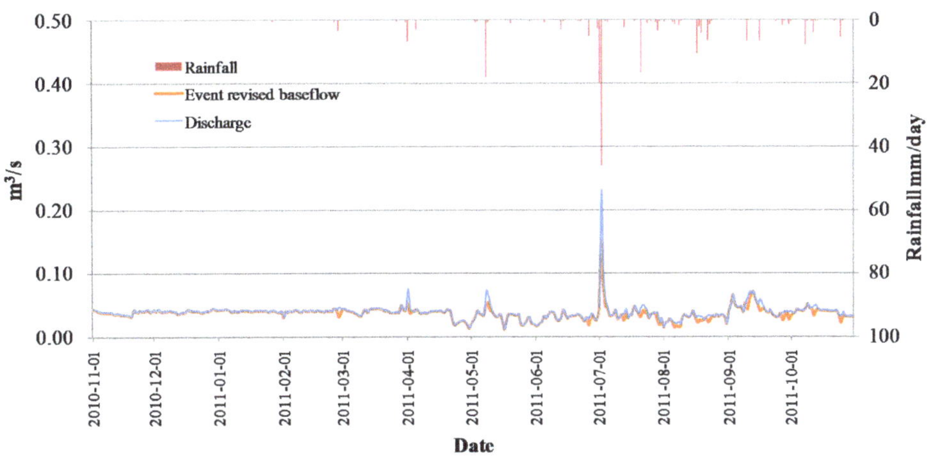

Figure 8. Relations among the observed discharge, separated baseflow, and the rainfall in the Bulang catchment.

The coefficient of determination of the computed and measured groundwater levels is as high as 0.99 indicating that the steady state model is capable to reproduce the measured groundwater levels with a very high accuracy. The optimized values of hydraulic conductivity are 8.59 m/day, 3.86 m/ay, and 1.86 m/day, respectively, in the three parameter zones. The computed discharge to the Bulang River and deep groundwater circulation are very close to the estimated values in the previous water balance computation (Table 3).

3.2.2. Calibration of the Transient Groundwater Model

The purpose of the transient model is to simulate seasonal variations of groundwater levels and stream flow caused by varying precipitation and evapotranspiration values. The simulation period is one year from November 2010 to October 2011. A stress period of 8 days was chosen in line with estimated ET values from remote sensing data. The computed groundwater levels by the steady state model were used as initial groundwater heads in the transient model. The average values of daily precipitation and evapotranspiration values in every 8 days were used to compute transient net recharge values for the transient model. The specific yield values of three zones in the first model layer were optimized using PEST so that computed groundwater level series fit well with the measurement series. Figure 9 shows the match of the computed groundwater levels to the measurements in four observation wells. The RMSE values are 0.26 m, 0.14 m, 0.15 m, and 0.82 m for the computed groundwater levels at well Blh, eco-site, well a, and well b, respectively. The optimized values of specific yield are 0.29, 0.25 and 0.1, respectively, in the three parameter zones.

Figure 10 shows the comparison of the computed groundwater discharge to the stream and separated baseflow with the stream discharge measurements. The RMSE is 0.0075 m^3/s for the computed groundwater discharge. In general, the transient model computes more smooth discharges than the results of baseflow separation. Large differences in summer months may be caused by irrigation water use diverted from the river which is not considered in the model.

3.2.3. Sensitivity of the Transient Groundwater Model

Since the model was calibrated based on the one year measurements of groundwater levels and stream discharge and there are uncertainties in upscaling ET and estimating hydraulic conductivities, sensitivity analysis was performed to test sensitivities of computed groundwater levels and stream discharges to uncertainties in ET and hydraulic conductivity. In order to evaluate effects of inter-annual variations of precipitation, sensitivity of precipitation is also investigated. The coefficient of variation of the annual precipitation is 0.265 in the area. Therefore, precipitation, ET, and hydraulic conductivities were alternately increased and decreased by 26.5%, the transient model was run to compute the changes of groundwater levels and stream discharges while the other variables were kept fixed. The results are shown in Figure 11. The stream discharges are more sensitive to uncertainties in hydraulic conductivities while groundwater levels are more sensitive to changes in precipitation and ET values.

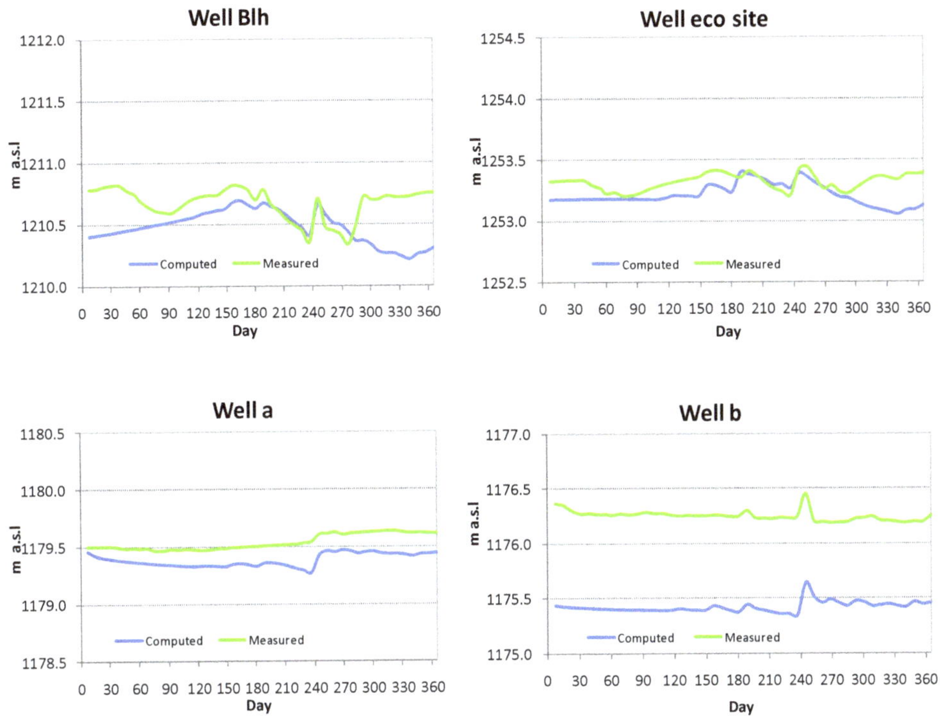

Figure 9. Fit of the computed groundwater levels to the measured ones in four wells in Bulang sub-catchment.

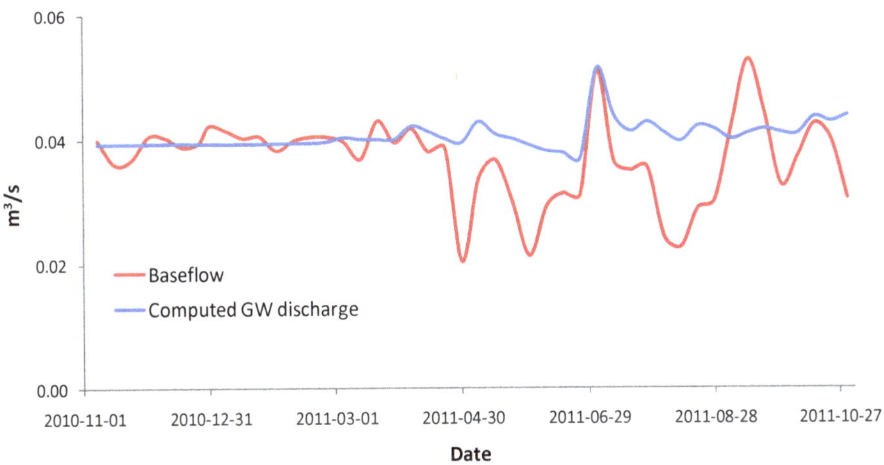

Figure 10. Comparison of simulated groundwater discharge to the Bulang River and the separated base flow from November 2010 to October2011.

The calibrated transient groundwater flow model was used to simulate the scenarios. For each scenario, the model computes groundwater level changes, groundwater discharge to the Bulang River, deep groundwater circulation to the Hailiutu River, and change of groundwater storage.

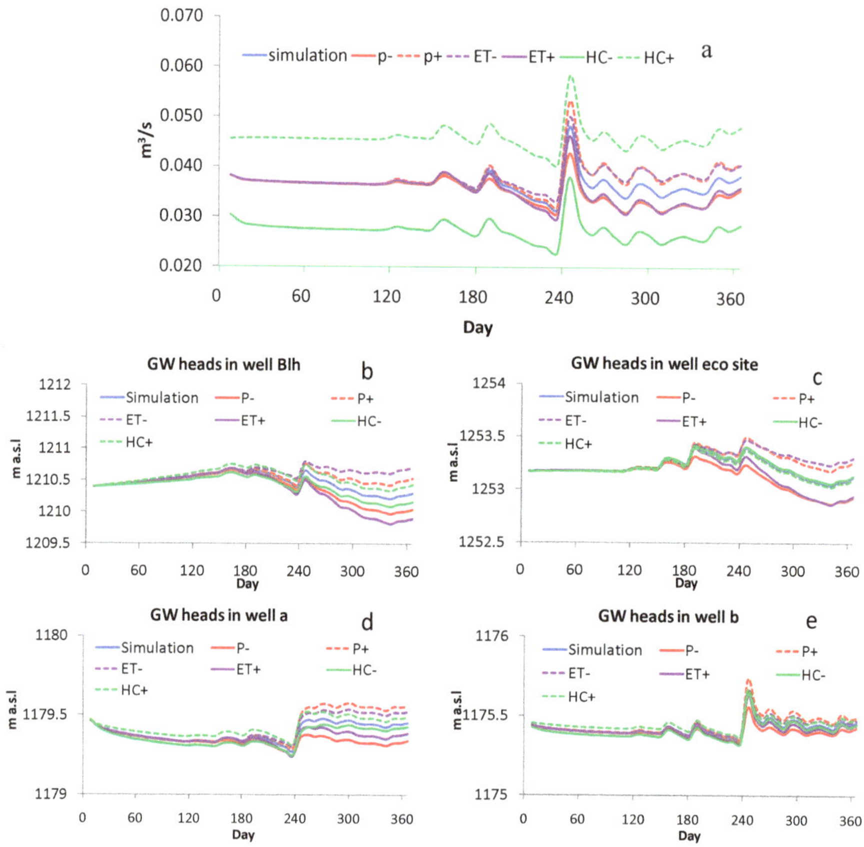

Figure 11. Sensitivity of computed stream discharge (**a**); and groundwater levels (**b–e**) to precipitation (P), evapotranspiration (ET), and hydraulic conductivity (HC); the sign (−) and (+) in the legend indicates the decreasing and increasing values by 26.5%.

3.3. Simulation of Land Use Change Scenarios

Annual water balance components for the four scenarios are summarized in Table 4. Time series of changes in discharge to the Bulang River and groundwater levels in four observation wells are plotted in Figure 12. The annual water balance computations show that net groundwater recharge increases significantly in scenario (1) and (4) when desert grasses were used for the protection of sand dunes.

Discharge to the Bulang River increases slightly with increasing net recharge, while deep groundwater circulation discharging to the Hailiutu River remains constant. The benefit of increasing net recharge is largely in increasing groundwater storage, especially in scenarios (1) and (4). Increase of discharge to the Bulang River and increase of groundwater levels occur in the growing period from mid April to mid September (Figure 12). In the winter months, river discharge and groundwater levels are stable since there is neither precipitation nor evapotranspiration.

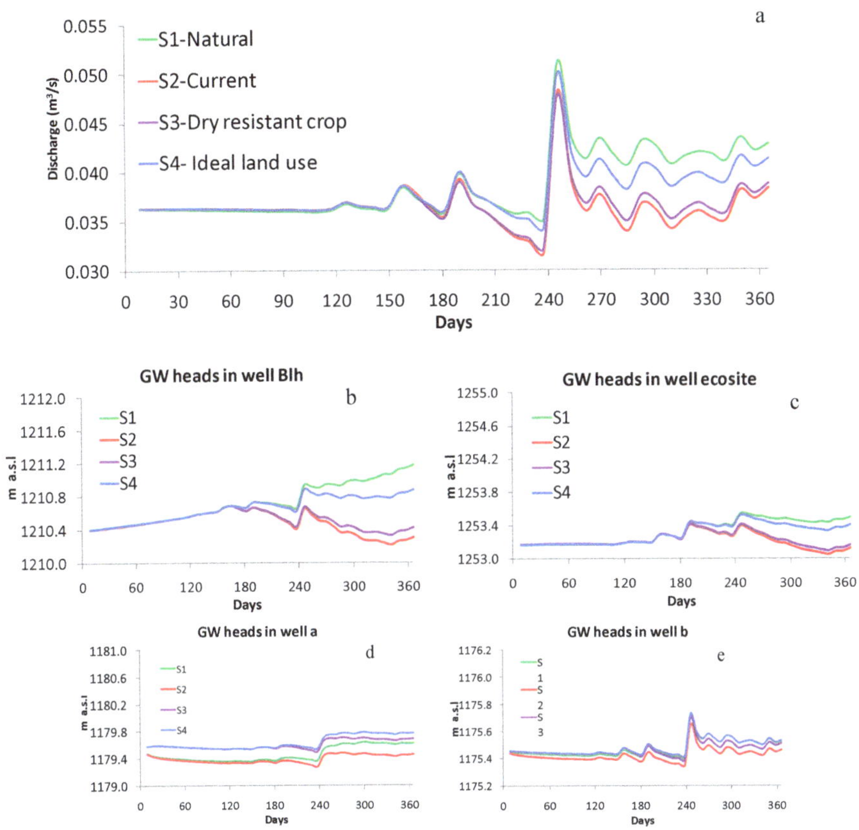

Figure 12. Simulated groundwater discharge to the Bulang River (**a**); and the calculated groundwater heads in well Blh (**b**); eco_site (**c**); well a (**d**); and well b (**e**) for four land use scenarios.

Annual water balance components for scenarios 2 and 4 under different hydrological years are summarized in Table 5. Time series of changes in discharge to the Bulang River and groundwater levels in four observation wells are plotted in Figure 13. The simulation results show that net groundwater recharge is increased significantly in normal and wet hydrological years as a result of increase of precipitation while evapotranspiration is only increased slightly comparing to the dry year. The increase of net recharge contributes largely to the increase of groundwater storage while discharge to the stream also increases. The increase of groundwater storage and stream discharges is more profound under the ideal land use scenario 4 compared to the current land use scenario 2. Under the ideal land use scenario, a healthy vegetation cover can be sustained in all years while water resources can be conserved for other social and economic uses.

Table 4. Comparison of simulated water balance for four scenarios (mm/year).

Annual Components	Precipitation	ET	Net Recharge	Discharge	Deep Circulation	Change of Storage
Scenario 1	214.8	102.7	112.1	13.4	8.1	90.1
Scenario 2	214.8	196.0	18.8	12.7	8.0	−4.1
Scenario 3	214.8	189.2	25.6	12.8	8.0	3.9
Scenario 4	214.8	120.9	93.9	13.2	8.1	72.4

Table 5. Comparison of simulated annual water balance components for scenarios 2 and 4 under different hydrological years (mm/year).

Annual Components		Precipitation	ET	Net Recharge	Discharge	Deep Circulation	Change of Storage
Scenario 2	Dry	214.8	196.0	18.8	12.7	8.0	−4.06
	Normal	340.0	218.4	121.6	13.6	8.2	99.24
	Wet	420.0	280.5	139.5	13.8	8.4	115.86
Scenario 4	Dry	214.8	120.9	93.9	13.2	8.1	72.4
	Normal	340.0	136.7	203.3	14.1	8.3	179.9
	Wet	420.0	166.0	254.0	14.5	8.4	228.6

In order to evaluate the long-term impact of land use scenarios on groundwater discharges and groundwater levels in consideration of inter-annual variability of climate in the Bulang catchment, the calibrated transient model was extended to multiple years from 2000 to 2009. The month was chosen as stress period so that the transient model simulates 120 months. The monthly net groundwater recharge was estimated from monthly precipitations at Wushenqi station and estimated ET values. The monthly ET values for the current land use and ideal land use scenarios were estimated using the ratio of monthly ET to precipitation in the annual simulation of dry, normal and wet years. The simulation results are presented in Figure 14. In general, groundwater discharges and groundwater levels exhibit inter-annual variations. Groundwater levels and discharges were increased during the wet years of 2001 and 2002, and decreased during dry years of 2005 and 2006, and recovered during subsequent normal years. The depleted groundwater storage during the dry years can be restored during the normal and wet years. The alternating dry, normal and wet years will not cause degradation of vegetations since groundwater provides a reliable resource to sustain the vegetation. In comparison to the current land use, under the ideal land use scenario, groundwater levels in 4 observation wells are increased which lead to the increase of groundwater discharge to the river.

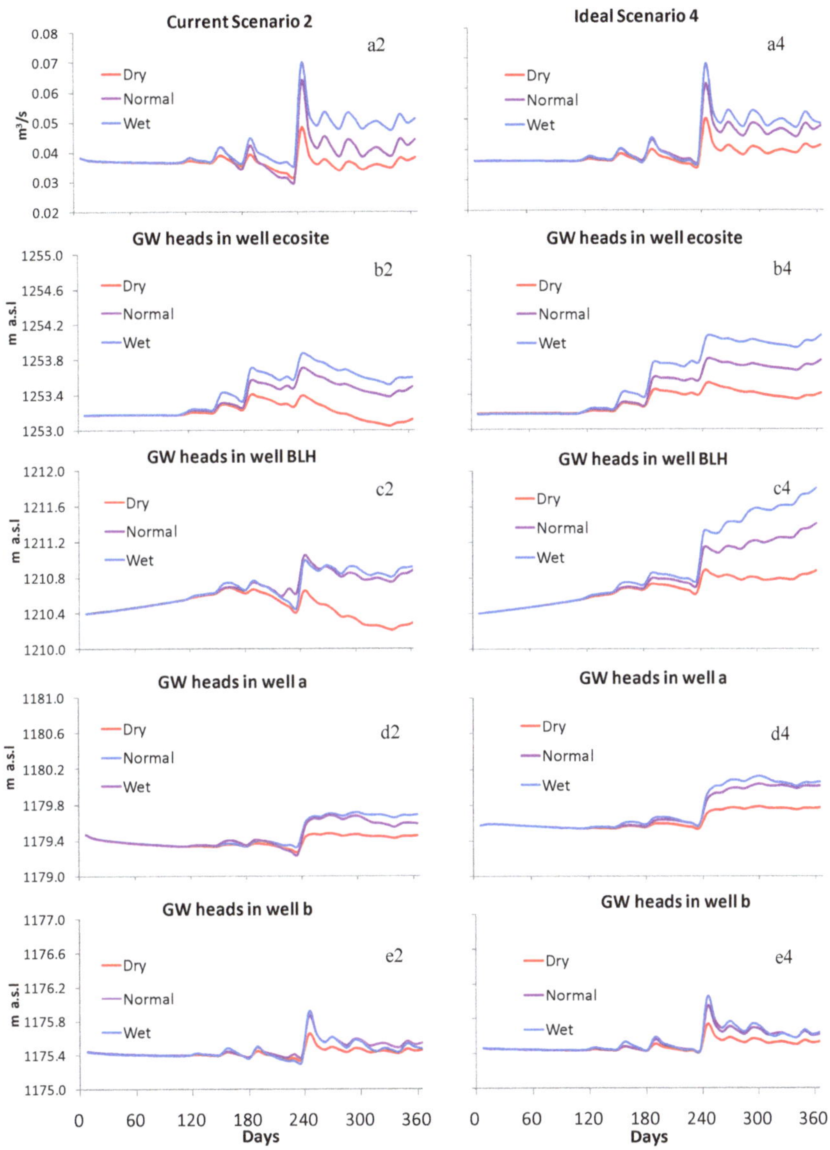

Figure 13. Simulated groundwater discharge to the Bulang River and calculated groundwater heads with current land use (**a2–e2**); and Ideal land use scenario (**a4–e4**) under dry, normal and wet hydrological years.

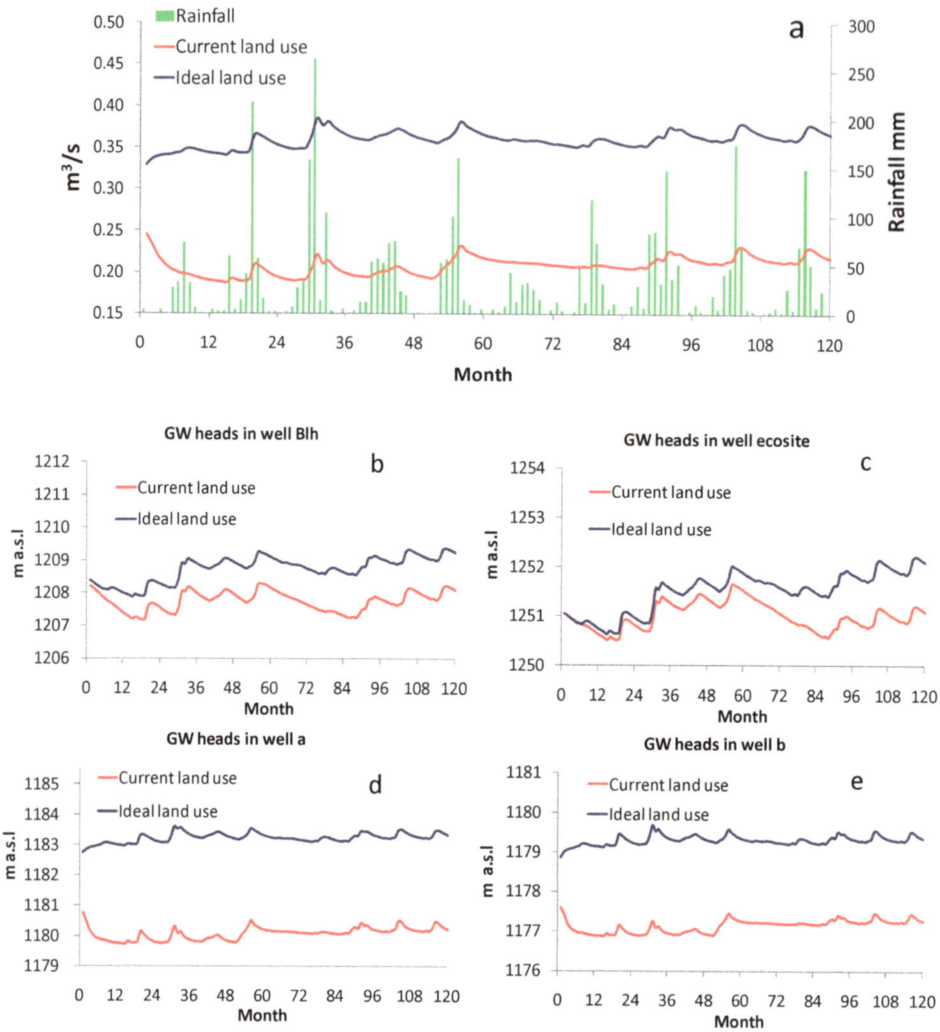

Figure 14. Simulated groundwater discharge to the Bulang River and the monthly rainfall at Wushenqi from 2000 to 2009 (**a**); the calculated groundwater heads in well Blh (**b**); eco_site (**c**); well a (**d**); and well b (**e**) for current and ideal land use scenarios from 2000 to 2009.

4. Discussions

Located in a semi-arid region, the Bulang catchment is covered by sand dunes. The hydrological processes are dominated by direct infiltration of precipitation and evapotranspiration. Direct surface runoff is limited and river discharge is maintained by groundwater discharge. Local authorities have implemented count mobilization measures against desertification by planting shrubs on unstable sand dunes, but large scale plantations of salix bushes to prevent desertification consumes too much water resources. The model simulation of current land use scenario in 2011 shows that in total 91% of the annual precipitation is consumed by the vegetation and crops in the catchment, the net groundwater recharge amounts to only 9% of the precipitation, which maintains stream discharge. However, for the ideal land use of planting desert grasses and dry resistant crops, only 56% of the annual precipitation would be consumed by dry resistant plants and crops and net groundwater recharge will be increased to 44% of the annual precipitation. In comparison with the result of scenario 3, the increase of net groundwater recharge is mainly contributed from planting desert grasses since the cropland area is limited to 3% of the land area in the Hailiutu River [60].

Inter-annual variations of precipitation have large impacts on the catchment water balance. The net groundwater recharge will be increased to 60.0% of the annual precipitation during normal and wet hydrological years for the ideal land use scenario. Groundwater storage and stream discharges are increased significantly during normal and wet years. The preliminary long-term simulations indicate that the depleted groundwater storage can be restored during normal and wet hydrological years while alternating dry, normal and wet hydrological years occur.

Although the study attempts estimating evapotranspiration rates using site measurements and upscaling with NDVI values, there are a number of limitations to be considered in the future research.

First, the site measurements of sap flow of salix bush, willow tree, and maize should be continued to obtain long-term reference evapotranspiration values for dry, normal and wet years. A field ecohydrological research site is under construction which includes large diameter lysimeters for measuring evapotranspiration of dominant plants and crops and net groundwater recharge at various groundwater table depths. Second, evapotranspiration of desert grasses (*Artemisia Ordosica*) and dry resistant crops (Sorghum and millet) should be measured. Eddy covariance technique [61] is most suited to measure evapotranspiration of grasses and crops at the plot scale. Third, the same NDVI map was used to compute the grid ET values for scenarios 2, 3, and 4 with the reference ET values of crops and plants. There are uncertainties in the estimated grid ET values with this upscaling method. Seasonal variations of NDVI values of crops and plants were not considered. ET values might change for different combinations of crops and plants in different scenarios. A more reliable relationship between seasonal NDVI and ET values could be established with long-term systematic measurements of plot scale ET values of dominant crops and plants with Eddy covariance technique.

The constructed groundwater model could be further improved by installing a number of monitoring wells in the recharge area in the northwestern part of the catchment. One more rain gauge at the northern boundary could provide better information about the spatial distribution of the rainfall input. Since most of the cropland is irrigated by means of groundwater abstraction, the investigation of the location and amount of groundwater abstraction should be implemented. Furthermore, the scattered small pools in local depressions with open water evaporation should be taken into account, where the evapotranspiration (ET) package of MODFLOW would be considered in the groundwater model.

The interaction between evapotranspiration and groundwater table depth was not simulated in the model. For scenarios with a significant increase of groundwater storage, groundwater table depth becomes shallower resulting higher evapotranspiration. This interaction can be partially simulated by using the ET package in MODFLOW.

The long-term simulations show that groundwater levels and discharges are relatively stable and fluctuate around long term means. A recent study by Yin *et al.* [62] simulated impacts of long-term climate variations on tree water use in the area. The result shows that trees did not suffer from water stress during the dry years because of the availability of groundwater for transpiration. The long-term monitoring of metrology, hydrology, and vegetation should be continued to ascertain the findings from model simulation.

5. Conclusions

Groundwater is the most important resource for local society and ecosystem protection in the semi-arid Bulang catchment. Groundwater maintains stream discharge and sustains vegetation growth. Net groundwater recharge, defined as precipitation minus evapotranspiration, is a good indicator of water resources availability in the area since dominant hydrological processes are direct infiltration of precipitation and evapotranspiration.

The simulation of the current land use in 2011 (being a dry year) indicates that nearly 91% of the annual precipitation is consumed by evapotranspiration of salix bushes and maize crops. Only 9% of the annual precipitation becomes net groundwater recharge which maintains a stable stream discharge. Although it is not possible to restore pristine land cover by desert grasses (scenario 1), an ideal land use (scenario 4) can be achieved by planting desert grasses for fixing sand dunes and dry resistant crops (sorghum and millet) for the local society. The simulation shows that the evapotranspiration consumes only 56% of the annual precipitation under the ideal land use scenario in dry year, and 40% in normal and wet years. The increase of net groundwater recharge is mainly contributed by planting desert grasses since the cropland area is limited.

The simulation of scenarios 2 and 4 under normal and wet hydrological years show significant increase of groundwater storage and slight increase of river discharges comparing to the dry year. The long-term simulations show that groundwater levels and discharges are relatively stable and fluctuate around long-term means. The depleted groundwater storage during the dry years can be restored during normal and wet years.

The results of this study have relevant implications for water and ecosystem management in the catchment. In order to maintain river discharges and sustain a healthy growth of groundwater dependent vegetations, future land use changes should aim introducing dry resistant crops and desert grasses for vegetating sand dunes.

Acknowledgments: This study was supported by the Honor Power Foundation (China) and UNESCO-IHE (Delft, The Netherlands) and its donors. Assistance during the field work by Xi'an Center of Geological Survey is gratefully acknowledged. The critical review of three anonymous referees helped for the further improvement of the manuscript.

Author Contributions: Zhi Yang and Yangxiao Zhou were primarily responsible for implementing the research, conducting groundwater model simulations, and preparing manuscript. Jochen Wenninger was involved in the experimental field measurement setup, the hydrograph separation calculations, and revision of the manuscript. Stefan Uhlenbrook and Li Wan supervised the research and critically reviewed the draft manuscript.

Conflicts of Interest: The authors declare to have no conflict of interest.

References

1. Snyman, H.; Fouché, H. Production and water-use efficiency of semi-arid grasslands of South Africa as affected by veld condition and rainfall. *Water SA* **1991**, *17*, 263–268.
2. Deng, X.P.; Shan, L.; Zhang, H.; Turner, N.C. Improving agricultural water use efficiency in arid and semiarid areas of China. *Agric. Water Manag.* **2006**, *80*, 23–40.
3. Mitchell, D.; Fullen, M.; Trueman, I.; Fearnehough, W. Sustainability of reclaimed desertified land in Ningxia, China. *J. Arid Environ.* **1998**, *39*, 239–251.
4. Bellot, J.; Sanchez, J.; Chirino, E.; Hernandez, N.; Abdelli, F.; Martinez, J. Effect of different vegetation type cover on the soil water balance in semi-arid areas of south eastern Spain. *Phys. Chem. Earth Part B: Hydrol. Oceans Atmos.* **1999**, *24*, 353–357.
5. Mermoud, A.; Tamini, T.; Yacouba, H. Impacts of different irrigation schedules on the water balance components of an onion crop in a semi-arid zone. *Agric. Water Manag.* **2005**, *77*, 282–295.
6. Brown, A.E.; Zhang, L.; McMahon, T.A.; Western, A.W.; Vertessy, R.A. A review of paired catchment studies for determining changes in water yield resulting from alterations in vegetation. *J. Hydrol.* **2005**, *310*, 28–61.
7. Jothityangkoon, C.; Sivapalan, M.; Farmer, D. Process controls of water balance variability in a large semi-arid catchment: Downward approach to hydrological model development. *J. Hydrol.* **2001**, *254*, 174–198.
8. Farmer, D.; Sivapalan, M.; Jothityangkoon, C. Climate, soil, and vegetation controls upon the variability of water balance in temperate and semiarid landscapes: Downward approach to water balance analysis. *Water Resour. Res.* **2003**, *39*.

9. Scott, R.L.; James Shuttleworth, W.; Goodrich, D.C.; Maddock, T., III. The water use of two dominant vegetation communities in a semiarid riparian ecosystem. *Agric. For. Meteorol.* **2000**, *105*, 241–256.
10. Li, Y.; Cui, J.; Zhang, T.; Okuro, T.; Drake, S. Effectiveness of sand-fixing measures on desert land restoration in Kerqin Sandy Land, northern China. *Ecol. Eng.* **2009**, *35*, 118–127.
11. Zhang, L.; Dawes, W.; Walker, G. Response of mean annual evapotranspiration to vegetation changes at catchment scale. *Water Resour. Res.* **2001**, *37*, 701–708.
12. Lørup, J.K.; Refsgaard, J.C.; Mazvimavi, D. Assessing the effect of land use change on catchment runoff by combined use of statistical tests and hydrological modelling: Case studies from Zimbabwe. *J. Hydrol.* **1998**, *205*, 147–163.
13. Bronstert, A.; Niehoff, D.; Bürger, G. Effects of climate and land-use change on storm runoff generation: Present knowledge and modelling capabilities. *Hydrol. Process.* **2002**, *16*, 509–529.
14. Niehoff, D.; Fritsch, U.; Bronstert, A. Land-use impacts on storm-runoff generation: Scenarios of land-use change and simulation of hydrological response in a meso-scale catchment in SW-Germany. *J. Hydrol.* **2002**, *267*, 80–93.
15. Yang, Z.; Zhou, Y.; Wenninger, J.; Uhlenbrook, S. The causes of flow regime shifts in the semi-arid Hailiutu River, Northwest China. *Hydrol. Earth Syst. Sci.* **2012**, *16*, 87–103.
16. Ahearn, D.S.; Sheibley, R.W.; Dahlgren, R.A.; Anderson, M.; Johnson, J.; Tate, K.W. Land use and land cover influence on water quality in the last free-flowing river draining the western Sierra Nevada, California. *J. Hydrol.* **2005**, *313*, 234–247.
17. Zampella, R.A.; Procopio, N.A.; Lathrop, R.G.; Dow, C.L. Relationship of land-use/land-cover patterns and surface-water quality in the mullica river Basin1. *J. Am. Water Resour. Assoc.* **2007**, *43*, 594–604.
18. Harbor, J.M. A practical method for estimating the impact of land-use change on surface runoff, groundwater recharge and wetland hydrology. *J. Am. Water Resour. Assoc.* **1994**, *60*, 95–108.
19. Scanlon, B.R.; Reedy, R.C.; Stonestrom, D.A.; Prudic, D.E.; Dennehy, K.F. Impact of land use and land cover change on groundwater recharge and quality in the southwestern US. *Glob. Chang. Biol.* **2005**, *11*, 1577–1593.
20. Salama, R.; Hatton, T.; Dawes, W. Predicting land use impacts on regional scale groundwater recharge and discharge. *J. Environ. Qual.* **1999**, *28*, 446–460.
21. Batelaan, O.; De Smedt, F.; Triest, L. Regional groundwater discharge: Phreatophyte mapping, groundwater modelling and impact analysis of land-use change. *J. Hydrol.* **2003**, *275*, 86–108.
22. Furukawa, Y.; Inubushi, K.; Ali, M.; Itang, A.; Tsuruta, H. Effect of changing groundwater levels caused by land-use changes on greenhouse gas fluxes from tropical peat lands. *Nutr. Cycl. Agroecosyst.* **2005**, *71*, 81–91.
23. Jeong, C.H. Effect of land use and urbanization on hydrochemistry and contamination of groundwater from Taejon area, Korea. *J. Hydrol.* **2001**, *253*, 194–210.

24. McLay, C.; Dragten, R.; Sparling, G.; Selvarajah, N. Predicting groundwater nitrate concentrations in a region of mixed agricultural land use: A comparison of three approaches. *Environ. Pollut.* **2001**, *115*, 191–204.
25. Molénat, J.; Gascuel-Odoux, C. Modelling flow and nitrate transport in groundwater for the prediction of water travel times and of consequences of land use evolution on water quality. *Hydrol. Process.* **2002**, *16*, 479–492.
26. Cole, M.L.; Kroeger, K.D.; McClelland, J.W.; Valiela, I. Effects of watershed land use on nitrogen concentrations and δ 15 nitrogen in groundwater. *Biogeochemistry* **2006**, *77*, 199–215.
27. Choi, W.J.; Han, G.H.; Lee, S.M.; Lee, G.T.; Yoon, K.S.; Choi, S.M.; Ro, H.M. Impact of land-use types on nitrate concentration and δ^{15} N in unconfined groundwater in rural areas of Korea. *Agric. Ecosyst. Environ.* **2007**, *120*, 259–268.
28. Krause, S.; Jacobs, J.; Bronstert, A. Modelling the impacts of land-use and drainage density on the water balance of a lowland–floodplain landscape in northeast Germany. *Ecol. Model.* **2007**, *200*, 475–492.
29. Leblanc, M.; Favreau, G.; Tweed, S.; Leduc, C.; Razack, M.; Mofor, L. Remote sensing for groundwater modelling in large semiarid areas: Lake Chad Basin, Africa. *Hydrogeol. J.* **2007**, *15*, 97–100.
30. Courault, D.; Seguin, B.; Olioso, A. Review on estimation of evapotranspiration from remote sensing data: From empirical to numerical modeling approaches. *Irrig. Drain. Syst.* **2005**, *19*, 223–249.
31. Waters, P.; Greenbaum, D.; Smart, P.L.; Osmaston, H. Applications of remote sensing to groundwater hydrology. *Remote Sens. Rev.* **1990**, *4*, 223–264.
32. Sener, E.; Davraz, A.; Ozcelik, M. An integration of GIS and remote sensing in groundwater investigations: A case study in Burdur, Turkey. *Hydrogeol. J.* **2005**, *13*, 826–834.
33. Münch, Z.; Conrad, J. Remote sensing and GIS based determination of groundwater dependent ecosystems in the Western Cape, South Africa. *Hydrogeol. J.* **2007**, *15*, 19–28.
34. Shaban, A.; Khawlie, M.; Abdallah, C. Use of remote sensing and GIS to determine recharge potential zones: The case of Occidental Lebanon. *Hydrogeol. J.* **2006**, *14*, 433–443.
35. Tweed, S.O.; Leblanc, M.; Webb, J.A.; Lubczynski, M.W. Remote sensing and GIS for mapping groundwater recharge and discharge areas in salinity prone catchments, southeastern Australia. *Hydrogeol. J.* **2007**, *15*, 75–96.
36. Bobba, A.; Bukata, R.; Jerome, J. Digitally processed satellite data as a tool in detecting potential groundwater flow systems. *J. Hydrol.* **1992**, *131*, 25–62.
37. Ben-Dor, E.; Goldshleger, N.; Braun, O.; Kindel, B.; Goetz, A.; Bonfil, D.; Margalit, N.; Binaymini, Y.; Karnieli, A.; Agassi, M. Monitoring infiltration rates in semiarid soils using airborne hyperspectral technology. *Int. J. Remote Sens.* **2004**, *25*, 2607–2624.
38. Kalma, J.D.; McVicar, T.R.; McCabe, M.F. Estimating land surface evaporation: A review of methods using remotely sensed surface temperature data. *Surv. Geophys.* **2008**, *29*, 421–469.

39. Groeneveld, D.P.; Baugh, W.M.; Sanderson, J.S.; Cooper, D.J. Annual groundwater evapotranspiration mapped from single satellite scenes. *J. Hydrol.* **2007**, *344*, 146–156.
40. Groeneveld, D.P. Remotely-sensed groundwater evapotranspiration from alkali scrub affected by declining water table. *J. Hydrol.* **2008**, *358*, 294–303.
41. Jin, X.; Wan, L.; Zhang, Y.; Xue, Z.; Yin, Y. A study of the relationship between vegetation growth and groundwater in the Yinchuan Plain. *Earth Sci. Front.* **2007**, *14*, 197–203.
42. Jin, X.M.; Schaepman, M.E.; Clevers, J.G.; Su, Z.B.; Hu, G. Groundwater depth and vegetation in the Ejina area, China. *Arid Land Res. Manag.* **2011**, *25*, 194–199.
43. Lv, J.; Wang, X.S.; Zhou, Y.; Qian, K.; Wan, L.; Eamus, D.; Tao, Z. Groundwater-dependent distribution of vegetation in Hailiutu River catchment, a semi-arid region in China. *Ecohydrology* **2013**, *6*, 142–149.
44. Zhou, Y.; Wenninger, J.; Yang, Z.; Yin, L.; Huang, J.; Hou, L.; Wang, X.; Zhang, D.; Uhlenbrook, S. Groundwater—Surface water interactions, vegetation dependencies and implications for water resources management in the semi-arid Hailiutu River catchment, China—A synthesis. *Hydrol. Earth Syst. Sci.* **2013**, *17*, 2435–2447.
45. Lubczynski, M.W.; Gurwin, J. Integration of various data sources for transient groundwater modeling with spatio-temporally variable fluxes—Sardon study case, Spain. *J. Hydrol.* **2005**, *306*, 71–96.
46. Bethenod, O.; Katerji, N.; Goujet, R.; Bertolini, J.; Rana, G. Determination and validation of corn crop transpirationby sap flow measurement under field conditions. *Theor. Appl. Climatol.* **2000**, *67*, 153–160.
47. Ford, C.R.; Hubbard, R.M.; Kloeppel, B.D.; Vose, J.M. A comparison of sap flux-based evapotranspiration estimates with catchment-scale water balance. *Agric. For. Meteorol.* **2007**, *145*, 176–185.
48. Stisen, S.; Jensen, K.H.; Sandholt, I.; Grimes, D.I. A remote sensing driven distributed hydrological model of the Senegal River basin. *J. Hydrol.* **2008**, *354*, 131–148.
49. Hendricks Franssen, H.; Brunner, P.; Makobo, P.; Kinzelbach, W. Equally likely inverse solutions to a groundwater flow problem including pattern information from remote sensing images. *Water Resour. Res.* **2008**, *44*.
50. Li, H.; Brunner, P.; Kinzelbach, W.; Li, W.; Dong, X. Calibration of a groundwater model using pattern information from remote sensing data. *J. Hydrol.* **2009**, *377*, 120–130.
51. Yang, Z.; Zhou, Y.; Wenninger, J.; Uhlenbrook, S. A multi-method approach to quantify groundwater/surface water-interactions in the semi-arid Hailiutu River basin, northwest China. *Hydrogeol. J.* **2014**, *22*, 527–541.
52. McDonald, M.G.; Harbaugh, A.W. *A Modular Three-Dimensional Finite-Difference Ground-Water Flow Model*; Scientific Publications Company: Reston, VA, USA, 1984.
53. Doherty, J.; Brebber, L.; Whyte, P. *PEST: Model-Independent Parameter Estimation*; Watermark Computing: Corinda, Australia, 1994.
54. Hou, L.; Wenninger, J.; Shen, J.; Zhou, Y.; Bao, H.; Liu, H. Assessing crop coefficients for Zea mays in the semi-arid Hailiutu River catchment, northwest China. *Agric. Water Manag.* **2014**, *140*, 37–47.

55. Xuejun, D.; Xinshi, Z.; Baozhen, Y. A preliminary study on the water balance for some sand land shrubs based on transpiration measurements in field condition. *Acta Phytoecol. Sin.* **1997**, *21*, 208–225.
56. Huang, J.; Zhou, Y.; Yin, L.; Wenninger, J.; Zhang, J.; Hou, G.; Zhang, E.; Uhlenbrook, S. Climatic controls on sap flow dynamics and used water sources of Salix psammophila in a semi-arid environment in northwest China. *Environ. Earth Sci.* **2015**, *73*, 289–301.
57. Yang, Z.; Li, X.; Sun, Y.; Liu, L.; Zhang, X.; Ma, Y. Characteristics of rainfall interception and stemflow for Salix Psammophila in Maowusu sandland, Northwest China. *Adv. Water Sci.* **2008**, *19*, 693–698.
58. Lin, D.J.; Zheng, Z.C.; Zhang, X.Z.; LI, T.X.; Wang, Y.D. Study on the effect of maize plants on rainfall redistribution processes. *Sci. Agric. Sin.* **2011**, *44*, 2608–2615.
59. Mu, Q.; Zhao, M.; Running, S.W. Improvements to a MODIS global terrestrial evapotranspiration algorithm. *Remote Sens. Environ.* **2011**, *115*, 1781–1800.
60. Yangxiao, Z.; Zhi, Y.; Danrong, Z.; Xiaomei , J.; Jun, Z. Inter-catchment comparison of flow regime between the Hailiutu and Huangfuchuan rivers in the semi-arid Erdos Plateau, Northwest China. *Sci. J.* **2015**, *60*, 688–705.
61. Zhao, X.; Jia, Z.; Liu, H.; Song, T.; Wang, Y.; Shi, L.; Song, C.; Wang, Y. Effects of the conversion of marshland to cropland on water and energy exchanges in northeastern China. *J. Hydrol.* **2008**, *355*, 181–191.
62. Yin, L.; Zhou, Y.; Huang, J.; Wenninger, J.; Zhang, E.; Hou, G. Interaction between groundwater and trees in an arid site: Potential impacts of climate variation and groundwater abstraction on trees. *J. Hydrol.* **2015**, *528*, 435–448.

Quantification of Environmental Flow Requirements to Support Ecosystem Services of Oasis Areas: A Case Study in Tarim Basin, Northwest China

Jie Xue, Dongwei Gui, Ying Zhao, Jiaqiang Lei, Xinlong Feng, Fanjiang Zeng, Jie Zhou and Donglei Mao

Abstract: Recently, a wide range of quantitative research on the identification of environmental flow requirements (EFRs) has been conducted. However, little focus is given to EFRs to maintain multiple ecosystem services in oasis areas. The present study quantifies the EFRs in oasis areas of Tarim Basin, Xinjiang, Northwest China on the basis of three ecosystem services: (1) maintenance of riverine ecosystem health, (2) assurance of the stability of oasis–desert ecotone and riparian (Tugai) forests, and (3) restoration of oasis–desert ecotone groundwater. The identified consumptive and non-consumptive water requirements are used to quantify and determine the EFRs in Qira oasis by employing the summation and compatibility rules (maximum principle). Results indicate that the annual maximum, medium, and minimum EFRs are 0.752×10^8, 0.619×10^8, and 0.516×10^8 m^3, respectively, which account for 58.75%, 48.36%, and 40.29% of the natural river runoff. The months between April and October are identified as the most important periods to maintain the EFRs. Moreover, the water requirement for groundwater restoration of the oasis–desert ecotone accounts for a large proportion, representing 48.27%, 42.32%, and 37.03% of the total EFRs at maximum, medium, and minimum levels, respectively. Therefore, to allocate the integrated EFRs, focus should be placed on the water demand of the desert vegetation's groundwater restoration, which is crucial for maintaining desert vegetation to prevent sandstorms and soil erosion. This work provides a reference to quantify the EFRs of oasis areas in arid regions.

Reprinted from *Water*. Cite as: Xue, J.; Gui, D.; Zhao, Y.; Lei, J.; Feng, X.; Zeng, F.; Zhou, J.; Mao, D. Quantification of Environmental Flow Requirements to Support Ecosystem Services of Oasis Areas: A Case Study in Tarim Basin, Northwest China. *Water* **2015**, *7*, 5657–5675.

1. Introduction

Water is an extremely crucial natural resource for the socioeconomic sustainable development of arid and semiarid regions worldwide. Oases, being the center of human survival and biodiversity, are a distinctive landscape in arid areas. The existence and stabilization of oases depend on water availability. Recently, the oasis ecosystem has been considered highly vulnerable and is suffering from severe threats because of the scarcity and excessive utilization of water resources [1–3]. In particular, the stabilization of the natural oasis ecosystem is seriously threatened under the water-use competition between artificial and natural oasis ecosystems [4]. Xinjiang, Northwest China, as one of the world's largest arid areas, is characterized by considerably fragile water resources and associated ecological and environmental challenges [5]. Its landscape is a typical mountain-oasis-desert ecosystem, and its oases are situated between the mountainous areas and among the desert plains; these oases are essential for human settlement, preventing desertification, and supporting vegetable cultivation [6,7], hence requiring a stable water supply. In recent years, however, the increase in water demand has been attributed to anthropogenic activities such as agricultural irrigation of artificial oasis areas in the lower reaches of the river basin, thereby increasing the competition for water resources between artificial and natural oasis ecosystems. To resolve the conflict for water resources, the stability and health of natural oasis ecosystems must receive the highest priority to maintain such arid regions [4].

Water demands associated with maintenance of the health of riverine ecosystems have been quantified in previous research through environmental flow requirements (EFRs), which refer to the amount of water needed to restore and maintain the stability and health of ecosystems [8]. Subsequently, several researchers indicated that the EFRs do not only include the water demand of the riverine ecosystem but also that of the eco-environment ecosystem outside the river [9–12]. Over the next few years, the environmental flows have been classified into consumptive and non-consumptive water requirements, which have been developed to quantify the water demands of various ecosystem functions in wetlands, estuaries, and rivers ecosystem services [13–19]. In the literature, most studies have estimated the EFRs of natural vegetation or riparian forests in arid areas [6,20]. These findings indicated that water supply is crucial to ensuring ecosystem stabilization in arid areas. However, quantification of the EFRs through the integration of multiple components is lacking in these areas, particularly in the oases of arid areas under multiple ecosystem services. Moreover, few standard methodologies are available for the identification of EFRs in the oases of arid areas because previous works cannot be used in a straightforward manner. This mismatch is ascribed to the differences in ecosystem services and data availability. Notably, the concentration on limited ecosystem functions of rivers, wetlands, or estuaries in ecosystem services leads to oversimplification of EFRs' identification [17,18]. To fill the gap in EFRs for the oasis in arid areas such as those of Tarim Basin, an appropriate approach should be developed to determine these EFRs.

This paper, on the basis of the characteristics of the oasis ecosystem in Tarim Basin, identifies the EFRs using three functions of the oasis ecosystem services:

(1) keeping the riverine ecosystem health to maintain freshwater habitat for biodiversity conservation, sediment transport for riverbed desalting, and salinity balance for salt equilibrium; (2) guaranteeing the stability of the oasis-desert ecotone and riparian forests for combatting desertification and soil erosion; and (3) restoring the groundwater of the oasis-desert ecotone to ensure desert vegetation growth. Notably, water consumption associated with irrigation for farmland ecosystems is not included in the oasis ecosystem services in terms of priority maintenance in the arid region. The Qira oasis case study is conducted to quantify the applicability and practicability of EFRs in the oasis areas of the Tarim Basin in accordance with the summation rule, which is applied to identify consumptive water requirements, as well as the compatibility rule, which is used to determine non-consumptive water requirements [17,18]. The present study first determined the maximum, medium, and minimum levels of EFRs in the Qira oasis and subsequently identified the most important periods for maintaining the EFRs. The proportions for allocation of EFRs in the oasis areas of the Tarim Basin are also discussed. Finally, the allocations of environmental flows and water demands of human activities are presented in this paper.

2. Materials and Methods

2.1. Study Area and Classification of Environmental Flow Requirements

The Tarim Basin, located in Xinjiang, Northwest China, is the world's largest inland basin and contains a large number of oases. Owing to the extremely arid climate, with an annual precipitation of below 100 mm and annual potential evaporation of approximately 2000 mm, the water supply in the oasis areas of the Tarim Basin depends only on river water [4,21]. Thus, river discharge is vital for maintaining natural oasis ecosystems (e.g., riparian forests and desert vegetation), as well as artificial oasis ecosystems (e.g., farmland and urban vegetation) in this region. However, with the significant increase of artificial oasis extensions in recent years, several dams have been built in the upstream and/or midstream so that river water can be extracted for anthropogenic activities (e.g., agricultural irrigation). The increase in water demand in such arid areas with severe water deficits has resulted in frequent drying up of the lower reaches of Tarim Basin's inland rivers. Many serious problems have been exposed, such as vanishing of the aquatic ecosystem, degradation of riparian forests, and decline of the groundwater table. Most importantly, the downstream of the natural oasis ecosystem, which acts as a natural barrier to prevent desertification and sandstorms, has been severely weakened and even nearly lost [4]. These cases are similar to the situations among oases in other arid regions, such as those of Central Asia [22].

On the basis of the ecosystem structures in the oases of arid areas, Abd El-Ghani [23] divided the Qara oasis of Egypt into three functional belts: desert vegetation belt around the desert region, shelter forest belt around the desert vegetation, and the crop belt at the center. The oasis structures in the Tarim Basin exhibit comparable characteristics. According to the oasis characteristics and ecosystem functions, the oasis space structures in the Tarim Basin are classified into three functional zones, namely, desert, shelter forest, and farmland [24]. However, the riverine ecosystems in the oases of arid areas receive little attention. River discharge serves not only as water supply for human and oasis ecosystem demands but also contributes to the water left in or released into the river channel for aquatic biodiversity and growth of riparian forests. Therefore, inland river ecosystems should also be included in the ecosystem services of oases in arid areas. In the present study, we divide the Tarim Basin oasis' structure into three circular zones and a strip (Figure 1). From the center, the first circular zone represents the farmland ecosystem, the second represents the shelter forest, and the third represents the desert vegetation ecosystem; the strip represents the inland river ecosystem.

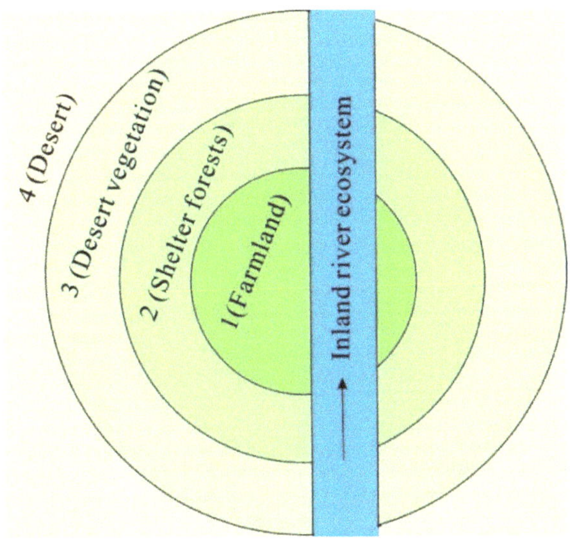

Figure 1. Schematic map of ecosystem functions in the oasis areas of Tarim Basin.

The water supply that is solely dependent on river water must meet multiple functions of ecosystem services in the oasis areas of the Tarim Basin. From the aforementioned classification, the riverine ecosystem and the natural oasis ecosystem are regarded as two ecosystem services in the region, whereas the artificial oasis ecosystem controlled by human activities is not considered as one of the ecosystem services among the maintenance priorities in such arid regions [25]. Therefore, given the different ecosystem functions, EFRs in oasis areas of the Tarim Basin are classified into various components (Table 1). Four types of components are identified, including water demands for river health, desert vegetation growth, maintenance of riparian and shelter forests, and groundwater restoration. These water requirements must ensure the integrity of ecosystem functions, including water quantity (e.g., maintenance of riverine habitat and desert vegetation) and water quality (e.g., salinity balance and nutrient transport).

Furthermore, on the basis of water consumption characteristics, the EFRs in the oasis areas of the Tarim Basin are classified into non-consumptive and consumptive water demands. Water volumes for ensuring the replacement of evapotranspiration by desert vegetation, riparian and shelter forests, groundwater restoration, and water losses of river channel are identified as consumptive EFRs. By contrast, water volumes for maintaining the riverine habitat and providing adequate transport of sediments and nutrients are considered as non-consumptive EFRs. Non-consumptive water requirements ensure the maintenance of river health and flow, as well as prevent drying up of the oasis areas in the Tarim Basin.

Table 1. Classification of environmental flow requirements (EFRs) in the oasis areas of Tarim Basin.

Ecosystem Type	Environmental Flow Requirements	Ecosystem Functions
Riverine ecosystem	EFRs for riverine ecosystem health	Maintenance of riverine habitat Sediment transport Salinity balance Nutrients transport
Nature oasis ecosystem	EFR for maintenance of desert vegetation	Combating desertification
	EFRs for maintenance of riparian and shelter forests	Preventing sandstorm
	EFR for groundwater restoration	Ensuring desert vegetation growth

2.2. Quantification of Environmental Flow Requirements in Oasis Areas of Tarim Basin

According to the classification of oasis ecosystem services above, water demands for the riverine ecosystem, desert vegetation ecosystem, and riparian and shelter forests ecosystem are considered prior to agricultural water conversion. At higher elevations before the mountain pass, the river is recharged by groundwater. In the downstream plain oasis of the Tarim Basin, little groundwater recharge or return flow occurs because the riverbed is higher than the surrounding land [4]. Groundwater of the oasis areas comes from overbank flow of river water. Therefore, the water supply in the oasis areas of Tarim Basin solely depends on river discharge, which originates from the glacier or snow melt and precipitation of Kunlun mountain ranges in the South, Tianshan mountain ranges in the North, and Pamir mountain ranges in the

West. The discharge then flows through the oasis areas and finally discharges into the extremely arid desert. The water requirements of different components in the EFRs demonstrate spatial difference. The health of riverine, riparian, and shelter forest ecosystems (mainly *Populus euphratica*) is maintained by perennially uninterrupted discharge. The desert vegetation ecosystem (mainly shrub vegetation) preventing desertification and soil erosion is fed by the seasonal flooding period (recruiting more shrubs and grassland) and groundwater (ensuring that the shrubs do not wither). Notably, the EFRs between desert vegetation and groundwater restoration do not overlap. The groundwater depth in the oasis areas can only support deep-rooted desert vegetation that does not wither throughout the year from surface water shortage. Desert vegetation (short-root plants and weeds) are fed by river discharge and can also recruit new shrubs and plants in the seasonal flooding period. The water requirements of EFRs' components exhibit spatial and temporal differences. However, in the present study, these water requirements are simply quantified using an annual average scale based on the previous literature. The water requirements (e.g., desert vegetation) are basically synchronous with river runoff variation on temporal scales [26]. Therefore, the temporal allocation for the EFRs is based on the proportion of natural river runoff on different temporal scales. Consequently, the EFRs' components are quantified in the oasis areas of the Tarim Basin as follows.

2.2.1. Environmental Flow Requirement for Base Flows

The environmental flow requirement for base flows is minimum river discharge, which is used to maintain the health of aquatic habitats, improve the self-purification ability and integrity of riverine ecosystem, and ensure continuous flows. In the current study, EFRs for base flows are determined by the annual average of minimum monthly river runoff during a period of little human disturbance [18]:

$$W_{bf} = \frac{1}{n} \sum_{i=1}^{n} min(Q_{ij}) \tag{1}$$

where W_{bf} is the EFR for river base flows (m^3), Q_{ij} is the *j*th monthly river runoff at the *i*th year (m^3), and n is the number of years. The runoff data (Q) resulted is obtained from the hydrological station, which presents a long-term daily mean runoff time series during 1960–2010 for hydrological studies upstream of the oasis areas of Tarim Basin.

2.2.2. Environmental Flow Requirement for Sediment Transport

Water demand for sediment transport requires that river discharge is applied to maintain the dynamic equilibrium of scour and erosion of river sediment transport. Therefore, EFRs for sediment transport are determined by the carrying capacity and quantity of sediment in the oasis areas of Tarim Basin. The expression is given as follows [17–19]:

$$W_{st} = S_t/C_t \tag{2}$$

where W_{st} refers to the EFR for sediment transport (m^3), S_t is the quantity of sediment in inland rivers of the oasis areas (kg), and C_t denotes the sediment carrying capacity (kg/m^3). Both S_t and C_t are determined from long-term records of sediment transport data at the hydrological station upstream of oasis areas of Tarim Basin.

Moreover, salinity balance and nutrient transport exert certain effects on the riverine ecosystem. Generally, the EFR for maintaining sediment transport can meet the requirement of water quality in inland rivers of arid areas [16]. Therefore, the water quality of riverine ecosystem is not discussed hereafter.

2.2.3. Environmental Flow Requirements for Evaporation of Surface Water

To maintain river water flow, the evaporation of surface water in the river channel should be considered. The EFR for water evaporation in the river channel in arid areas is calculated as follows [27]:

$$W_e = \varepsilon \cdot l \cdot b \cdot E_0 \tag{3}$$

where W_e is EFR for the evaporation of surface water of river channel (m^3); ε is reduction coefficient of evaporation; l and b are length and width of river channel, respectively; and E_0 is the annual evaporation based on records of an evaporating dish with a 20 cm diameter.

2.2.4. Environmental Flow Requirements for Different Vegetation Types

Water requirements for different vegetation types (*i.e.*, riparian and shelter forests and desert vegetation) are crucial for wind prevention and sand resistance. The EFRs for plant communities represent the summation of all water demands [27–29]. The formula is expressed as follows:

$$W_f = \sum_{i=1}^{2} S_i R_i \tag{4}$$

where W_f is the EFRs for riparian and shelter forests, as well as desert vegetation (m^3); S_i is the area of the ith vegetation types (km^2); R_i is the ecological water quota of ith vegetation types estimated via evapotranspiration results using lysimeter method (m^3/km^2). In 2003, Wang *et al.* [28] estimated the ecological water quotas for high-, medium-, and low-coverage grass to be 2343.8, 1486.7, and 629.7 m^3/hm^2, respectively, and that for riparian and shelter forests to be 3405 m^3/hm^2 in the oasis areas of Tarim Basin.

2.2.5. Environmental Flow Requirements for Groundwater Restoration

Water requirements for groundwater restoration are water volumes restoring to the target water table to maintain desert vegetation growth. Therefore, EFRs for groundwater restoration are calculated as follows [27]:

$$W_{gr} = M \cdot \Delta H \cdot S \cdot n \quad (5)$$

where W_{gr} is the EFR for groundwater restoration (m^3); M is the saturation deficiency of the zone of fluctuation of the water table (cm^3/g). Song *et al.* [27] estimated the saturation deficiency of the fluctuation belt of water table (M) to be 0.16 (cm^3/g), given the aquifer saturation rate of 25.8%; ΔH is the change in height of ground water (m); S is the area of desert vegetation cover (km^2); and n is the bulk density of soil (g/cm^3), which is determined as 1.36 (g/cm^3) by $n = 35.72e^{-0.185H}$ (H is the groundwater depth).

2.2.6. Integration of Environmental Flow Requirements for Various Ecosystem Functions

According to the classification of consumptive and non-consumptive water requirements, the rule of summation is used to obtain consumptive EFRs, whereas the rule of maximum is applied to calculate non-consumptive ones. Therefore, EFRs in the oasis areas of Tarim Basin are integrated by the following expression:

$$W = W_e + W_f + W_{gr} + max(W_{bf}, W_{st}) \quad (6)$$

where W is the total environmental flow requirement in the oasis areas of Tarim Basin (m^3), and max (a, b) refers to the maximum value of parameters a and b.

2.2.7. Temporal Allocation of Environmental Flow Requirements

EFRs generally show temporal variability at different scales (e.g., annual and monthly scales), but the water supply for EFRs (the river discharge) is easily influenced by anthropogenic activities (e.g., extensive irrigation or water diversion project). Consequently, the water requirements for multiple ecosystem services cannot be satisfied by the natural oasis areas of Tarim Basin. To maintain the stability and health of riverine and natural oasis ecosystems, allocation of temporal EFRs should be considered under the temporal variability of natural river discharge.

To ensure a natural flow regime in the oasis of Tarim Basin, the temporal pattern of natural river runoff is considered as the indicator of the temporal variability of EFRs. Therefore, EFRs are allocated on the basis of the proportion of natural river runoff at different temporal scales. The allocation proportion of EFRs is expressed as follows:

$$P_i = \sum_{j=1}^{n} Q_{ji} / \sum_{j=1}^{n} Q_j \tag{7}$$

where P_i is the proportion of the monthly river runoff in ith month to the annual runoff during 1960–2010, Q_{ji} is the river runoff in ith month of jth year (m^3) from 1960 to 2010, and Q_j is the annual runoff in jth year (m^3) from 1960 to 2010.

2.3. Case Study Illustrating the Application of Environmental Flow Requirements in Qira Oasis

Oases in the Tarim Basin present similar structures and characteristics. Hence, in this research, the Qira oasis located in the south rim of Tarim Basin in Xinjiang is selected for the case study of EFRs by virtue of the available data. The Qira oasis lies in the lower reaches of the Qira River Basin (latitudes 36°54′N–37°09′N and longitudes 80°37′E–80°59′E) (Figure 2). Its existence mainly depends on the availability of Qira River, which is generated from the high altitude valley of Kunlun Mountains, then flows through the Qira oases, and finally discharges into the extremely arid desert. Given an annual precipitation of approximately 39 mm and a strong evaporation of 2700 mm which is estimated by the measured data of a 20 cm diameter evaporating dish during 1960–2010, the water supply in the Qira oasis area solely depends on river discharge [30–32]. On the basis of the meteorological data (Qira station) in the Qira oasis and the hydrological data obtained by a hydrological station 19.5 km from the Qira county (Figure 2), the annual mean temperature, annual evaporation, and precipitation increased during the period of 1960–2010 by rates of 0.28 °C/10a, 90.98 mm/10a, and 17.20 mm/10a, respectively. By contrast, the annual runoff declined from 1960 to 2010 by a rate of -0.03×10^8 $m^3/10a$ (Figure 3). Although the climate of Xinjiang has begun to transform from warm and dry conditions to warm and wet conditions [2,33], the decrease of runoff caused by regional climate change (less rainfall and strong evaporation) leads to more pressure being placed on integrated water resource management for eco-environmental protection and economic sustainable development in the Qira oasis.

The Qira oasis consists of four types of ecosystems, *i.e.*, desert vegetation ecosystem, shelter forest ecosystem, farmland ecosystem, and riverine ecosystem. The farmland ecosystem is controlled by human activities and, hence, not included within the ecosystem services under maintenance priority in the region. With regard to the desert vegetation ecosystem, the remote sensing data from Landsat™ imagery in 2010 are compiled to obtain desert vegetation coverage in the Qira oasis. On the basis of the TM spectral characteristics, the classification maps of land cover are assessed with the scale of 1:47,000, and the classification accuracy is determined using the Kappa coefficient [34]. According to the classification results, desert vegetation in the Qira oasis is composed of high-coverage grass (69.34 km^2) (60%–90% vegetation coverage), medium-coverage grass (22.54 km^2) (20%–60% vegetation coverage), and low-coverage grass (20.28 km^2) (5%–20% vegetation coverage). From the evaluation of classification precision, the Kappa coefficients of classification data in 2010 was found to account for 93%. Therefore, these classification data can reflect the changes of land cover in the study region with relative high accuracy [34]. The riparian and shelter forests (mainly *Populus euphratica*) in the Qira oasis accounted for 50 km^2, which comprises 30.86% of the total coverage areas (Table 2). Considering the uncertainty in compiling and classifying the desert vegetation, the actual area of each type grass is quantified by classifying three levels (*i.e.*, maximum, medium, and minimum), respectively [19]. According to water requirement quota estimated by the results from the experiment [28], the water demand for each type of grass is determined by the water requirement quota multiplied by the corresponding actual area. Moreover, at upstream of the Qira oasis, long-term records of runoff and sediment discharge from 1960 to 2010 are available in the Qira hydrological station (1557 m above sea level).

Table 2. Vegetation types, vegetation cover indices, and areas occupied by different vegetation types. Maximum, medium, and minimum denote three levels of water demands by vegetation.

Vegetation Type	Vegetation Cover Index	Total Area (km^2)	Area (km^2)		
			Maximum	Medium	Minimum
Desert vegetation	High-coverage grass (60% to 90%)	69.34	62.41	52.01	41.60
	Medium-coverage grass (20% to 60%)	22.54	13.52	9.02	4.51
	Low-coverage grass (5% to 20%)	20.28	4.06	2.54	1.01
Riparian and shelter forests	-	50	50	50	50
Total	-	162.16	129.99	113.56	97.13

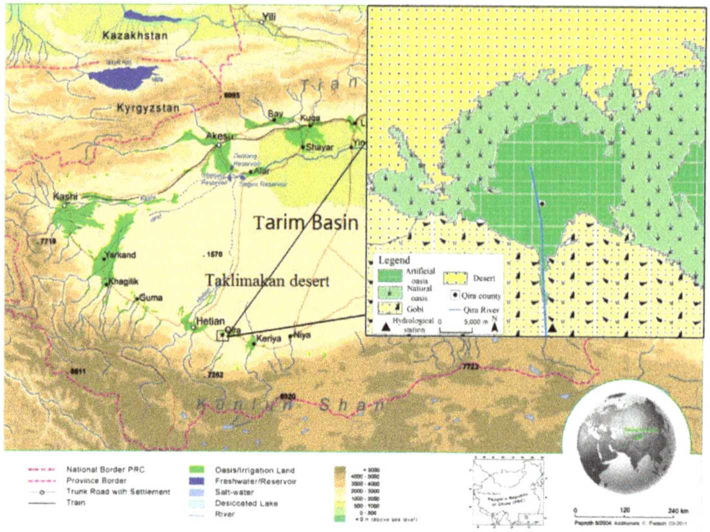

Figure 2. Location and topography of the study area, adapted from the project SuMaRiO [35].

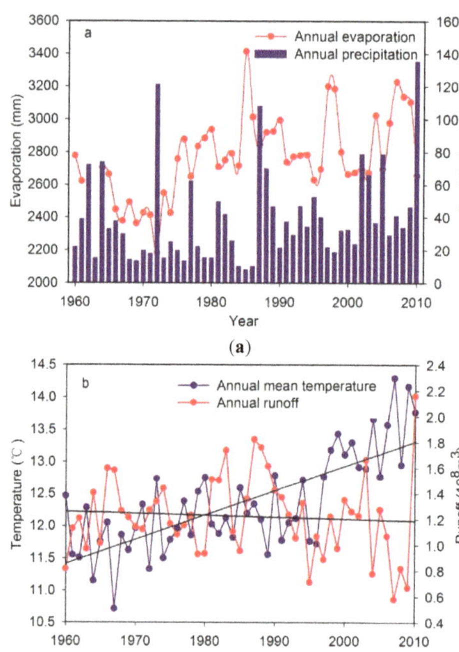

(a)

(b)

Figure 3. Variations in annual precipitation and evaporation (**a**); annual mean temperature, and annual runoff (**b**) in the Qira oasis.

3. Results and Discussion

3.1. Quantification of Environmental Flow Requirements

According to different ecosystem functions of the riverine ecosystem in the Qira oasis area, the EFRs for base flows, sediment transport, and water evaporation of river channel are calculated using Equations (1)–(3), respectively. Considering there was little human disturbance during the period from 1960 to 1979, the riverine eco-environment maintained its natural state in the absence of the establishment of water conservation measures and severe utilization of river water in the Qira oasis area. Therefore, the EFRs for base flows are determined on the basis of the average of annual minimum monthly river runoff between 1960 and 1979. The EFR of base flow is estimated to be 0.016×10^8 m^3, which accounts for 1.25% of natural river discharge. For sediment transport, the quantity of sediment in Qira oasis area is 1,810,000 kg, and the sediment carrying capacity is 13.5 kg/m^3. As a result, the EFR for sediment transport is determined to be 0.014×10^8 m^3, which represents 1.09% of total river discharge. Moreover, river length and river width are 115 km and 23.88 m, respectively, and annual evaporation and reduction coefficient of evaporation are 2505 mm and 0.5, respectively. Therefore, the EFR for water evaporation of river channel is calculated to be 0.034×10^8 m^3, which accounts for 2.66% of total river discharge. The results imply that the EFR for water evaporation of the river channel has the largest water demand, which represents 53.10% of the EFRs for all ecological functions in the riverine ecosystem. The conclusions are different from the observations on the EFRs in wetlands and estuaries, in which maintaining the sediment transport and salinity balance requires higher EFRs than other components [16,17,19]. The contrasting results are attributed to very strong evaporation (approximately 2700 mm) and small runoff (3.9 m$^3\cdot$ s^{-3}) in the Qira oasis area, so that the EFR for water evaporation of river channel is greater than that for base flow and sediment transport.

The EFRs for different ecosystem functions of the natural oasis ecosystem (including desert vegetation and riparian and shelter forests) in the Qira oasis area are calculated using Equation (4). Given that desert vegetation in the Qira oasis consists of high-, medium-, and low-coverage grass, based on corresponding water quota, the EFRs for desert vegetation are estimated to be 0.169×10^8, 0.137×10^8, and 0.105×10^8 m^3 at the maximum, medium, and minimum levels, respectively. Moreover, the EFRs of riparian and shelter forests (mainly comprised *Populus euphratica*) are calculated to be 0.170×10^8 m^3, which accounts for a larger proportion than that of desert vegetation (Table 3). According to Zhang (2003) [36], the stability of desert vegetation and riparian and shelter forests mainly depended on groundwater depth. The most suitable groundwater depth for desert vegetation and forests is approximately 4–7 m. When the groundwater depth is greater than 7 m, desert vegetation and forests begin to degenerate [36]. Given the severe water utilization in the upstream of the Qira oasis, groundwater depth has reached 8.45 m. Based on those estimated groundwater tables, the EFR for groundwater restoration is calculated using Equation (5). The maximum (groundwater depth restored to 4 m), medium (groundwater depth restored to 6 m) and minimum (groundwater depth restored to 7 m) water requirements are 0.363×10^8, 0.262×10^8, and 0.191×10^8 m^3, respectively. The EFRs of the components (including desert vegetation, riparian and shelter forests, and groundwater restoration) are found to require large water supply in the natural oasis ecosystem of the Qira oasis area [37]. Particularly, the EFRs for groundwater restoration are considerably higher than other water demand components in the natural oasis ecosystem. In fact, with the Qira oasis comprising the most important ecosystem services, satisfying the EFRs has become crucial for combating desertification and soil erosion in the Qira oasis and other oasis areas of Tarim Basin.

Through classification of non-consumptive and consumptive water volumes, the total EFRs in the Qira oasis area are determined using Equation (6). The EFRs at the maximum, medium, and minimum levels are determined to be 0.752×10^8, 0.619×10^8 and 0.516×10^8 m^3, respectively, which respectively account for 58.75%, 48.36%, and 40.29% of the total river discharge. The calculated water volumes are essential to maintain the stability and health of the Qira oasis ecosystem. Moreover, the temporal pattern of the total EFRs at the different time scales in the Qira oasis area is calculated using Equation (7). The ecological objective for ecosystem services is determined by the monthly proportion of the total EFRs during 1960–2010 in the Qira oasis. The monthly allocation proportion of the total EFRs from April to October represents the most important periods to maintain the EFRs in the Qira oasis (Figure 4).

The river discharge during this period ensures greater than 93% of the total EFRs. The allocation of the EFRs between October and April of the subsequent year is very low at different levels, whereas that between April and October has become crucial to support the water demands of multiple ecosystem services in the Qira oasis (Figure 5). In previous studies, the EFRs are classified by consumptive and non-consumptive water demand to meet various ecosystem functions in river, wetland, or estuary ecosystem services [16,18,19]. Under multiple ecosystem services, the EFRs in the oasis areas of Tarim Basin are also divided into consumptive water (including EFRs for groundwater restoration, riparian and shelter forests, desert vegetation, and evaporation of surface water) and non-consumptive water (including EFRs for base flows and sediment transport). Maintaining the total EFRs in the Qira oasis area requires >50% of the natural river flows at the maximum level. The results are consistent with the data obtained from wetlands or estuaries [16,19]. Moreover, at a monthly scale, the allocation of the EFRs is concentrated mainly on the flooding period, which is the crucial time for supporting and recruiting the desert vegetation and forests in oasis–desert ecotone [38]. Therefore, ensuring the river discharge at temporal scale performs a key function to maintain the stability and health of natural oasis ecosystems.

Table 3. Environmental flow requirements for natural oasis ecosystem based on different ecosystem functions.

Ecological Functions	Water Quota (m^3/hm^2)	Maximum (10^8 m^3)	Medium (10^8 m^3)	Minimum (10^8 m^3)
High-coverage grass	2343.8	0.146	0.122	0.098
Medium-coverage grass	1486.7	0.020	0.013	0.007
Low-coverage grass	629.7	0.003	0.002	0.001
Riparian and shelter forests	3405	0.170	0.170	0.170
Total	-	0.339	0.307	0.276

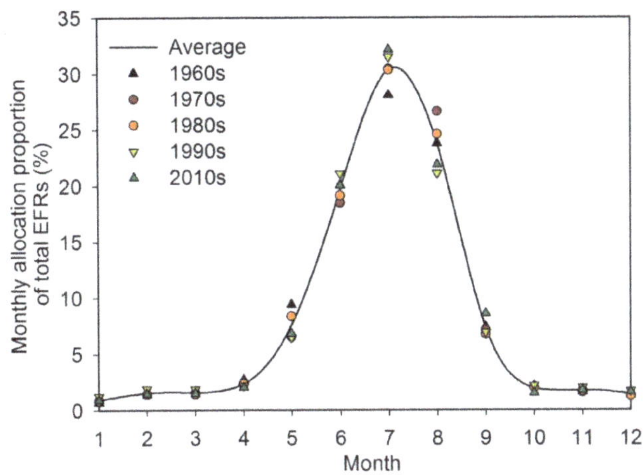

Figure 4. Proportion of temporal allocation for the total EFRs during different periods in Qira oasis.

3.2. Proportion of Environmental Flow Requirements for Multiple Ecosystem Functions

The EFRs of various ingredients for multiple ecosystem services require different water demands in oasis areas. According to classification in Table 2, three levels of each component are calculated by corresponding equation (Table 3), and the proportion of each EFR at different levels is determined by water demand of each component dividing by total EFRs. In non-consumptive water, EFRs for base flows and sediment transport at the maximum level represent 2.09% and 1.83% of the total EFRs, respectively (Figure 6). Similarly, the EFRs for sediment transport are higher than those for base flows at the medium and minimum levels, implying that the EFR for sediment transport is sufficient to satisfy the water requirement for base flows.

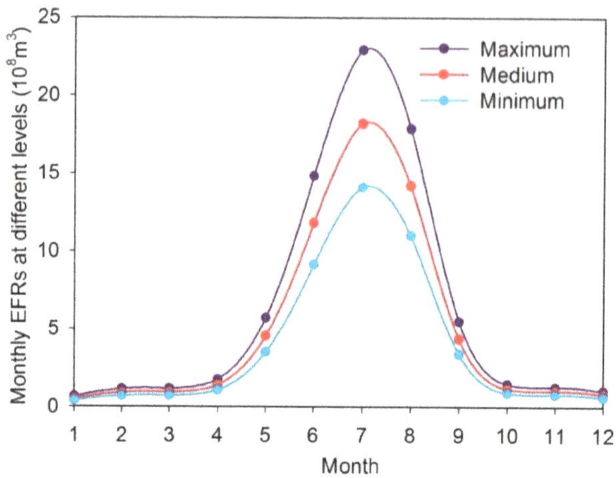

Figure 5. Monthly EFRs for multiple ecosystem functions at different levels (Maximum, Medium, and Minimum) in Qira oasis.

For consumptive water systems, the riparian and shelter forests, desert vegetation, and groundwater restoration require more water volumes at the maximum level and account for 22.19%, 22.06%, and 47.39% of the total environmental flow requirements, respectively (Figure 6). Among the consumptive water systems, the environmental flow requirements for groundwater restoration of oasis–desert ecotone comprise the largest proportion, which account for 48.27%, 42.32%, and 37.03% of the total environmental flow requirements at the maximum, medium, and minimum levels, respectively.

According to the above EFR analysis, the proportion of the EFR of each ingredient in natural oasis ecosystem is larger than that in the riverine ecosystem. In extremely arid areas, the EFRs for natural oasis ecosystem undoubtedly account for a larger proportion [6]. Moreover, in all components of the EFRs, the EFR for groundwater restoration requires the greatest water supply. In the past few years, groundwater depth in the oasis areas of Tarim Basin has rapidly declined because of the increase in water demand for human activities (particularly agricultural irrigation) [38], thereby increasing pressure for groundwater restoration in the oasis area. Groundwater, once depleted, is very difficult to restore [39]. Therefore, in future research, more focus should be given to groundwater management in the Qira oasis and other oasis areas of Tarim Basin.

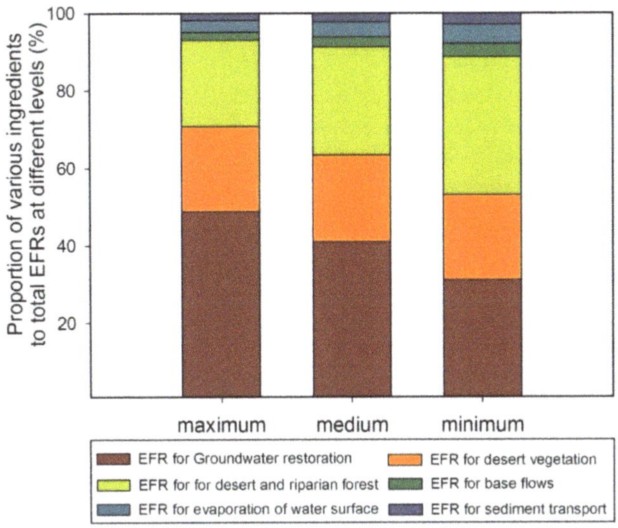

Figure 6. Proportion of various ingredients to the total EFRs for multiple ecosystem functions at different levels (Maximum, Medium, and Minimum) in Qira oasis.

3.3. *Response of Environmental Flow Requirements to Natural River Runoff*

Figure 7 illustrates the proportion of annual total EFR to annual river discharge from 1960 to 2010 in the Qira oasis. The natural river runoff can almost satisfy EFRs at the maximum levels in dry years, such as in 1995, 2007, and 2009. Given that the river runoff started to change abruptly and remarkably and then decreased in 1994, the proportion of annual total EFR to annual river discharge in the pattern of the graph changes more considerably. Thus, a major challenge is to meet water requirements for both human activities and natural ecosystems in the time allocation, especially during dry years.

In the Qira oasis, water used for anthropogenic activities, especially agricultural irrigation, is mainly extracted from river discharge. Therefore, considering the EFRs with priority, a water volume that meets the demands of the EFRs is crucial to support human activities. Figure 8 indicates the annual water volume to ensure EFRs in the Qira oasis from 1960 to 2010, *i.e.*, the water volume for annual runoff subtracting the total EFRs. The water volumes are essential to satisfy the water demands in socioeconomic sustainable developments. The annual average water volumes available for human activities are respectively determined to be 0.765×10^8, 0.662×10^8, and 0.529×10^8 m^3 at the maximum, medium, and minimum levels while ensuring EFRs are met in the Qira oasis. Meeting the EFR requirements has been a recommendation for water use of an artificial oasis (such as farmland) in water resource management.

In previous studies, the EFRs for multiple ecosystem services are proposed with higher priority in river, wetland, or estuary ecosystem studies to coordinate the conflicting water demands between ecosystem services and socioeconomic sustainable developments [16–19]. Similarly, particular focus in the Qira oasis and other oasis areas of Tarim Basin should be placed on the EFRs, preferentially on the basis of different ecosystem services. However, approximately 95% of water resources extracted from river and groundwater is used for anthropogenic activities, especially agricultural irrigation [4]. Thus, the water demands for human activities and the natural oasis ecosystem in Qira and other oasis areas of Tarim Basin leads to competition for water extracted from river discharge. Considering the severe water utilization through human activities such as agricultural irrigation, the oasis–desert ecotone has recently experienced a substantial shrink and is frequently eroded by windblown sand in Qira and other oasis areas of Tarim Basin [37].

Consensus action on supplying water for human activities while ensuring EFRs has been achieved [40–43]. However, prioritizing EFRs probably results in economic losses in oasis farmlands owing to shortages in the water used for irrigation in such extremely arid regions. Water trade-off between human activities and eco-environmental requirements can pose greater challenges in Qira and other oasis areas of Tarim Basin, particularly during dry periods. Under such circumstances, economic compensation may be an effective measure to relieve conflicts between supply and demand [44–46]. Therefore, to ensure that the EFRs of natural oasis ecosystems are met, optimal allocation of water resources of artificial oasis and integrated water resource management should be a focus.

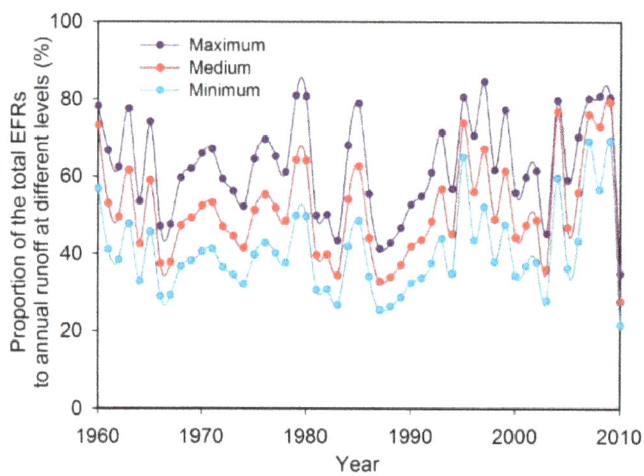

Figure 7. Proportion of the total EFRs to annual runoff at different levels (Maximum, Medium, and Minimum) in Qira oasis.

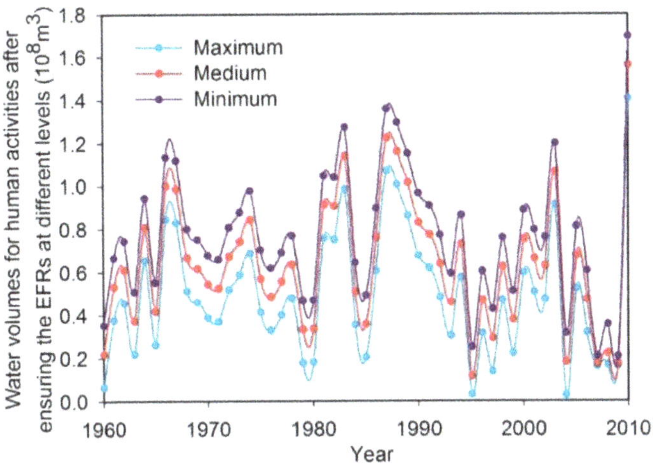

Figure 8. Water volumes needed to support human activities after ensuring the EFRs at different levels in Qira oasis.

4. Conclusions

This study identifies and quantifies the EFRs to support the three ecosystem services in the Qira oases of Tarim Basin in accordance with the summation and compatibility rules (maximum principle). The results indicate that EFRs for supporting various ecosystem services are 0.752×10^8, 0.619×10^8, and 0.516×10^8 m^3 at the annual maximum, medium, and minimum levels, which account for 58.57%, 48.36%, and 40.29% of the natural river runoff in the Qira oasis, respectively. The results imply that natural river discharge, which should be discharged within the Qira oasis, should provide at least the abovementioned water volumes to ensure the stability and health of oasis ecosystems in the Qira oasis. The months spanning from April to October are identified as the most crucial periods to maintain the EFRs. River discharge during this period ensures more than 93% of the annual total EFRs.

In the integrated environmental flow allocation, the EFR for groundwater restoration of the oasis–desert ecotone comprises the largest proportion, which accounts for 48.27%, 42.32%, and 37.03% of the total EFRs at the maximum, medium, and minimum levels, respectively. Therefore, more focus should be given to the water demands of groundwater restoration of desert vegetation, which is vital to preventing desertification and occurrence of sandstorms. Annual water volumes available for human activities to satisfy EFRs in the Qira oasis are 0.765×10^8, 0.662×10^8, and 0.529×10^8 m^3 at the maximum, medium, and minimum levels, respectively. EFRs serve as the foundation for socioeconomic sustainable developments to meet the water demands of ecosystem services.

Although this work quantifies the EFRs by considering multiple ecosystem services in oasis areas; however, estimation of the EFRs related to data uncertainty (such as error of coverage area classification in TM image compilation) and methodological drawbacks (e.g., error of empirical parameters in the equation) may result in a large uncertainty. In future research, focus will be given to parameter calibration with experimental data and uncertainty analysis to quantify the EFRs in the oasis areas.

Acknowledgments: This work was financially supported by the National Natural Science Foundation of China (41471031), the Thousand Youth Talents Plan Project (Y472241001), and the project of Science and Technology Service Network Initiative (STS) of Chinese Academy of Sciences. The authors would like to acknowledge two anonymous reviewers for their valuable comments on the paper.

Author Contributions: All authors contributed to the design and development of this manuscript. Jie Xue carried out the data analysis and prepared the first draft of the manuscript; Jiaqiang Lei, Dongwei Gui and Ying Zhao are the graduate advisor of Jie Xue and contributed many ideas to the study; Xinlong Feng, Fanjiang Zeng, Jie Zhou, and Donglei Mao provided some important advices on the concept of methodology and writing of the manuscript.

Conflicts of Interest: The authors declare no conflict of interest.

References

1. Boehmer, K.; Memon, A.; Mitchell, B. Towards sustainable water management in Southeast Asia—Experiences from Indonesia and Malaysia. *Water Int.* **2000**, *25*, 356–377.
2. Chen, Y.N.; Xu, Z.X. Plausible impact of global climate change on water resources in Tarim River Basin. *Sci. China Ser. D* **2005**, *48*, 65–73.
3. Ling, H.B.; Zhang, Q.Q.; Xu, H.L.; Shi, W. Regional climate change and its effects on the runoff of Manas River, Xinjiang, China. *Environ. Earth Sci.* **2011**, *64*, 2203–2213.
4. Rumbaur, C.; Thevs, N.; Disse, M.; Ahlheim, M.; Brieden, A.; Cyffka, B.; Duethmann, D.; Feike, T.; Fror, O.; Gartner, P.; *et al.* Sustainable management of river oases along the Tarim River (SuMaRiO) in Northwest China under conditions of climate change. *Earth Syst. Dynam.* **2015**, *6*, 83–107.
5. Chen, Y.N. *Water Resources Research in Northwest China*; Springer: Berlin, Germany, 2014; pp. 3–18.
6. Ye, Z.X.; Chen, Y.N.; Li, W.H. Ecological water demand of natural vegetation in the lower Tarim River. *J. Geogr. Sci.* **2010**, *20*, 261–272.
7. Ling, H.B.; Xu, H.L.; Fu, J.Y.; Fan, Z.L.; Xu, X.W. Suitable oasis scale in a typical continental river basin in an arid region of China: A case study of the Manas River Basin. *Quatern. Int.* **2013**, *286*, 116–125.
8. Covich, A. *Water and Eeosystems*; Oxford University Press: NewYork, NY, USA, 1993; pp. 40–52.
9. Petts, G.E. Water allocation to protect river ecosystems. *Regul. Rivers Res. Manag.* **1998**, *12*, 353–365.

10. Prat, N.S.; Munné, A. Water use and quality and stream flow in a Mediterranean stream. *Water Res.* **2000**, *34*, 3876–3881.
11. Richter, B.D.; Mathews, R.; Harrison, D.L.; Wigington, R. Ecologically sustainable water management: Managing river flows for ecological integrity. *Ecol. Appl.* **2003**, *13*, 206–224.
12. Cai, X.; Rosegrant, M.W. Optional water development strategies for the Yellow River Bas in: Balancing agricultural and ecological water demands. *Water Resour. Res.* **2004**, *40*, W08S04.
13. Flannery, M.S.; Peebles, E.B.; Montgomery, R.T. A percent-of-flow approach for managing reductions of freshwater inflows from unimpounded rivers to the southwest Florida Estuaries. *Estuaries* **2002**, *25*, 1318–1332.
14. Tharme, R.E. A global perspective on environmental flow assessment emerging trends in the development and application of environmental flow methodologies for rivers. *River Res. Appl.* **2003**, *19*, 397–441.
15. Benetti, A.D.; Lanna, A.E.; Cobaichini, M.S. Current practices for establishing environmental flows in Brazil. *River Res. Appl.* **2004**, *20*, 427–444.
16. Yang, Z.F.; Sun, T.; Qin, X.S. Calculating methods for quantifying environmental flows in estuaries: A case study of Haihe River Basin, China. *J. Environ. Inf.* **2005**, *6*, 72–79.
17. Sun, T.; Yang, Z.F.; Cui, B.S. Critical environmental flows to support integrated ecological objectives for the Yellow River Estuary, China. *Water Resour. Manag.* **2008**, *22*, 973–989.
18. Yang, Z.F.; Sun, T.; Cui, B.S.; Chen, B.; Chen, G.Q. Environmental flow requirements for integrated water resources allocation in the Yellow River Basin, China. *Commun. Nonlinear Sci.* **2009**, *14*, 2469–2481.
19. Sun, T.; Yang, Z.F.; Shen, Z.Y.; Zhao, R. Ecological water requirements for the source region of China's Yangtze River under a range of ecological management objectives. *Water Int.* **2012**, *37*, 236–252.
20. Thomas, F.M.; Foetzki, A.; Arndt, S.K.; Bruelheide, H.; Gries, D.; Li, X.; Runge, M. Water use by perennial plants in the transition zone between river oasis and desert in NW China. *Basic Appl. Ecol.* **2006**, *7*, 253–267.
21. Tang, Q.C.; Chen, H.Y. Water resources and oasis construction in Tarim Basin. *Chin. Geogra. Sci.* **1992**, *2*, 173–182.
22. CAWA. Regional Research Network-Central Asian Water. Available online: http://www.cawa-project.net/ (accessed on 5 March 2015).
23. Abd El-Ghani, M.M. Flora and vegetation of Gara oasis, Egypt. *Phytocoenol* **1992**, *21*, 1–14.
24. Hu, S.J.; Song, Y.D.; Tian, C.Y.; Li, Y.T.; Li, X.C.; Chen, X.B. Suitable scale of Weigan River plain oasis. *Sci. China Ser. D* **2007**, *50*, 56–64.
25. Malano, H.M.; Davidson, B.A. Framework for assessing the trade-off between economic and environmental uses of water in a river basin. *Irrig. Drain.* **2009**, *58*, 133–147.
26. Wang, F.; Liang, R.L.; Yang, X.L.; Chen, M.J. A study of ecological water requirements in northwest China: Theoretical analysis. *J. Nat. Resour.* **2002**, *17*, 1–8. (in Chinese).
27. Song, Y.D.; Fan, Z.L.; Lei, Z.D.; Zhang, F.W. *Research On Water Resources And Ecology Of Tarim River, China*; Volksverlag Xinjiang: Urumqi, China, 2000; pp. 400–406.

28. Wang, R.H.; Lu, X.M.; Song, Y.D.; Fan, Z.L.; Ma, Y.J. Patterns and characteristics of ecological water demand in west arid zone of China—A case study of green corridor in the lower reaches of Tarim River. *Chin. J. Appl. Ecol.* **2003**, *14*, 520–524. (in Chinese).
29. Fan, Z.L.; Ma, Y.J.; Zhang, H.; Wang, R.H.; Zhao, Y.J.; Zhou, H.F. Research of eco-water table and rational depth of groundwater of Tarim river Drainage Basin. *Arid Land Geogr.* **2004**, *27*, 8–13. (in Chinese).
30. Liu, B.; Zhao, W.Z.; Chang, X.X.; Li, S.B.; Zhang, Z.H.; Du, M.W. Water requirements and stability of oasis ecosystem in arid region, China. *Environ. Earth Sci.* **2010**, *59*, 1235–1244.
31. Wu, J.L.; Yang, F.X.; Zhou, J.; Zhang, Z.W. Desert types and characteristics in the Qira River Basin. *Arid Land Geogr.* **2013**, *36*, 803–811. (in Chinese).
32. Wu, Y.W.; Yang, F.X.; Hua, T. Landforms and their effects on ecological pattern in the Qira River Basin. *Arid Zone Res.* **2011**, *28*, 355–362. (in Chinese).
33. Shi, Y.F.; Shen, Y.P.; Kang, E.S.; Li, D.L.; Ding, Y.J.; Zhang, G.W.; Hu, R.J. Recent and future climate change in Northwest China. *Clim. Chang.* **2007**, *80*, 379–393.
34. Congalton, R.G.; Green, K. *Assessing the Accuracy of Remotely Sensed Data: Principals and Practices*; CRC Press: London, UK, 1999; pp. 137–145.
35. The project SuMaRiO. Sustainable Management of River Oases Along the Tarim River. Available online: http://www.sumario.de/ (accessed on 15 March 2015).
36. Zhang, F. *Water Resource Variation and Its Effects on landscape changes Research in Qira Oasis*; Xinjiang University: Urumqi, China, 2003.
37. Bruelheide, H.; Jandt, U.; Gries, D.; Thomas, F.M.; Foetzki, A.; Buerkert, A.; Wang, G.; Zhang, X.M.; Runge, M. Vegetation changes in a river oasis on the southern rim of the Taklamakan Desert in China between 1956 and 2000. *Phytocoenologia* **2003**, *33*, 801–818.
38. Chen, Y.N.; Li, W.H.; Xu, H.L.; Liu, J.Z.; Zhang, H.F.; Chen, Y.P. The influence of groundwater on vegetation in the lower reaches of Tarim River, China. *Acta Geogr. Sin.* **2003**, *58*, 542–549.
39. Cui, Y.L.; Shao, J.L. The role of ground water in arid/semiarid ecosystems, Northwest China. *Groundwater* **2005**, *43*, 471–477.
40. Andrea, F.; Acreman, M.; Dunbar, M.J. Defining environmental river flow requirements: A review. Hydrol. *Earth Syst. Sci.* **2004**, *8*, 861–876.
41. Yates, D.N.; Sieber, J.; Purkey, D.; Huber, A. WEAP21—A demand-, priority-, and preference-driven water planning model part 1: Model characteristics. *Water Int.* **2005**, *30*, 487–500.
42. Wei, Y.P.; Davidson, B.; Chen, D.L.; White, R.E. Balancing the economic, social and environmental dimensions of agro-ecosystems: An integrated modeling approach. *Agr. Ecosyst. Environ.* **2009**, *131*, 263–273.
43. Pang, A.P.; Sun, T.; Yang, Z.F. A framework for determining recommended environmental flows for balancing agricultural and ecosystem water demands. *Hydrol. Sci. J.* **2014**, *59*, 890–903.
44. Ran, B.Q.; Xu, F.R.; Zhang, C.L. Review of research and practice of river basin ecological compensation. *J. Hydraul. Eng.* **2008**, *39*, 1220–1225.

45. Liu, Y.L.; Hu, P. Ecological compensation standard for Xinjiang River basin based on Pareto optimization. *J. Hydraul. Eng.* **2009**, *40*, 703–713.
46. Sisto, N.P. Environmental flows for rivers and economic compensation for irrigators. *J. Environ. Manag.* **2009**, *90*, 1236–1240.

Spatiotemporal Distribution of Soil Moisture and Salinity in the Taklimakan Desert Highway Shelterbelt

Yuan Huang, Yongdong Wang, Ying Zhao, Xinwen Xu, Jianguo Zhang and Congjuan Li

Abstract: Salinization and secondary salinization often appear after irrigation with saline water. The Taklimakan Desert Highway Shelterbelt has been irrigated with saline ground water for more than ten years; however, soil salinity in the shelterbelt has not been evaluated. The objective of this study was to analyze the spatial and temporal distribution of soil moisture and salinity in the shelterbelt system. Using a non-uniform grid method, soil samples were collected every two days during one ten-day irrigation cycle in July 2014 and one day in spring, summer, and autumn. The results indicated that soil moisture declined linearly with time during the irrigation cycle. Soil moisture was greatest in the southern and eastern sections of the study area. In contrast to soil moisture, soil electrical conductivity increased from 2 to 6 days after irrigation, and then gradually decreased from 6 to 8 days after irrigation. Soil moisture was the greatest in spring and the least in summer. In contrast, soil salinity increased from spring to autumn. Long time drip-irrigation with saline groundwater increased soil salinity slightly. The soil salt content was closely associated with soil texture. The current soil salt content did not affect plant growth, however, the soil in the shelterbelt should be continuously monitored to prevent salinization in the future.

Reprinted from *Water*. Cite as: Huang, Y.; Wang, Y.; Zhao, Y.; Xu, X.; Zhang, J.; Li, C. Spatiotemporal Distribution of Soil Moisture and Salinity in the Taklimakan Desert Highway Shelterbelt. *Water* **2015**, *7*, 4343–4361.

1. Introduction

In arid and semiarid regions, water scarcity is a serious and chronic environmental problem threatening the ecosystem [1]. To overcome this shortage, lower quality water is widely used [2,3]. For instance, saline groundwater is applied in the ecological shelterbelt in the Taklimakan Desert, the largest mobile desert in China. The desert has a harsh environment with little precipitation and strong evaporative potential [4]. To access oil fields, a highway was built from north to south across the Taklimakan Desert in 1997. A shelterbelt was planted along both sides of the highway in 2003, to limit sand drift into the highway [5]. The shelterbelt, which is sometimes called "Great Green Wall", consists of several drought- and salt-tolerant plants including *Tamarisk* L., *Haloxylon* Bunge, and *Calligonum* L. [6]. The moving dunes have been successfully stabilized on both sides of the highway for more than ten years. The highway shelterbelt is irrigated with underground water (containing 2.8–29.7 $g \cdot L^{-1}$ salt). The water comes from 108 wells. Each well irrigates 3.12×10^3 m^2. The water salinity averages 5 $g \cdot L^{-1}$ [6].

Saline water irrigation can result in an increase in salinity at the soil surface. This process, which is known as secondary salinization, is a major environmental hazard [7,8] that threatens about 6% of the total global land area [9]. Salt accumulation in soil not only affects physical, chemical, and biological soil processes [10], but also reduces soil productivity [11], soil sustainability [12,13], and vegetation growth [14]. Soil salinity in the irrigated areas of Xinjiang has increased 40% from 1983 to 2005 [8]. Previous studies in the Taklimakan Desert Highway Shelterbelt indicated that long-term drip irrigation with saline water irrigation did not significantly increase soil salinity among lateral roots but did significantly increase soil salinity at the soil surface (0–10 cm) [15]. Furthermore, Zhang *et al.* [6] indicated this salt accumulation was a major threat to the shelterbelt. Drip irrigation with saline water within the shelterbelts may increase soil nutrient content and improve soil structure [15].

Many factors can influence the distribution of soil moisture and salinity. Sumner *et al.* [16] showed that the salinity at which vegetation growth is negatively affected as soil clay content increases. Furthermore, they observed that soils with high clay content had greater water retention capacity, soil salinity and osmotic potential. Li *et al.* [17] reported significant differences between the inter-dunes and dune-tops, indicated that different topographic characteristics between plots result in the differences in soil organic carbon (SOC), available nitrogen (AN), and available phosphorus (AP), whereas the heterogeneity of soil pH and electrical conductivity (EC) arise from plant species and their distribution. In the arid desert environment, the aerial parts of vegetation for nutrients retardation are beneficial to soil evolution. Li *et al.* [15] showed that irrigation salinity water significantly affected soil EC but not soil pH. Salt accumulation in soil increased as irrigation water salinity increased. Low salinity (3.6–15.5 $g \cdot L^{-1}$) in irrigation water increased SOC, total nitrogen (TN) and total phosphorus (TP). In addition, saline irrigation is beneficial for the development of Aeolian sandy soils, and there has been no salinity hazard to plants with saline irrigation for seven years. There was no obvious profile differentiation because of the short time since shelterbelt establishment, however, the color of the Aeolian sandy soil changed slightly [18].

However, these studies are often limited to point-scale. Information about the spatial patterns of soil moisture content and salinity on the field scale are lacking in this region. Soil salinity and moisture are highly variable over time and space [19–22], especially in harsh environments. To understand the factors governing hydrological process, information is needed about the spatiotemporal evolution of soil moisture and salinity under saline drip irrigation. Deriving conclusions from soil moisture and salinity measurements at only a few locations may result in large uncertainties because soil moisture and salinity can be highly variable. The prediction of soil moisture and salinity at un-sampled locations creates even more uncertainty. This requires knowledge of soil moisture and salinity magnitude, temporal dynamics, and spatial variability [23]. Since the 1990s, the study of spatiotemporal distribution of soil moisture and salinity was promoted by geostatistical techniques which reduce and quantify these uncertainties [24]. Currently, research interest is growing in mapping soil EC as a surrogate for soil salinity [25]. Zheng *et al.* [26] characterized the spatiotemporal variability of soil EC in a drip-irrigated field during the cotton growing and determined the factors that influence the spatiotemporal variability of soil salinity. Many findings have demonstrated that spatial and temporal information about soil EC can be used to improve the overall management of irrigated fields [27].

Soil moisture and salinity are essential factors restricting the sustainable development of shelterbelts. Usually, salinity tends to concentrate in shallow soils. To understand soil and plant development along the Taklimakan Desert Highway Shelterbelt, it is important to learn more about the spatiotemporal distribution of soil salinity within irrigation cycles and with seasons. In this study, our objectives were: (1) to characterize the spatiotemporal distribution of soil moisture and salinity either in one single drip-irrigation cycle, or in different growing seasons (spring, summer and autumn); and (2) to determine the factors that influence the spatiotemporal variability of soil moisture and salinity.

2. Materials and Methods

2.1. Study Area

The study was conducted in the Taklimakan Desert Highway Shelterbelt (Figure 1), which is located between 37–42° N and 82–85° E. The region has annual precipitation < 30 mm and annual potential evaporation > 3800 mm; Annual average temperature is 12.7 °C, with the highest temperature of 43.2 °C in August and the lowest temperature of −19.3 °C in January. It has more than 130 days for sand-shifting. The annual average wind speed is 2.5 $m \cdot s^{-1}$, with wind activity intensity index above 40,000 [5]. Most of the areas along the highway are covered with mobile shifting sands, mainly consisting of fine and very fine sand particles [4]. There are very few natural vegetation and residential animals [5].

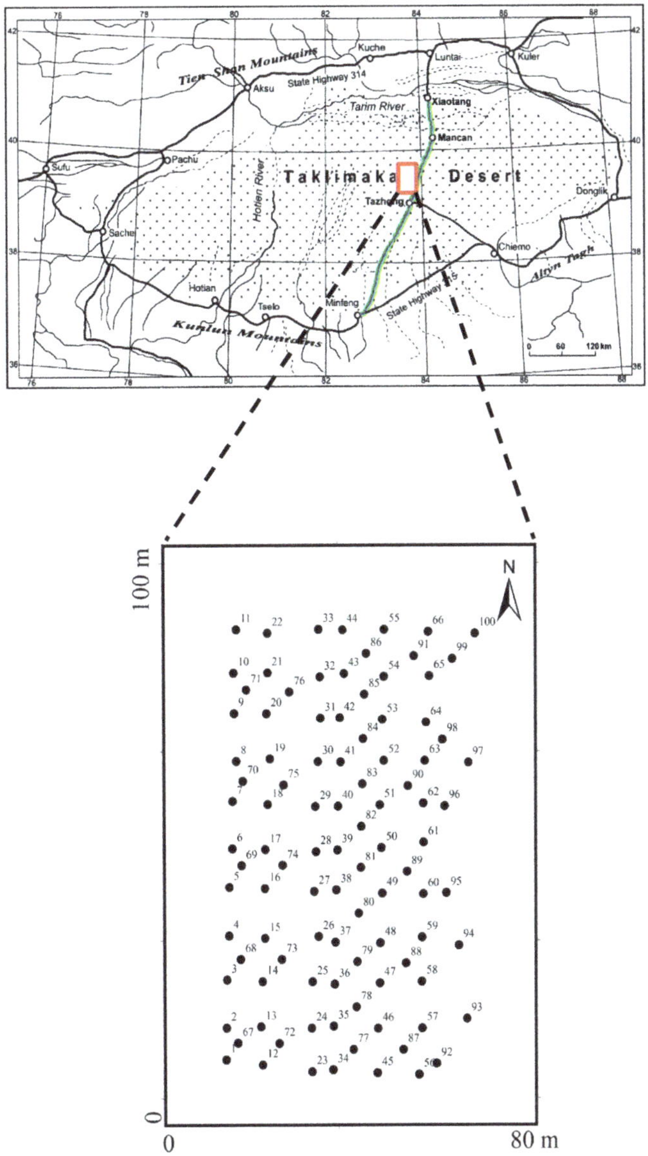

Figure 1. Locations of sampling sites at the study area. The green line shows the Taklimakan Desert Highway Shelterbelt. The red box shows the sampling site.

2.2. Sampling Design and Soil Analysis

Similar ground landscape, vegetation planting pattern, vegetation species, and irrigation measurements are developed along the highway shelterbelt. The ground landscape is mainly high mobile dunes combined with large complex dune chains.

The soil is mainly shifting Aeolian sandy soil. The shelterbelt was planted with a spacing of according to 1 m × 2 m intervals. The shelterbelt is irrigated twice per month, from March to May, and from September to October. Irrigation is applied three times per month, from June to August [6]. The only difference is the salinity of irrigation water, and the salinity around 5 g· L^{-1} represents most salinity of the irrigation water.

Since the spatial arrangement of shelterbelt controlled by groundwater irrigation is very similar from place to place along the highway shelterbelt, we mainly investigated the spatial distribution of soil water and salt of the shelterbelt irrigated by the most representative salinity, *i.e.*, 4.80 g L^{-1}. The geostatistical non-uniform grid sampling method was used, and 66 points in the shelterbelt, 25 points between the shelterbelts and 9 points in shifting sandy land were selected. In 2014, shallow soil samples (0–10 cm) were collected at each point on 23 April and 8 October. These dates were the middle dates during the irrigation cycle. Soil samples were also collected at each point every two days during one irrigation cycle (10 to 20 July). Light rainfall occurred early in the morning and again one or two hours before sampling on 20 July. In total, about 700 soil samples were taken. Five soil cores (*i.e.*, five replicates) were taken at each grid point which was about 30-cm from the dripper. Each point was recorded with GPS.

Each soil sample was divided into two parts. One part was used to measure soil moisture by the oven-drying method. The second part was air-dried and passed through a 1.0 mm sieve for soil chemical analyses. Topographical and vegetation conditions were noted at each sampling point. Soil SOC, total nitrogen (TN), total phosphorus (TP), total potassium (TK), pH, and particle size distribution were only measured at first sampling time in April. Soil pH and EC were measured in a soil–water suspension with a 1:5 soil: water ratio. The SOC content was determined using the $K_2Cr_2O_7$–H_2SO_4 oxidation method of Walkley–Black [28]. Total soil N was determined using the Kjeldahl procedure (UDK140 Automatic Steam Distilling Unit, Automatic Titroline 96, Usmate Velate, Italy). Total soil P was determined by flow injection analysis [28]. Total soil K was determined by flame photometry [29]. Particle size distribution was determined by the pipette method in a sedimentation cylinder, using sodium hexametaphosphate as the dispersing agent [30].

2.3. Descriptive Statistical Analysis

Descriptive statistics including mean, maximum, minimum, standard deviation (SD), coefficient of variation (CV), skewness, and distribution type (DT) among soil properties were calculated using SPSS 13.0. Analysis of variance (ANOVA) analysis was conducted to test the differences of moisture and EC among five sampling times during an irrigation cycle and among three seasons at a significant level of 0.05.

Person's Correlation Analysis was performed to identify the relationship among soil moisture, salinity, other selected physical, chemical factors, and terrain.

2.4. Geostatistics Analysis

Geostatistics were used to explore the variation and spatial dependency of spatially distributed data. The variogram function can be used for Kriging interpolation to generate predictions for unsampled locations within the sampling extent. The GS^+ software (version, 9.0; Gamma Design Software 2002) was used to create semivariograms for variance structure of soil moisture and salinity of topsoil [31]. Then, the ordinary Kriging interpolation method by Arcgis 9.3 was adopted to map spatial distribution of soil moisture and salinity [19,32].

A semivariogram was calculated for each soil property as follows [33]:

$$\gamma(h) = \frac{1}{2N} \sum_{i=1}^{N} \{z(x_i) - z(x_i + \mathrm{h})\}^2 \quad (1)$$

where $z(x_i)$ is a sample z at location x_i, and N(h) is the number of data pairs separated by h. A variogram function is fitted to the experimental variogram to obtain geostatistics, including nugget variance (C_0), structured variance (C), sill (C_0 + C), and range (A_0). To explore the degree of spatial dependency, the ratio of nugget to sill (*i.e.*, the nugget ratio) was calculated. A nugget ratio < 25% indicates a strong spatial dependency; a nugget ratio > 75% indicates no spatial dependency, otherwise, the spatial dependency is moderate [20].

The most common variogram functions are bounded models with a fixed range (*i.e.*, circular or spherical), or models that approach the sill asymptotically (exponential or Gaussian models). Semivariogram $\gamma(h)$ for soil salinity during the irrigation cycle and growing period in different seasons were calculated, and the scatterplot of $\gamma(h)$ *versus* h were generated. Then, different theoretical semivariance models were used to fit the calculated values and the model with the best-fitting value, which were based on R^2.

3. Results and Discussion

3.1. Descriptive Statistics of Soil Properties

Table 1 showed the basic descriptive statistics of soil physical and chemical properties in the shelterbelt. Based on skewness values, SOC, TN, TP, TK, pH, sand, and silt content were normally distributed, while soil moisture, EC, and clay content were log-normally distributed.

The large CV values were obtained for soil moisture (196.99%) and EC (107.00%), indicating a strong spatial variation. The CV values of TN, TP and TK ranged from

22.1% to 54.7%, corresponding to a moderate degree of variability. These results were consistent with other studies [34–36]. Corwin *et al.* [36] reported that CV values for soil nutrient all exceeded 20%. Zhang *et al.* [37] indicated that soil SOC, TN, TP, TK were in moderate variability in a subtropical Karst forest, which were mainly related to soil parent materials. The lowest CV values were observed for sand content (4.2%). Notably, a relatively small CV value (6.3%) was observed for pH, similar to the surface layer of an agricultural soil [38]. Others have also reported similar findings at various scales [34,39]. The SOC content of all sampling sites ranged from 0.16 to 1.52 $g \cdot kg^{-1}$, with a mean of 0.65 $g \cdot kg^{-1}$, indicating that SOC was still in a low level. The CV value of SOC is 49.8%, which could be mainly attributed to planting pattern of the shelterbelt.

Table 1. Statistical summary for selected soil physical and chemical properties.

Variable	Descriptive Statistics [a]					DT [b]
	Mean	Min.–Max.	SD	CV (%)	Skewness	
Moisture (%)	1.19	0.00–10.00	2.34	197.0	2.53	LN
EC ($mS \cdot cm^{-1}$)	1.56	0.09–11.46	1.66	107.0	3.82	LN
SOC ($g \cdot kg^{-1}$)	9.46	1.68–27.48	0.32	49.8	0.58	N
TN ($g \cdot kg^{-1}$)	0.06	0.01–0.19	0.03	54.7	0.79	N
TP ($g \cdot kg^{-1}$)	0.35	0.17–0.70	0.08	23.9	0.68	N
TK ($g \cdot kg^{-1}$)	7.27	4.60–8.80	0.81	22.1	−0.95	N
pH	8.32	7.08–9.48	0.53	6.3	−0.15	N
Sand (%)	89.74	82.93–98.26	3.72	4.2	−0.02	N
Silt (%)	10.24	1.74–17.07	3.69	36.1	−0.05	N
Clay (%)	0.03	0.00–0.57	0.11	440.2	4.80	LN

Notes:[a] Min., minimum; Max., maximum; SD, standard deviation; CV, coefficient of variation; [b] DT, distribution type; N, normal distribution; LN, log-normal distribution.

3.2. Spatiotemporal Distribution of Soil Moisture and Salinity

3.2.1. Spatiotemporal Distribution of Soil Moisture and Salinity during the Irrigation Cycle

To describe the spatial variation of moisture and salinity in one irrigation cycle, isotropic semivariograms for all studied variables were calculated and the best fitted models were selected based on R^2 (Figure 2). Moistures at 6th and 8th day were best fitted by the spherical model, whereas others were best fitted with the Gaussian model.

The nugget/sill ratio has been used extensively to define spatial properties. Table 2 shows that the nugget/sill ratios ranged from 0.03% to 27.31%, indicating that moisture and salinity exhibits strong spatial dependence excluding soil moisture in the 6th day. Their spatial correlation may be controlled by both intrinsic variations of soil properties and extrinsic factors such as topography or human-induced activities [39]. The nugget/sill ratio of moisture of the 6th day was in moderate spatial correlation, implying the interaction of moisture and salt itself, combined the small scale of spatial variation of topography and the influence of vegetation growing on the soil. The sill variance of moisture and salinity decreased during the irrigation cycle except the last day with raining. Except the ranges of moisture on the 6th and 8th day were larger than 50 m, other ranges were all about 6 m (Table 2), confirming that our samples spacing (10 m) is reasonable.

Table 2. Parameters of semivariogram models for soil moisture and salinity during one irrigation cycle.

Index	Time (d)	Mathematical Model	Range(A_0)/m	Nugget/C_0	Sill/(C_0 + C)	C_0/(C_0 + C)%	R^2
Moisture	2	Gaussian	6.10	0.01	7.77	0.13	0.63
	4	Gaussian	6.90	0.01	1.21	0.83	0.83
	6	Spherical	51.90	0.22	0.79	27.31	0.56
	8	Spherical	57.00	0.03	0.10	3.22	0.38
	10	Gaussian	6.80	0.00	0.21	0.47	0.68
Salinity	2	Gaussian	6.40	0.01	3.36	0.21	0.48
	4	Gaussian	6.00	0.00	3.07	0.03	0.45
	6	Gaussian	6.90	0.01	3.60	0.28	0.67
	8	Gaussian	6.40	0.00	1.13	0.08	0.67
	10	Gaussian	6.10	0.01	4.06	0.25	0.56

Soil moisture and EC maps obtained by ordinary Kriging interpolation are displayed in Figure 3. The values and spatial patterns of soil moisture and salinity were clearly affected by irrigation time and raining. In general, soil moisture decreased gradually during the 10-day irrigation cycle (Figures 3 and 4). Soil moisture was the highest in the 2th day after watering, and then decreased quickly with time, and reached the smallest value in the 8th day. The three smoothed contour maps of moisture distribution during 2-day, 4-day, and 6-day after irrigation display quite similar spatial patterns, with high soil water contents in the southern and eastern sections and low values in the northern and western parts. Furthermore, the area of high moisture content in the southern and eastern sections was decreased during the irrigation and presented a diminishing trend from west to east. This could be ascribed to the east side of the study area closed to quicksand where plants were planted sparser than west, and the plants growth was not as good as the west. If dividing the protection forest area into two parts, projection area of tree crown of west and east part were 0.31% and 0.89%, respectively. Thus, due to the weak canopy shadow, the effect of evaporation in east was higher, and thus soil moisture of east section was lower. Consequently, the locations with relatively higher and lower moisture remained about the same over time, which somewhat revealed the temporal stability of soil moisture spatial patter similar to Zhao *et al.* [40].

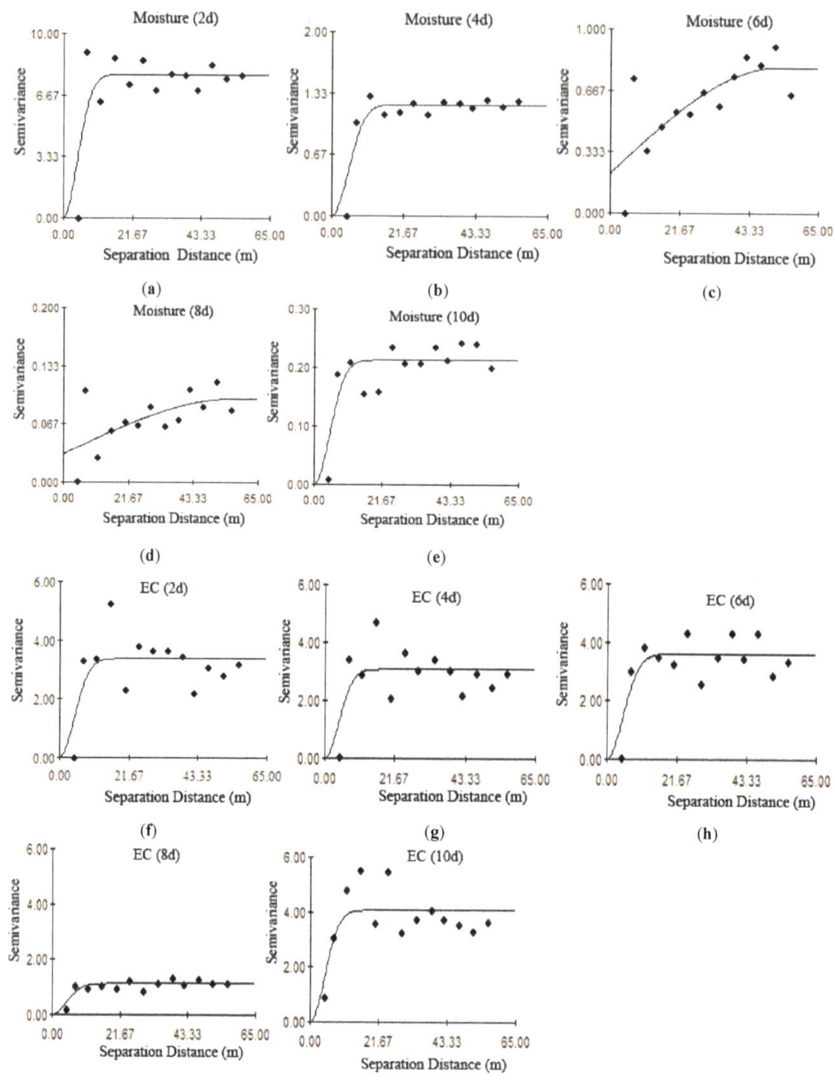

Figure 2. Values of moisture (**a**–**e**) and EC (**f**–**j**) variograms of all plots during one irrigation cycle in July. 2d, 4d, 6d, 8d, 10d represent 2nd, 4th, 6th, 8th, and 10th days after watering.

Eight days after irrigation, soil moisture reached stability, and the whole area had no significant differences on 18 and 20 July. This is mainly ascribed to infiltration and evaporation dynamics from topsoil layer. Due to the large water potential difference between soil under the dripper and the fringes of wetting area, moisture spread around at a large rate of the infiltration flux after the irrigation ended. Subsequently, with the decreasing of soil moisture under the dripper and the enlargement of wetting area, suction gradients and unsaturated hydraulic conductivity of the soil decreased, and the rate of migration of moisture also decreased. Generally, the typical distribution of soil moisture was presented at the second day during the drip irrigation cycle (Figure 3). Soil moisture maps of the 4th day can be explained that within the desert highway shelterbelt, the shallow soil was saturated after irrigation, then water evaporated, and water moved up through the capillary rise effectively [41]. At the 6th, 8th, and 10th day after the irrigation, soil moisture content continuously decreased due to strong atmospheric evaporation in desert areas, characterized by point source flowing sideways. Note that there are two kinds of soil water infiltration, *i.e.*, vertical and lateral infiltration. Vertical infiltration dominated when soil is less saturated, otherwise, lateral seepage is dominant. Although sand has strong permeability, lateral seepage processes would not occur until soil on the top-surface gradually saturated the soil below, with continuous drip irrigation. Furthermore, numerous small amounts of dripped water leached salt out by lateral seepage around the dripper, and gradually push the salt outward. Consequently, salt is concentrated on the soil surface and the edge of wetting-front, associated with the strong evaporation and capillary effect. In contrast, there is less salt near the dripper, corresponding to a lower thickness of salt crust and soil EC.

Based on the soil salinity classification system of FAO, five salinity classes were established using the EC values as follows: (1) very strongly saline, >16 mS·cm^{-1}; (2) strongly saline, 8–16 mS·cm^{-1}; (3) moderately saline, 4–8 mS·cm^{-1}; (4) slightly saline, 4–2 mS·cm^{-1}; and (5) non-saline, 0–2 mS·cm^{-1} [41]. The EC values indicated that the salt concentration over the irrigation cycle was non-saline to slight (Figure 5). In the 10-day irrigation cycle, soil moisture content was decreasing during the 2nd and 8th days and finally increased at the 10th day due to the unexpected rainfall before measurement. With different patterns to soil moisture, EC values increased from the 2nd to the 6th days, gradually decreasing at the 8th day and then increasing at the 10th day (Figure 4). Compared with soil moisture, the spatial distribution of the salinity presented plaques shape, indicating a local secondary salinization. During drip irrigation, soil salt content differs with different distances from the dripper. Note that each of 100 soil cores was a composite of sub-samples taken from five points, with one under the dripper and the other four centered from dripper, spaced 30 cm apart. Wang *et al.* [42] found that with saline drip-irrigation, the soil EC changing with a distance from the dripper of 30 cm was first increasing then decreasing. Meanwhile, three mechanisms including convection, diffusion, and mechanical dispersion affect salt transport. After saline irrigation, a series of complex chemical reactions, such as dissolution, precipitation, adsorption, desorption, and ion exchange occurred between the different elements of the soil solution and soil solid phase. With evaporation happened on the soil surface, salt in the soil solution can move up and become a precipitate. Consequently, with strong evaporation affected by soil capillary action, salts accumulate on the soil surface and form a salt crust, inhibiting soil evaporation, in turn [6,41]. Overall, non-secondary salinization was dominant at the irrigation cycle and accounted for 86%, 75%, 54%, 71%, and 63% at the 2th, 4th, 6th, 8th and 10th day, respectively (Figure 3).

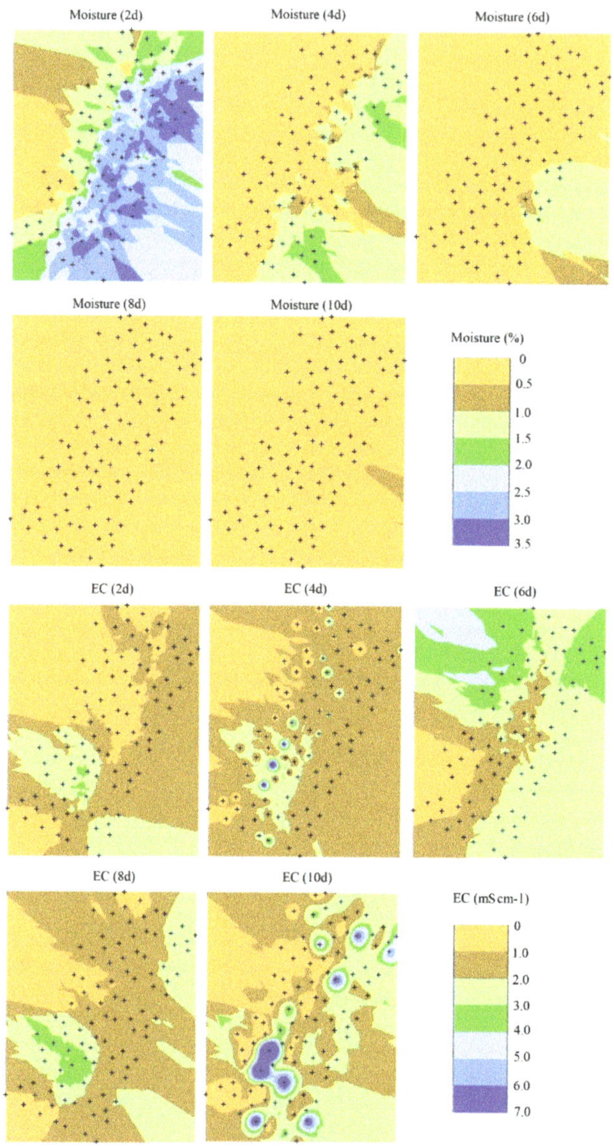

Figure 3. Ordinary Kriged maps of soil moisture and EC during the irrigation cycle in July. 2d, 4d, 6d, 8d, 10d represent 2nd, 4th, 6th, 8th, 10th days after watering.

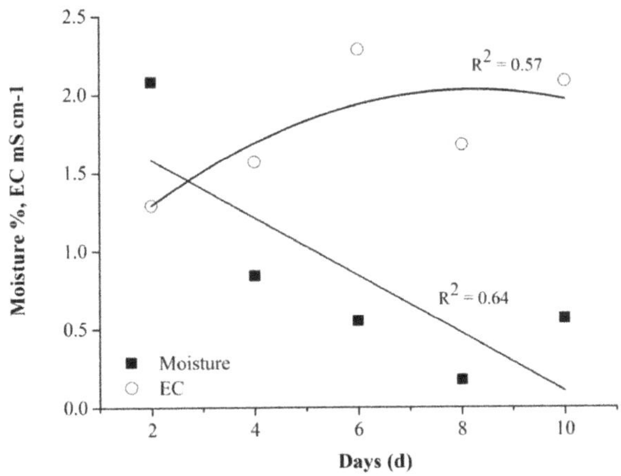

Figure 4. Functional diagram of averaged moisture and salt changes during the irrigation cycle.

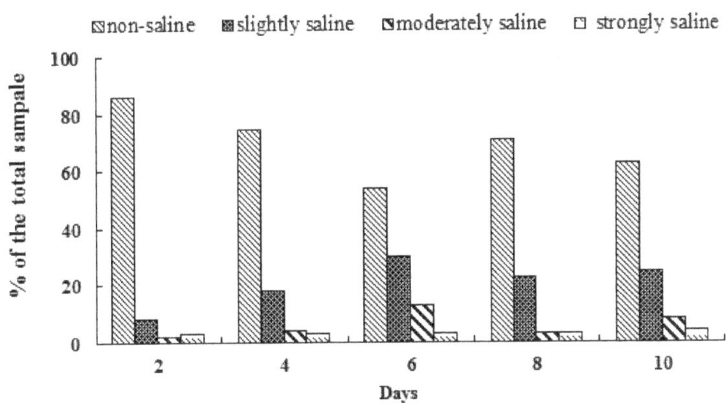

Figure 5. Distribution of soil salinity classes during the irrigation cycle.

Notably, both moisture and salinity increased at the 10th day (Figure 4). Due to the rainfall at the 10th day, soil salt content could be decreased due to the rainfall input. However, our case showed that this amount of rain actually was not enough to leach out salt from the topsoil. Conversely, it could be very effective to wash the salinity from the branches and leaves to soil surface. Pereira *et al.* [43] also pointed out that the increased salinization in the surface soil may be attributed to poly-salt or salt excretion from halophytes, indicating that halophytes can gather salt on their surface. There are salt-secreting plants in the shelterbelt, such as *Chinese tamarisk*, which the salt can be reasonably leached out by the stemflow to the topsoil. Consistent with our findings, Li *et al.* [44] also demonstrated the chemical elements in soil were formed by the effects of stemflow in the nearby Gurbantunggut Desert. The comparison of the chemical properties of stemflow and bulk precipitation revealed a higher content of chemical elements in the stemflow.

3.2.2. Spatiotemporal Distribution of Soil Moisture and Salinity in Different Seasons

In 2014, soil samples (0–10 cm) at each point were collected in two days before irrigation during the spring (23 April), summer (18 July), and autumn (8 October), respectively, to investigate the seasonal dynamics of soil moisture and salt contents. Soil moisture in spring and salt in autumn reached the highest sill and the comparably high nugget (Figure 6 and Table 3). According to R^2, salt in spring did not show an organized spatial pattern; a best fit to the data is accomplished by using a linear model with the strong nugget effects. Nugget/sill ratio of moisture ranged from 0.03% to 3.22%, and salt ranged from 0.08% to 100%, indicating that moisture was the strong spatial dependence and kept relatively stable with the season, whereas salt demonstrated seasonal-dependent spatial heterogeneity. The nugget/sill ratio of salt was in moderate spatial dependence in autumn, strong in summer, and weak in spring (Table 3).

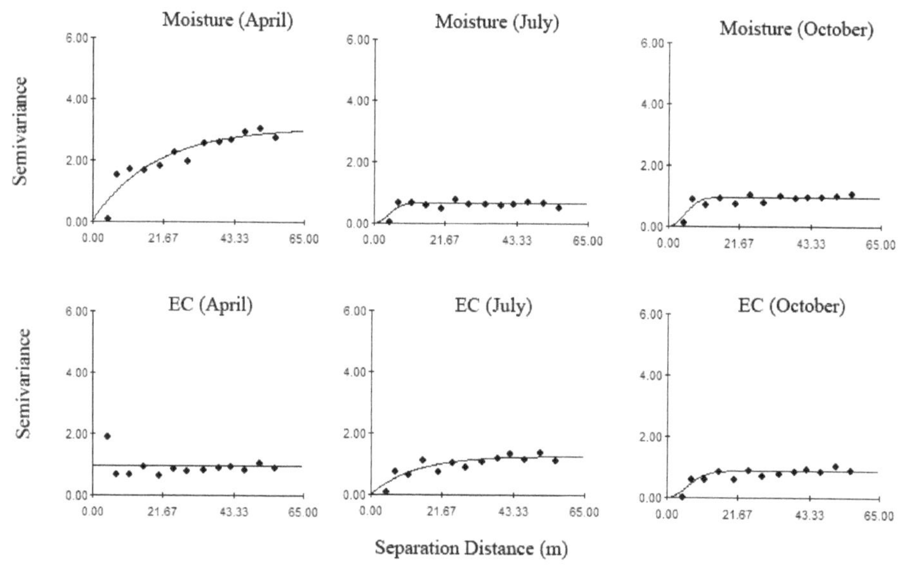

Figure 6. Variograms of soil moisture and EC among different seasons.

There were strong seasonal variations of soil moisture and salinity (0–10 cm). Soil moisture was the highest in spring and decreased in the later season due to the strong evaporative demand, whereas soil salinity showed the relatively smaller change (Figure 7). Although the amount of irrigation water was the same in different seasons, the accumulative characteristics of irrigation were not. The amount of intensity of water loss was mainly controlled by atmospheric evaporation capacity and soil moisture condition. Note that, although the frequency of irrigation in summer was higher than in spring and autumn, the evaporative demand was the highest in summer. Therefore, soil moisture content due to irrigation was the lowest in the summer. According to local meteorological data, the average temperature in autumn is lower than in summer, but higher than in spring. Figure 8 was indicative of the average moisture content in different seasons, which demonstrated soil moisture in the autumn was higher than in the summer, but lower than in the spring. Accordingly, it may imply that temperature is the main factor controlling soil water content.

Table 3. Parameters of semivariogram models for soil moisture and salt among different seasons.

Index	Month	Mathematical Model	Range(A_0)/m	Nugget/C_0	Sill/(C_0 + C)	C_0/(C_0 + C)%	R^2
Moisture	April	Exponential	57.3	0.05	3.07	1.60	0.86
	July	Spherical	57.00	0.03	0.10	3.22	0.38
	October	Exponential	33.90	0.00	0.30	0.03	0.72
Salinity	April	linear	55.83	0.95	0. 95	100	0.00
	July	Exponential	13.40	0.00	1.27	0.08	0.75
	October	Gaussian	52.30	2.43	5.88	41.33	0.86

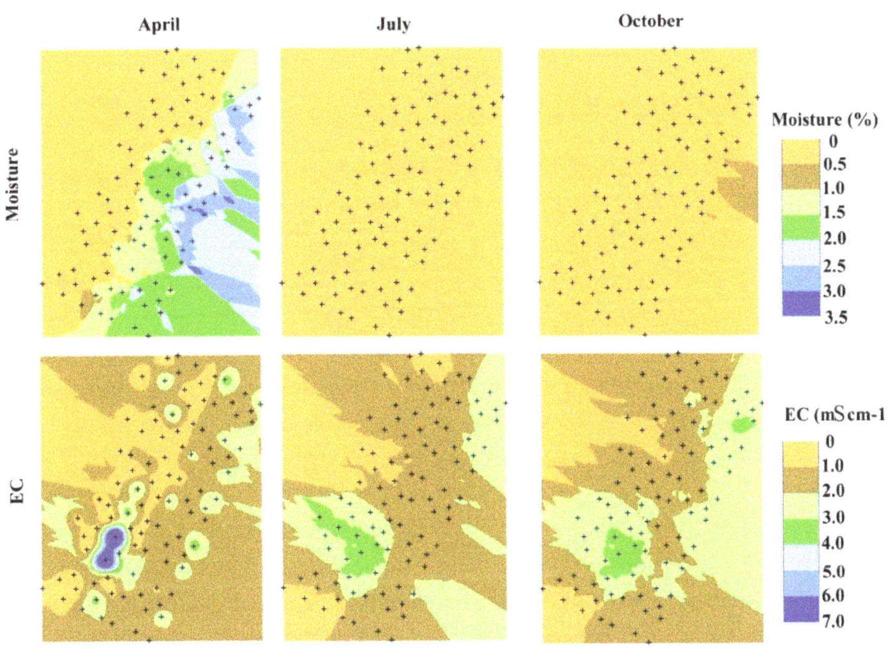

Figure 7. Ordinary kriged maps of soil moisture and EC during different seasons.

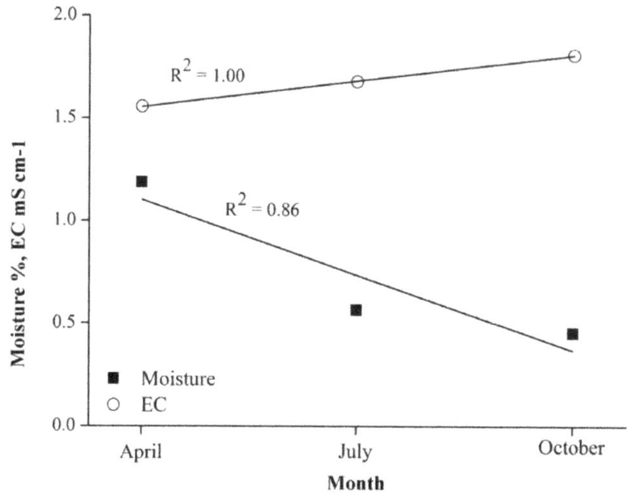

Figure 8. Functional diagram of soil moisture and EC changes in different months.

Under saline water irrigation, soil salinity slightly increased in summer, as compared to that in spring (Figure 8). This phenomenon was in accordance with the idea that, with continuing irrigation, salt accumulation and salt leaching occur simultaneously [35]. Similarly, Li *et al.* [15] reported that soil salinization increased near the surface soil throughout the year.

3.3. Correlation between Selected Parameters

To characterize the relationship between soil moisture, EC, and selected soil properties (SOC, TN, TP, TK, pH, sand, silt, clay and terrain), Pearson's correlation coefficient was calculated for each variable (Table 4). The results indicated that soil moisture is significantly positively correlated with TN and silt content, and negatively correlated with pH and sand content; whereas EC strongly positively correlated with TN, TP and silt content, and negatively with sand content. Remarkably, it seemed that both soil moisture and salt had significant correlation with soil texture and TN. For a finer-textured soil, the water retention capacity is greater than a coarse-textured soil, therefore at a given EC and matric potential, the osmotic potential of the soil solution is lower in the coarse-textured soil [16]. Due to saline water irrigation, a series of biochemical, radiation, and evaporation processes occur, which may input the nutrient components into soil. The artificial shelterbelts could be beneficial for the soil nutrient accumulation due to the vegetative litter decomposition, root growth, decomposition, and other biogeochemical cycles [24]. In addition, the shelterbelt growth and continuation is beneficial for Aeolian sandy soil development [7]. It is well know that sand content reduction corresponding with silt and clay content increase is the key improvements of soil quality. Brantley *et al.* [45] proved that ions tended to be adsorbed on the clayed soil and could not be extracted easily by water. Therefore, compared with moisture, salinity was significantly intercepted by the more clayed soil. Consequently, as water is absorbed by plants or evaporated, salt remains on the topsoil during early irrigation. In general, soil moisture and nutrients have positive effects on the growth of the shelterbelt, whereas EC has a negative effect on the growth of the shelterbelt. However, with saline drip-irrigation, the soil EC of shelterbelt increased [7], which may be somewhat beneficial to soil development, particularly taking soil texture into effect, causing a trade-off effect for promoting factors and inhibiting factors of plants growth. Wang *et al.* [46] indicated that clay content determined the moisture and salinity of soil, emphasizing that moisture of the topsoil layer was mainly affected by a single surface factor; ions tended to be adsorbed on the clay complexly and could not be extracted easily by water [45]. As a result, compared with moisture, salinity was more related to soil clay content.

Table 4. Correlation coefficients between soil salinity and environmental factors.

Index	Moisture (%)	EC ($ms\cdot cm^{-1}$)	SOC ($g\cdot kg^{-1}$)	TN ($g\cdot kg^{-1}$)	TP ($g\cdot kg^{-1}$)	TK ($g\cdot kg^{-1}$)	pH	Sand (%)	Silt (%)	Clay (%)	Terrain (m)
Moisture (%)	1										
EC ($mS\cdot cm^{-1}$)	0.12	1									
SOC ($g\cdot kg^{-1}$)	0.11	0.10	1								
TN ($g\cdot kg^{-1}$)	**0.34 ****	**0.36 ****	0.10	1							
TP ($g\cdot kg^{-1}$)	−0.06	**0.22 ***	0.10	**0.21 ***	1						
TK ($g\cdot kg^{-1}$)	0.13	−0.09	0.07	0.08	**−0.31 ****	1					
pH	**−0.28 ****	−0.19	−0.04	**−0.31 ****	−0.06	**0.25 ***	1				
Sand (%)	**−0.23 ***	**−0.40 ****	**−0.39 ****	**−0.40 ****	−0.14	−0.15	**−0.22 ***	1			
Silt (%)	**0.23 ***	**0.41 ****	**0.38 ****	**0.40 ****	0.15	0.15	**0.22***	**−1.00 ****	1		
Clay (%)	−0.11	−0.20	0.19	−0.05	**−0.22 ***	−0.02	0.06	**−0.27 ****	**0.24 ***	1	
Terrain (m)	−0.03	−0.05	0.06	−0.04	−0.19	0.12	0.04	**0.21 ***	**−0.22 ***	0.06	1

Notes: * Correlation is significant at the 0.05 level; ** Correlation is significant at the 0.01 level.

Table 4 indicates that SOC had positive correlation with silt content ($R^2 = 0.38$, $p < 0.01$), negative correlation with sand content ($R^2 = -0.39$, $p < 0.01$), and no significant correlation with clay content, since soil clay component was rare. Present studies suggest that the change of soil texture is firstly caused by wind erosion, desertification, and then by changes of soil organic matter and nutrient content, which finally results in soil impoverishment [47]. In the process of desertification, soil coarsens, and soil organic matter and nutrient content decrease. It is well known that loss of fine soil particles could result in loss of soil organic matter since organic matter is often combined with fine soil particles. Zhao *et al.* [48] addressed the idea that content of soil silt, clay, and SOC are the key factors for soil texture and fertility.

4. Conclusions

To obtain more information for further understanding of soil and plant development along the Taklimakan Desert Highway Shelterbelts, a better knowledge of spatial variation of soil moisture and salt are essential. The spatial pattern of soil moisture and salt content in the harsh ecosystem displayed considerable spatial correlation, largely because of the relative static controls of soil properties in dry climates. During the irrigation cycle, soil moisture gradually reduced, which was lower at the northwest than at the southeast. In contrast, soil salinity showed a tendency to increase with the local plaque distribution. Soil moisture was the highest in spring, and summer and autumn had relatively low soil moisture. Salinity was gathered in local area in different seasons. Correlation analysis showed that both soil moisture and salinity were significantly positively correlated with silt content and TN, and negative correlated with sand content. Soil with high moisture and nutrient content and low salt content is more suitable for shelterbelt growth, which pointed out soil moisture and salinity was a contradictory relation in this area. For further understanding of soil and plant development along the Taklimakan Desert Highway Shelterbelt, secondary salinization and soil properties of shelterbelt should be considered. For salt-leaching on top-soil and water supply to the shelterbelt in the northwest of the study area, long-term monitoring for the shelterbelt in the local area and high-water discharge drip irrigation on a regular basis are necessary. Meanwhile, the effects of stemflow to increase soil salinity should not be ignored.

Acknowledgments: This research was financially supported by a grant from the Natural Science Foundation of China (41271341, 41030530, 31300449), Thousand Youth Talents Plan Project (Y472241001) and the Western Light Project of the Chinese Academy of Sciences (XBBS201205). We thank the staff of the Tazhong Station of Desert Research for the field and laboratory assistances.

Author Contributions: All authors contributed to the design and development of this manuscript. Yuan Huang carried out the data analysis and prepared the first draft of the manuscript; Xinwen Xu, Yongdong Wang and Ying Zhao are the graduate advisor of Yuan Huang and contributed many ideas to the study; Jianguo Zhang and Congjuan Li provided some important advices on the concept of methodology and writing of the manuscript. All authors read and approved the final manuscript.

Conflicts of Interest: The authors declare no conflict of interest.

References

1. Chartzoulakis, K.S. Salinity and olive: Growth, salt tolerance, photosynthesis and yield. *Agric. Water Manag.* **2005**, *78*, 108–121.
2. Beltran, J.M. Irrigation with saline water: Benefits and environmental impact. *Agric. Water Manag.* **1999**, *40*, 183–194.
3. Feikema, P.M.; Morris, J.D.; Connell, L.D. The water balance and water sources of a eucalyptus plantation over shallow saline groundwater. *Plant Soil* **2010**, *332*, 429–449.
4. Lei, J.Q.; Li, S.Y.; Jin, Z.Z.; Fan, J.L.; Wang, H.F.; Fan, D.D.; Zhou, H.W.; Gu, F.; Qiu, Y.; Xu, B. Comprehensive eco-environmental effects of the shelter-forest ecological engineering along the Tarim Desert Highway. *Chin. Sci. Bull.* **2009**, *53*, 190–202.
5. Li, B.; Xu, X.; Lei, J.; Qiu, Y.; Xu, B.; Zhou, H.; Wang, Q.; Wang, B.; Su, W. Site type classification for the shelter-forest ecological project along the Tarim Desert Highway. *Chin. Sci. Bull.* **2009**, *53*, 31–40.

6. Zhang, J.; Xu, X.; Lei, J.; Sun, S.; Fan, J.; Li, S.; Gu, F.; Qiu, Y.; Xu, B. The salt accumulation at the shifting aeolian sandy soil surface with high salinity groundwater drip irrigation in the hinterland of the Taklimakan Desert. *Chin. Sci. Bull.* **2009**, *53*, 63–70.
7. Homer-Dixon, T.F. Environmental scarcities and violent conflict: Evidence from cases. *Int. Secur.* **1994**, *19*, 5–40.
8. Wang, Y.; Xiao, D.; Li, Y.; Li, X. Soil salinity evolution and its relationship with dynamics of groundwater in the oasis of inland river basins: Case study from the Fubei Region of Xinjiang Province, China. *Environ. Monit. Assess.* **2008**, *140*, 291–302.
9. Salvati, L.; Ferrara, C. The local-scale impact of soil salinization on the socioeconomic context: An exploratory analysis in Italy. *Catena* **2015**, *127*, 312–322.
10. Karlen, D.L.; Tomer, M.D.; Neppel, J.; Cambardella, C.A. A preliminary watershed scale soil quality assessment in north central Iowa, USA. *Soil Tillage Res.* **2008**, *99*, 291–299.
11. Talaat, N.B.; Ghoniem, A.E.; Abdelhamid, M.T.; Shawky, B.T. Effective microorganisms improve growth performance, alter nutrients acquisition and induce compatible solutes accumulation in common bean (phaseolus vulgaris l.) plants subjected to salinity stress. *Plant Growth Regul.* **2014**, *75*, 281–295.
12. Garg, N.; Singla, P. Naringenin- and *Funneliformis mosseae*-mediated alterations in redox state synchronize antioxidant network to alleviate oxidative stress in *Cicer arietinum* L. Genotypes under salt stress. *J. Plant Growth Regul.* **2015**, 1–16.
13. Zhang, H.J.; Dong, H.Z.; Li, W.J.; Zhang, D.M. Effects of soil salinity and plant density on yield and leaf senescence of field-grown cotton. *J. Agron. Crop Sci.* **2012**, *198*, 27–37.
14. Shannon, M.C.; Grieve, C.M. Tolerance of vegetable crops to salinity. *Sci. Hortic.* **1999**, *78*, 5–38.
15. Li, C.; Lei, J.; Zhao, Y.; Xu, X.; Li, S. Effect of saline water irrigation on soil development and plant growth in the taklimakan desert highway shelterbelt. *Soil Tillage Res.* **2015**, *146*, 99–107.
16. Sumner, M.; Rengasamy, P.; Naidu, R. Sodic soils: A reappraisal. In *Sodic Soil: Distribution, Management and Environmental Consequences*; Sumner, M., Naidu, R., Eds.; Oxford University Press: New York, NY, USA, 1998; pp. 3–17.
17. Li, C.J.; Li, Y.; Jian, M.A.; Fan, L.L.; Wang, Q.X. Spatial heterogeneity of soil chemical properties between haloxylon persicum and haloxylon ammodendron populations. *J. Arid Land* **2010**, *2*, 257–265.
18. Gu, F.X.W.; Wen, Q.K.; Pan, B.R.; Hu, Y.L.; Xu, H.L. Research on soil enzyme activities of aeolian soil under artificial plantation in Taklimakan Desert heartland. *J. Desert Res.* **2000**, *20*, 293–297.
19. Stoyan, H.; De-Polli, H.; Böhm, S.; Robertson, G.P.; Paul, E.A. Spatial heterogeneity of soil respiration and related properties at the plant scale. *Plant Soil* **2000**, *222*, 203–214.
20. Hu, W.; Shao, M.; Han, F.; Klaus, R.; Jing, T. Watershed scale temporal stability of soil water content. *Geoderma* **2010**, *158*, 181–198.
21. Wilczek, A.; Szypłowska, A.; Skierucha, W.; Cieśla, J.; Pichler, V.; Janik, G. Determination of soil pore water salinity using an FDR sensor working at various frequencies up to 500 MHz. *Sensors (Basel)* **2012**, *12*, 10890–10905.

22. Skierucha, W.; Wilczek, A.; Szypłowska, A.; Sławiński, C.; Lamorski, K. A TDR-based soil moisture monitoring system with simultaneous measurement of soil temperature and electrical conductivity. *Sensors (Basel)* **2012**, *12*, 13545–13566.
23. Douaik, A.; van Meirvenne, M.; Tóth, T. Statistical methods for evaluating soil salinity spatial and temporal variability. *Soil Sci. Soc. Am. J.* **2007**, *71*, 1629–1635.
24. Schneider, K.; Huisman, J.A.; Breuer, L.; Frede, H.G. Ambiguous effects of grazing intensity on surface soil moisture: A geostatistical case study from a steppe environment in Inner Mongolia, PR China. *J. Arid Environ.* **2008**, *72*, 1305–1319.
25. Eigenberg, R.A.; Doran, J.W.; Nienaber, J.A.; Ferguson, R.B.; Woodbury, B.L. Electrical conductivity monitoring of soil condition and available N with animal manure and a cover crop. *Agric. Ecosyst. Environ.* **2002**, *88*, 183–193.
26. Zheng, Z.; Zhang, F.; Ma, F.; Chai, X.; Zhu, Z.; Shi, J.; Zhang, S. Spatiotemporal changes in soil salinity in a drip-irrigated field. *Geoderma* **2009**, *149*, 243–248.
27. Corwin, D.L.; Lesch, S.M. Apparent soil electrical conductivity measurements in agriculture. *Comput. Electron. Agric.* **2005**, *46*, 11–43.
28. Nelson, D.W.; Sommers, L.E. Total carbon, organic carbon, and organic matter. *Methods Soil Anal.* **1996**, *3*, 961–1010.
29. Egner, H.; Riehm, H.; Domingo, W.E. Untersuchungen über die chemische Bodenanalyse als Grundlage für die Beurteilung des Nährstoffzustandes der Böden. II. Chemische Extraktionsmethoden zur Phosphor-und Kaliumbestimmung. *Lantbrukshogskolans Unnaler* **1960**, *26*, 199–215.
30. Day, P.R. Particle fractionation and particle-size distribution. *Methods Soil Anal.* **1965**, *9*, 548–549.
31. Western, A.W.; Blöschl, G.; Grayson, R.B. Geostatistical characterisation of soil moisture patterns in the Tarrawarra catchment. *J. Hydrol.* **1998**, *205*, 20–37.
32. Augustine, D.J.; Frank, D.A. Effects of migratory grazers on spatial heterogeneity of soil nitrogen properties in a grassland ecosystem. *Ecology* **2001**, *82*, 3149–3162.
33. Goovaerts, P. Geostatistical tools for characterizing the spatial variability of microbiological and physico-chemical soil properties. *Biol. Fertil. Soils* **1998**, *27*, 315–334.
34. Hu, K.; Li, H.; Li, B.; Huang, Y. Spatial and temporal patterns of soil organic matter in the urban–rural transition zone of Beijing. *Geoderma* **2007**, *141*, 302–310.
35. Zhang, J.G.; Chen, H.S.; Su, Y.R. Spatial variability and patterns of surface soil moisture in a field plot of karst area in southwest China. *Plant Soil Environ.* **2011**, *57*, 409–417.
36. Corwin, D.L.; Lesch, S.M.; Oster, J.D.; Kaffka, S.R. Monitoring management-induced spatio–temporal changes in soil quality through soil sampling directed by apparent electrical conductivity. *Geoderma* **2006**, *131*, 369–387.
37. Zhang, Z.; Hu, B.; Hu, G. Spatial heterogeneity of soil chemical properties in a subtropical karst forest, southwest China. *Sci. World J.* **2014**, *2014*, 9.
38. McBratney, A.B.; Minasny, B.; Cattle, S.R. From pedotransfer functions to soil inference systems. *Geoderma* **2002**, *109*, 41–73.
39. Liu, Z.P.; Shao, M.A.; Wang, Y.Q. Spatial patterns of soil total nitrogen and soil total phosphorus across the entire loess plateau region of China. *Geoderma* **2013**, *197*, 67–78.

40. Zhao, Y.; Peth, S.; Wang, X.Y.; Lin, H.; Horn, R. Controls of surface soil moisture spatial patterns and their temporal stability in a semi-arid steppe. *Hydrol. Processes* **2010**, *24*, 2507–2519.
41. Abrol, I.P.; Yadav, J.S.P.; Massoud, F.I. Salt-affected soils and their managemen. In *FAO Soils Bull*; United Nations: Rome, Italy, 1988.
42. Wang, X.J.; Xu, X.W.; Lei, J.Q. Spatiotemporal distribution of salt crust in a shelter-forest belt under drip-irrigation with salt water. *Arid Zone Res.* **2006**, *23*, 399–404.
43. Pereira, L.S.; Oweis, T.; Zairi, A. Irrigation management under water scarcity. Agricultural water management. *Agric. Water Manag.* **2002**, *57*, 175–206.
44. Li, C.; Li, Y.; Ma, J. Spatial heterogeneity of soil chemical properties at finescales induced by haloxylon ammodendron (chenopodiaceae) plants in a sandy desert. *Ecol. Res.* **2011**, *26*, 385–394.
45. Brantley, S.L. Understanding soil time. *Science* **2008**, *321*, 1454–1455.
46. Wang, G.; Garcia, D.; Liu, Y.; de Jeu, R.; Johannes Dolman, A. A three-dimensional gap filling method for large geophysical datasets: Application to global satellite soil moisture observations. *Environ. Model. Softw.* **2012**, *30*, 139–142.
47. Lowery, B.; Swan, J.; Schumacher, T.; Jones, A. Physical properties of selected soils by erosion class. *J. Soil Water Conserv.* **1995**, *50*, 306–311.
48. Zhao, H.L.; Zhou, R.L.; Zhang, T.H.; Zhao, X.Y. Effects of desertification on soil and crop growth properties in Horqin sandy cropland of Inner Mongolia, north China. *Soil Tillage Res.* **2006**, *87*, 175–185.

Chapter 3:
Management of Agricultural Water Use in Drylands

Geospatial Techniques for Improved Water Management in Jordan

Jawad T. Al-Bakri, Sari Shawash, Ali Ghanim and Rania Abdelkhaleq

Abstract: This research shows a case from Jordan where geospatial techniques were utilized for irrigation water auditing. The work was based on assessing records of groundwater abstraction in relation to irrigated areas and estimated crop water consumption in three water basins: Yarmouk, Amman-Zarqa and Azraq. Mapping of irrigated areas and crop water requirements was carried out using remote sensing data of Landsat 8 and daily weather records. The methodology was based on visual interpretation and the unsupervised classification for remote sensing data, supported by ground surveys. Net (NCWR) and gross (GCWR) crop water requirements were calculated by merging crop evapotranspiration (ETc), calculated from daily weather records, with maps of irrigated crops. Gross water requirements were compared with groundwater abstractions recorded at a farm level to assess the levels of abstraction in relation to groundwater safe yield. Results showed that irrigated area and GCWR were higher than officially recorded cropped area and abstracted groundwater. The over abstraction of groundwater was estimated to range from 144% to 360% of the safe yield in the three basins. Overlaying the maps of irrigation and groundwater wells enabled the Ministry of Water and Irrigation (MWI) to detect and uncover violations and illegal practices of irrigation, in the form of unlicensed wells, incorrect metering of pumped water and water conveyance for long distances. Results from the work were utilized at s high level of decision-making and changes to the water law were made, with remote sensing data being accredited for monitoring water resources in Jordan.

Reprinted from *Water*. Cite as: Al-Bakri, J.T.; Shawash, S.; Ghanim, A.; Abdelkhaleq, R. Geospatial Techniques for Improved Water Management in Jordan. *Water* **2016**, *8*, 132.

1. Introduction

Drylands are limited by soil moisture due to low rainfall and high evaporation. They show a gradient of increasing primary productivity in the order of hyper-arid, arid, and semiarid to dry sub-humid areas [1]. Drylands represent fragile ecosystems that are highly susceptible to environmental changes. Nevertheless, they provide important ecosystem services in terms of land and water, particularly in developing countries [2]. The limited surface water resources of drylands makes water management a challenge for planners and decision-makers, as groundwater will be pumped for different uses to compensate for the limited surface water.

The indiscriminate use of groundwater, particularly for irrigation, may result in aquifer depletion and salinization of soil and water. The problems resulting from groundwater depletion are usually aggravated in countries with high population growth and limited water resources [3].

Among the countries of the Middle East, Jordan represents a challenging case in terms of dryland water management, as the country's water resources are limited and its population is growing at high rates [3]. In terms of water availability, Jordan is among the four poorest countries in the world, with a per capita share of less than 146 m^3 per year, which is far below the international water poverty line of 500 m^3 per year [4]. Drylands dominate most of Jordan, where more than 90% of the country's land is arid and receives low rainfall amounts (Figure 1). Therefore, agricultural production in the country depends on irrigation using both developed surface and groundwater resources. At present, agriculture consumes more than 60% of the developed water resources. This share is expected to decrease in the near future as the population is increasing at high rates that reached an average of 2.9% during 2000–2012 [3,5]. According to Jordan's Department of Statistics [6], the population of Jordan increased from 5 million in 2002 to 6.7 million in 2014. The problem of population growth during 2002–2012 was aggravated by the influx of 0.45 and 1.5 million refugees from Iraq and Syria, respectively [7,8]. Such conditions of scarce resources and high population would put more stresses on water use and management. In addition, they might create environmental threats resulting from the use of marginal water resources [9].

Managing water resources under the conditions of limited rainfall and increased competition among sectors requires updated information on water consumption by the different sectors. For domestic and industrial sectors, this can be achieved by the use of direct metering and information from pumping stations operated by the Government. For agricultural water use, the information on water consumption becomes more difficult when groundwater is used for irrigation. In Jordan, most of the figures on groundwater consumption for irrigation are inaccurate due to the lack of information on the actual irrigated area and crop water consumption. Information on the spatial distribution of irrigation with accurate areas for the different irrigated crops is still needed by the Ministry of Water and Irrigation (MWI) to assess and improve water management at the country level. The use of ground surveys to map irrigation is impractical, particularly when information is annually needed for water budget calculations and for revising plans of water management. Alternatively, remote sensing data and geospatial techniques can be used for this purpose. The improvements in spatial, spectral and temporal resolution of remote sensing data and the access to different sources of earth observation systems justify the adoption of geospatial techniques for managing water resources. The most important application in this regard is the use of medium and high resolution data to map crops and to estimate their evapotranspiration, the major component of crop water requirements or consumption use [10–14].

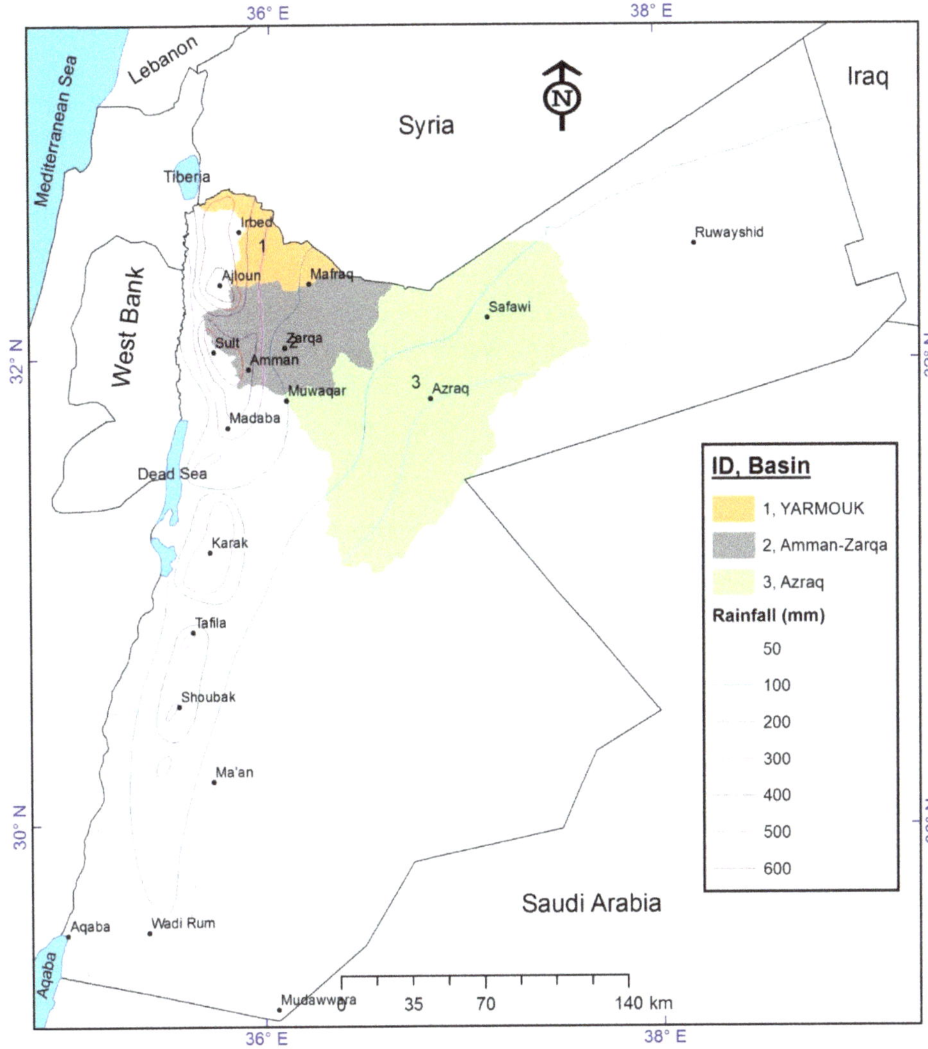

Figure 1. Location of the groundwater basins of Yarmouk, Amman-Zarqa and Azraq.

The use of remote sensing and geospatial techniques in water management took a new turn as these techniques were utilized in new approaches of assessment, generally called water accounting systems [15]. At the water basin level, the major components of these systems are the resource base, which represents water supply, crop evapotranspiration and water withdrawal, which represent water depletion, and the biomass production, which represents water productivity [16–18]. The most important inputs provided by remote sensing to the abovementioned components are the crop evapotranspiration and the land use maps [19]. The contribution of remote sensing and geospatial techniques will become more important in ungauged and poorly gauged basins as inputs will be provided with low cost and at reasonable accuracy [19]. Whether water accounting systems are implemented or not, remote sensing and geospatial techniques will continue to support water managers and decision makers with information needed for improved water management.

Within Jordan's efforts to manage its scarce water resources, MWI is in the stage of adopting geospatial techniques to improve water management at the country level. Recently, the MWI implemented a study in two pilot areas (Azraq and Mafraq) to map irrigated crops and to estimate agriculture water use using remote sensing data [20]. Initial results of the study showed that remote sensing maps were more accurate than estimates and records obtained by field crews of MWI. The difference between actual and estimated areas was important in terms of water consumption. Although the differences between both estimates could be attributed to the non-extensive ground survey that was limited by the available resources, they could, however, indicate possible violations of illegal drilling of wells and inaccurate metering of pumped water [21]. Therefore, the MWI adopted remote sensing for crop mapping and for auditing agricultural water consumption in the main irrigated areas in highlands. The areas included Jordan's parts of Yarmouk, Amman-Zarqa and Azraq basins (Figure 1). The overall aim of crop mapping using remote sensing data, presented in this work, is to provide MWI with information on the gap between official records and remote sensing estimates so that decisions can be taken to manage water resources in the three basins. Other objectives include the utilization of remote sensing data to map irrigated crops and the capacity building for MWI staff in using geospatial techniques for managing water resources. Despite their importance, spatial distribution of irrigated lands and their areas in Jordan were based on estimates, as the MWI lacked the capacity needed to fully utilize geospatial techniques. The inaccurate estimate of irrigated lands is also a worldwide problem and still remains uncertain [22]. Furthermore, the political nature of irrigation in countries that share water resources across borders often sets the stage for under-reporting of water use [23]. Therefore, this work provides proposed methods for mapping irrigated crops in Jordan's drylands and similar environments and will thus help in downscaling irrigation mapping to the basin level and provide important information that can be used for auditing records of water abstraction.

2. Study Areas and Data Collection

2.1. Study Areas

The first basin considered in this study was Yarmouk (Figure 1). About 30% (1393 km^2) of the basin's land is located in Jordan, while 70% is located in Syria. The basin has a semiarid Mediterranean climate in the west and an arid climate in the east. A rainfall gradient is obvious across the east-west direction, where the mean annual rainfall is 600 mm in the west and 150 mm in the east. The rainy season starts in November and ends by early May. The mean annual minimum and maximum temperatures in the west are 12.3 °C and 23.1 °C, respectively. The mean annual minimum and maximum temperatures in the east are 9.3 °C and 24.0 °C, respectively. The western parts of the basin are mainly rainfed areas cultivated with field crops, olives and vegetables, while the eastern parts extend to the low rainfall zone of the country where irrigation occurs using groundwater. Irrigation also occurs around Ramtha and in the northern parts of Jordan valley where citrus and vegetable crops are cultivated.

The second study area where irrigation was mapped is the Amman-Zarqa basin, also known as the Zarqa River basin. The basin has an area of 3600 km^2, where 95% of its area is inside Jordan and 5% is in the south of Syria. The basin has an arid climate in the east and the southeast, while the western parts have typical Mediterranean climates that are semiarid in Amman (Capital of Jordan) and dry sub-humid in Ajloun, where rainfall exceeds 560 mm. The western parts are mountainous and characterized by cool temperature in winter and mild temperature in summer. The annual rainfall ranges from more than 500 mm in the northwest to less than 100 mm in the east, with an average annual precipitation of 250 mm.

The third study area is Azraq basin (Figure 1), which has a total area 11,742 km^2. In this basin, water drains from all directions to the center (depression), which used to be a permanent oasis in Jordan's desert. The climate of Azraq Basin is arid with hot and dry conditions, leading to very high evapotranspiration rates. The mean annual temperature ranges between 7.4 and 24.5 °C in the west (Mafraq area) and between 2.6 to 36.6 °C in Azraq. The maximum air temperature in 2014 reached 42 °C. The mean annual rainfall is about 150 mm in the west and decreases to 70 mm in the east and south and to less than 50 mm in the center of the basin (Azraq depression). Over-pumping of groundwater for irrigation and drinking is noticed in this basin [24,25] and has resulted in the loss of wetlands surrounding the depression.

In terms of surface water resources, both the Yarmouk and Amman-Zarqa basins have perennial rivers that drain to Al-Wehdah and King Talal Dams (KTD), respectively. The collected water in both dams is used to irrigate different crops in Jordan Valley. Surface water resources in Azraq are mainly small desert dams that are used to irrigate small farms of olives and barley. Among the three basins, the Amman-Zarqa aquifer system is the most important groundwater supply in Jordan. This basin hosts more than half of Jordan's population [5] and more than 85 % of the industries in Jordan [26]. Groundwater resources in Yarmouk and Azraq basins, on the other hand, are also used for drinking and irrigation. Other sources of water include the treated wastewater from different plants in the three basins. Part of the effluent from treated wastewater plants (WWTP) is used for restricted agriculture, according to agreements between MWI and local farmers. The annual amount of treated wastewater effluent in Amman-Zarqa basin is 98 million cubic meters (MCM). About 15 MCM are used for irrigating forage crops in the basin, while the remaining is mixed with stored rainfall water in KTD and used to irrigate crops in Jordan Valley. The total amount of treated wastewater allocated for irrigation in Yarmouk is 3.8 MCM. Table 1 summarizes available water resources in the three basins and their use in 2014. All of the above figures were used in calculating gross groundwater amounts used for irrigation. In addition to over-pumping, other threats to water resources in the three basins are the trends of climate change, droughts and desertification [3,27–29].

Table 1. Summary of water resources in the study areas and their use in year 2014 (Source: MWI report of water budget [25]).

Basin	Groundwater Safe Yield (MCM)	Number of Agricultural Wells	Groundwater Use (MCM)		Surface Water Used in Irrigation (MCM)
			Agricultural	Non-Agricultural	
Yarmouk	40	129	36.4	9.4	3.8
Amman-Zarqa	88	590	63.9	92.4	15
Azraq	24	488	37.6	21.0	-

2.2. Remote Sensing Data

The main data used to map crop types were the images of the Operational Land Imager (OLI) of Landsat 8. The images have medium spatial resolution of 30 m. A higher resolution data of RapidEye, with 5-m pixel size, was used to delineate parcels of irrigated fields. The OLI data covered the period October 2013–September 2014, corresponding to a one-year period for calculating the water budget by the MWI. The data of RapidEye included two sets of cloud free images acquired in May and August 2014. Specifications of Landsat and RapidEye bands (B) used in the study are shown in Table 2.

2.3. Ground Data

Ground surveys were conducted during May–October 2014 to collect data on the different irrigated crops in the three basins. During these surveys, information was collected on the types of crops, growth stage, planting and harvesting dates, in addition to irrigation systems and agricultural management followed for the different crops. This information was critical for verifying crop maps and for calculating water consumption for irrigated crops in the three basins. In total, field surveys covered 310 farms in the three basins.

Table 2. Specification of Landsat OLI and RapidEye images used in crop mapping for the three basins.

Satellite Data	Path/Row	Band (Wavelength in μm)	Use	Processing
Landsat 8 OLI	174/37 for Yarmouk, 174/37 and 174/38 for Amman-Zarqa, and 173/38 for Azraq.	B1 (0.43–0.45), B2 (0.45–0.51), B3 (0.53 –0.59), B4 (0.64–0.67), B5 (0.85–0.88), B6 (1.57–1.65), B7 (2.11–2.29).	Crop type identification	1. Atmospheric correction 2. Vegetation index
RapidEye	Mosaic datasets covering the three basins.	B1 (0.44–0.51), B2 (0.52–0.59), B3 (0.63 –0.68), B4 (0.69–0.73), B5 (0.76–0.85).	Delineation of fields	1. Geometric correction. 2. Visual interpretation

Other ground data were collected for atmospheric correction of OLI images. Since remote sensing data of Landsat 8 were provided in digital numbers (DN) corresponding to radiance values at the top of atmosphere, it was important to convert the data into reflectance values at ground level. For this purpose, a handheld multispectral radiometer was used to collect ground measurements for target objects with standard or reference reflectance. The target objects were newly paved car parks (Dark objects) in front of Azraq Syrian Refugees Camp, dry desert mud flat, fields of parsley and tomatoes with full vegetation cover and bright surfaces (floors of abandoned quarries and desert surfaces). The selected locations were large enough so that they appeared in the form of several pixels on the images. Collection of ground data with the handheld radiometer was carried out within one hour of the satellite overpass on the 30 June 2014. The Landsat 8 image acquired for the day in which measurements were made was LC81730382014181LGN00.

2.4. MWI Data

Data of air temperature, relative humidity, wind speed and solar radiation were provided by the MWI. The data included daily records for the stations of Azraq, Safawi Um Ejjmal in Mafraq, Khirbet Samra, KTD, Amman, Ramtha and Irbed. The data were arranged in a spreadsheet and processed to interpolate missing records before being used for calculating crop water requirements. In addition, the MWI provided maps of groundwater wells, with attributes that included amounts of monthly abstraction, in addition to data on irrigated crops and crop management practices in Azraq and Amman-Zarqa basin. The data, originally collected from members of the highland water forum [30], were used for calculating crop water requirement and for verifying crop maps.

3. Methodology

Geospatial techniques and remote sensing data can be used for mapping irrigated lands at local, regional and global scales, according to objective, cost and time. The methods for crop mapping at local scales would include the visual interpretation or a digital classification method or a combination of both [10]. The improvement in remote sensing data encouraged researchers to implement and adopt digital classification methods, which are cost effective and consume less time when compared with the visual interpretation method [31]. The most commonly used approaches of digital classification include the parametric methods of supervised and unsupervised clustering [32]. These methods depend on the use of a single image acquired during the period of peak growth of crops. Due to multi-cropping and varying cropping calendars, a single image may not adequately characterize irrigated areas [33]. More advanced techniques may implement decision tree classifications, which are based on ground data, in addition to knowledge and experience of the classifier [10,34,35]. Further, the use of multi-temporal images of vegetation indices will increase the accuracy of mapping, as temporal resolution may compensate for the poor or moderate spatial resolution [10,36,37]. Among vegetation indices, the normalized difference vegetation index (NDVI) is considered a useful remote sensing product for mapping irrigation at local scales [36,37]. In this study, crop identification was based on the outputs from digital classification techniques of multi-temporal images of Landsat 8 (Figure 2), while boundaries of irrigated fields were digitized from very high resolution images of Google Earth (GE) and RapidEye (Figure 3), as described in the following subsections.

3.1. *Processing of Remote Sensing Data*

The set of downloaded images was processed to identify and exclude images with high cloud cover. A cloud mask was applied to each image using open source software [38]. Using the capabilities of a geographic information system (GIS), each image derived from the cloud mask was decoded using reclassification functions. The output images of cloud cover were displayed and processed in GIS to interpolate cloudy pixels in each image from previous and subsequent images. The total number of OLI images was 23 per year. However, the images with good quality and minimum cloud cover were only ten for Azraq and twelve for the two other basins. For the period of irrigation during April–September, most of images were free of clouds.

Remote sensing data of OLI were corrected to obtain reflectance values at ground level. This was carried out using measurements from the handheld multispectral radiometer (with similar spectral regions to OLI). The data of the handheld radiometer were used for correcting one image (Day 181 of year 2014). This absolute atmospheric correction was based on empirical line calibration [39] derived from linear equations that correlated ground measurements of spectral reflectance with DN for locations of target objects that had unchangeable spectral reflectance among the set of OLI images. The multiple-date image normalization, using linear regression equations, was carried out to correct other images [40]. This relative atmospheric correction method was based on the use of invariant features that had unchangeable spectral reflectance among the set of OLI images. Both methods of absolute and relative atmospheric corrections were considered as standard methods that would not require information about atmospheric conditions at the time of image acquisition [41]. The correction was carried out for the red and near infrared bands of OLI. The corrected bands were then used to derive NDVI, based on the following equation:

$$NDVI = \frac{B5 - B4}{B5 + B4}, \tag{1}$$

where B4 and B5 are the red and near infrared bands, respectively. Both bands represent spectral reflectance at ground level. The steps of processing OLI images to derive NDVI are summarized in Figure 2.

3.2. Extraction and Analysis of NDVI Profiles

Observations collected during ground surveys were transferred into a GIS layer, representing points with different types of irrigated crops. The map was intersected with the layer of NDVI images to assign NDVI values for each crop type in the form of a profile that represented temporal changes in this index during the year. Examples of these profiles for Azraq are shown in Figure 4. The data of NDVI profiles were exported to spreadsheets and NDVI values were plotted for growing seasons. Statistical functions were applied in the spreadsheets to identify ranges of NDVI and to aid in building a decision tree for creating crop maps. Analysis of NDVI profiles showed that separation of the irrigated crops was possible with the use of multi-temporal images. NDVI profiles showed that alfalfa was characterized by the highest NDVI that reached its maximum theoretical value (1.0), with fluctuations resulting from crop cutting. Olives, on the other, had stable NDVI profiles (with a nearly constant value that was slightly above 0.50 for old plantations). Younger olives (3–4 years) had a straight line NDVI that was in the range of 0.20 to 0.30. The profiles for vegetables varied according to crop type and growing season, with peak NDVI values occurring during the midseason for each vegetable crop.

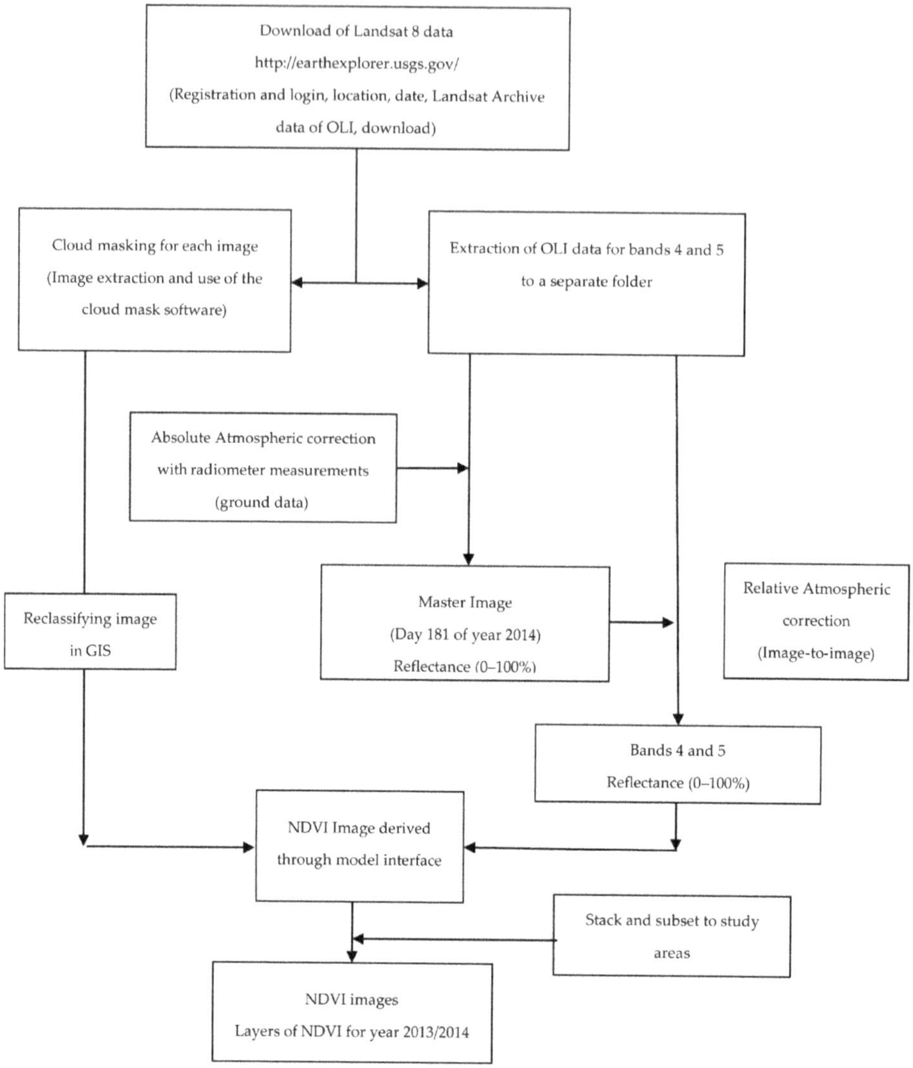

Figure 2. Flowchart of the image processing steps followed to derive NDVI images.

Figure 3. Use of high resolution images for digitizing parcels of irrigated lands and identifying ground cover for the different crops (Source: Google Earth).

Following the stage of NDVI extraction and analysis, decisions were developed to identify the NDVI ranges (bounds) for each crop and for each time of the season. These decisions were mainly developed for Azraq and Amman-Azraq basins, while for Yarmouk, the unsupervised classification method was used to separate rainfed from irrigated areas. The NDVI range for each vegetable crop was based on the mean and standard deviation for NDVI image, as profiles of NDVI showed normal distribution patterns during the growing seasons. For alfalfa and olives, the range was modified according to the degree of agreement between classified pixels and ground observations. The NDVI profiles were also used to identify the length of the growing season, which was crucial for calculating crop evapotranspiration. Based on ground observations and analysis of NDVI, a classification scheme (Table 3) was adopted for mapping the different irrigated crops. The scheme included the different classes and sub-categories that included all irrigated crops and growing seasons in the three basins.

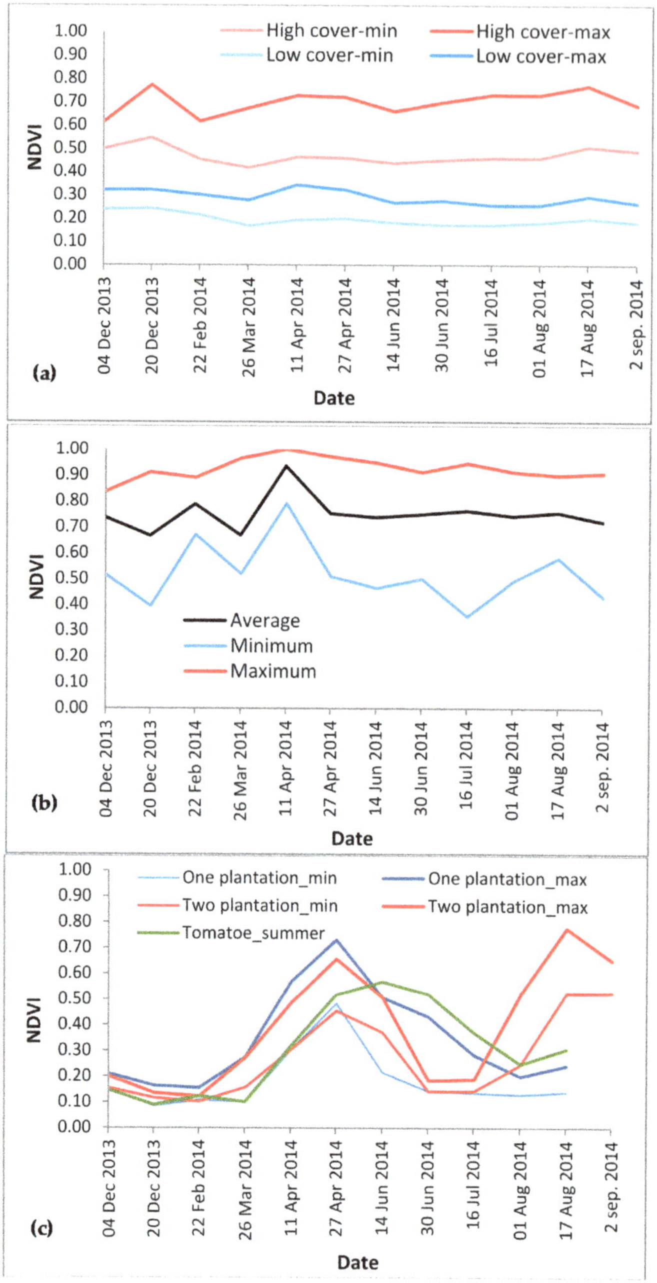

Figure 4. NDVI profiles of olives (**a**); Alfalfa (**b**) and vegetable crops (**c**) in Azraq area.

Table 3. Classification scheme of irrigated crops in Azraq and Mafraq.

Crop Category	Sub-Categories and Description
Olives	Farms of irrigated olive trees with different ground cover, depending on age and spacing. The class included three levels of cover that were grouped in the final map: low (<40%); medium (40%–60%); and high (>60%).
Fruit trees	Farms of deciduous orchards of peaches, apricots, stone fruits and table grapes grown on trellis. The sub-category of this class includes citrus in the northwest of Yarmouk.
Alfalfa and forage crops	Open spaces and farms cultivated with alfalfa under center pivot and solid sets of sprinkler irrigation systems in Azraq and fields of alfalfa and forage crops in the two other basins. The class includes barley fields cultivated under sprinkler irrigation.
Mixed cropping	Farms with two or more crop types, found mainly in Azraq. The fields include combinations of date palm, olives and alfalfa in the small landholdings near urban areas. The class is characterized by 100% ground cover for farms that included alfalfa in combination with olives, date palm and fruit trees.
Vegetables (Open fields)	Vegetables grown during February-June, March-July, April-August, June-September, July-October and September-December. The main irrigated crop is tomato. Other irrigated crops are melon, water melon, eggplant, zucchini, cauliflower, pepper and lettuce.
Vegetables and nursery plantations (Plastic houses)	Vegetables grown in plastic houses in the area of Mafraq and near Muwaqar in the middle west of the basin. The main crops are tomato and watermelon in Mafraq, and tomato and other vegetables in Muwaqar. The other sub-category of this class includes nurseries the lower part of Zarqa River between the WWTP and KTD and in the area of Baqa'a where some of the plastic houses are used as nurseries.

3.3. *Mapping Irrigated Fields*

Parcels of irrigated fields were mapped using an on-screen digitizing process with high resolution images of RapidEye and images of GE as backgrounds for delineation. The GE images were very useful as they had very high spatial resolution and horizontal accuracy that could reach 1 m [42]. RapidEye images, on the other hand, provided recent data that covered irrigated lands in year 2014, while GE images were mainly used in separating tree crops from vegetables. Identification of crop type, on the other hand, was based on outputs from the unsupervised classification of the multi-temporal images of Landsat 8 OLI. For Azraq and Amman-Zarqa basins, crop identification was carried out by the application of models that included decisions based on NDVI ranges derived from the multi-temporal images. The decisions were based on conditional statements that identified the range of NDVI for each crop and for each date in the set of the multi-temporal images of OLI. The models were applied in a sequence that prioritized crops according to their proportions, as observed during ground surveys. Subsequently, the sequence of models started with olives, followed by vegetables, fruit trees and other crops.

In Yarmouk basin, initial results showed some mixing between rainfed and irrigated areas. Therefore, principal component analysis (PCA) was used to transfer NDVI images of Yarmouk into three principal components. Applying the PCA to NDVI images resulted in transferring these images into three principal components that accounted for all seasonal changes in NDVI, as indicated by previous research [43]. Results from the unsupervised classification of PCA showed six distinguished spectral classes. The PCA images were then incorporated into unsupervised classifications with the ISODATA algorithm to derive spectral classes of irrigated and rainfed areas in this basin. Visual inspection of the output image showed that irrigated areas were distributed among three out of the six classes. In order to increase accuracy of mapping, models of knowledge based classifiers were used to refine the map of irrigated crops by incorporating layers of digital elevation model (DEM), rainfall and images of minimum NDVI values in these models. The DEM was used to exclude areas with high altitudes, where Jordan's forests were located. A rainfall map was used to exclude areas with high rainfall, as rainfed agriculture was dominant in these locations. The map of minimum NDVI separated irrigated areas from natural vegetation in the low rainfall zone. Accuracy assessment was made for the output maps using the confusion-matrix method [44], which compared remote sensing results with observations collected during ground surveys.

3.4. Assessment of Groundwater Abstraction Records

Assessment of groundwater abstraction for irrigation would require the use of ground measurements to compare amounts of abstraction with crop water consumption. Ground measurements of crop water requirements, however, are time consuming and become very expensive for large geographical areas. Therefore, crop water requirements can be calculated using crop maps and meteorological data. Remote sensing techniques can also provide spatial estimates of crop evapotranspiration (ETc), which is the main component of net crop water requirements (NCWR). This can be achieved by the surface energy balance models that utilize remote sensing data to derive the main factors controlling water evaporation from soil and plants and thus calculate ETc [45]. These models, however, require validation before being adopted for calculating ETc [46]. Therefore, crop maps were utilized in this study to calculate the crop ETc using the standard method of Food and Agricultural Organization [47]. The other component included in calculations of NCWR was the salt leaching requirement, which depended on water salinity and crop type. Rainfall in Yarmouk and Amman basin was also considered in these calculations. The FAO56, based on Penman-Monteith method, was used to calculate ETc as follows [47]:

$$\mathrm{ETc} = \mathrm{ETo} \times \mathrm{Kc}, \tag{2}$$

where ETc is the actual crop evapotranspiration (mm); Kc is the mean monthly crop coefficient; and ETo is the grass reference evapotranspiration. The NCWR, in million cubic meters (MCM), was calculated by multiplying the seasonal ET of the crop with its corresponding area, taking rainfall into consideration.

The FAO56 method calculates the theoretical crop water requirements and assumes standard conditions of crop spacing, ground cover and growing periods. These assumed conditions might not be valid in all fields in the three basins. Therefore, crop spacing and cover were considered in the calculations by adjusting

Kc values based on ground cover fraction [48]. Ground cover was derived from the high resolution images of GE (Figure 3) and the data collected during ground surveys. The length of each growing stage was adjusted from NDVI profiles (Figure 4), which showed some differences between the values reported by the FAO56 method and the actual growing periods. Following this step, daily weather records of MWI were used to calculate ETo using an ETo calculator, developed by FAO [49]. The outputs of daily ETo were summed for each month to calculate monthly ETc. Values of Kc were adjusted using NDVI profiles and data collected during ground surveys. The adjustment was made according to ground cover and used a linear interpolation to derive monthly Kc. Irrigation efficiency was used to convert the NCWR to gross crop water requirements (GCWR). The average irrigation efficiency in the three basins was about 80% [50]. The calculated GCWR were compared with the records of groundwater abstraction for irrigation purposes to assess the compliance between the recorded abstraction in relation to crop water requirements and safe yields in the three basins. For vegetables cultivated in plastic houses and nurseries, the GCWR were given an average value of 800 m^3 per year, based on the average for tomato [51,52] for two cultivations per year.

4. Results and Discussion

4.1. Ground Surveys

Observations from ground surveys showed that olives, fruit trees and vegetables were the main irrigated crops in the three basins. Irrigated vegetables included tomato as the main crop and other crops of melon, water melon, eggplant, zucchini, cauliflower and lettuce. Fruit trees included peaches, apricots, table grapes, pomegranate and date palm (in Azraq only). The size of some irrigated farms reached 400 ha (equivalent to 4000 dunums in the locally used unit), while the size of vegetable farms (open fields) was in the range of 20–30 ha and reached 70 ha in some areas, indicating considerable investment in irrigated agriculture. In the area of Mafraq in Amman-Zarqa basin, many of the drylands were turned into agricultural fields by clearing soil surface from basalt rocks. This practice, shown in Figure 5, had been conducted since the 1980s, when irrigation started in the drylands of Mafraq [53]. Ground observations also showed different planting and harvesting dates for vegetables, which were confirmed by the NDVI profiles (Figure 4).

Figure 5. Examples on irrigated crops and irrigation practices in the three basins including fields of tomatoe under drip irrigation (**a**) and alfalfa under center pivot irrigation (**b**); watermelon at harvest stage (**c**); fields of cauliflower cultivated during March–July (**d**); an irrigated farm of table grapes (**e**) and a piece of land in Mafraq (inside Amman-Zarqa basin) with the soil surface partially cleared from basalt rocks to cultivate tomatoes (**f**).

4.2. *Crop Maps and Irrigated Areas*

4.2.1. Irrigation in Yarmouk

The crop map of Yarmouk showed that the total irrigated area was 5.9 thousand ha, distributed in different parts of this basin (Figure 6). The main irrigated crops were tree crops (olives and fruits) and vegetables. Analysis of the crop map (Table 4) showed that vegetables constituted about half of the irrigated area, while the other half included olives and fruit trees (48%), with a small proportion of forage crops and nurseries. Most irrigation was practiced in the drylands of Mafraq, followed by the north middle areas. In addition, irrigation was practiced in the western parts of Ramtha, where the total irrigated area in that part constituted 21% of irrigation taking place in the basin. Irrigation was limited in the southwestern parts, as the area was mountainous and lacked irrigation infrastructure (dams and groundwater wells). The absence of irrigation to the west and north of Irbid would be mainly attributed to the high rainfall that supported the rainfed agriculture, which included olives and field crops. Further, urbanization was encroaching into these areas and was competing with other land uses.

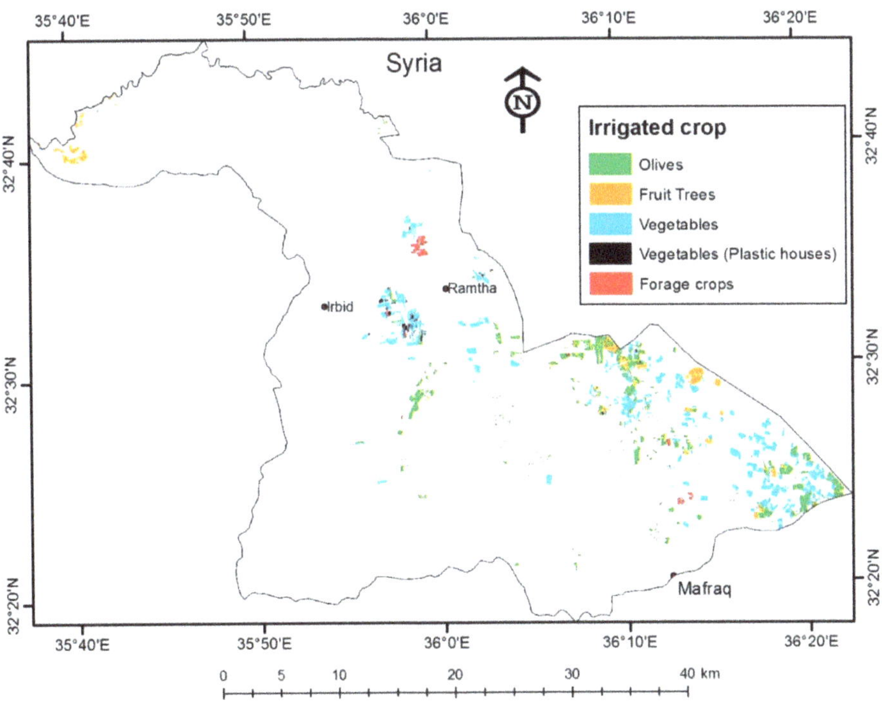

Figure 6. Distribution of irrigated crops in Yarmouk basin, as derived from remote sensing data.

Table 4. Analysis of irrigated areas in the three basins, as mapped from remote sensing data for the period October 2013–September 2014.

Class	Yarmouk		Amman-Zarqa		Azraq	
	Area (ha)	%	Area (ha)	%	Area (ha)	%
Olives	2018	34.1	4986	27.5	3050	39.0
Fruit trees	837	14.1	4118	22.7	1286	16.5
Alfalfa and forage crops	187	3.2	638	3.5	587	7.5
Mixed cropping	-	-	439	2.4	425	5.4
Vegetables (Open fields)	2763	46.7	7737	42.7	2465	31.6
Vegetables and nursery plantations (Plastic houses)	115	1.9	210	1.2	-	-
Total	**5920**		**18,128**		**7811**	

In terms of irrigated areas, this study provided detailed maps of irrigation in Yarmouk basin. Previous work [28,29] showed similar spatial distribution of irrigation in the middle and eastern parts of this basin without details on crop types. Therefore, this study contributed to the governmental efforts in managing water resources by providing more detailed maps for irrigation, particularly in the western parts of the basin. The best estimate of MWI for irrigated lands in the basin was 2.1 thousand ha, while initial results obtained from the digital classification of PCA images of NDVI showed that the irrigated area was more than 5.1 thousand ha [21,54]. The reason behind these counterintuitive findings could be the non-extensive ground survey carried out by MWI, which was limited by time and cost. Such results, therefore, encouraged MWI to take more actions toward the adoption of remote sensing technology for mapping irrigation in the basin and to audit data of groundwater abstraction.

4.2.2. Irrigation in Amman-Zarqa

The total irrigated area in Amman-Zarqa basin was 18.1 thousand ha. Vegetables constituted 43% of this area, while olives and fruit trees constituted more than half of the irrigated lands. Irrigation was practiced in two areas: Mafraq-Hallabat and on both sides of Zarqa River (Figure 7). This distribution could be attributed to the sources of irrigation water, which were groundwater in the Mafraq-Hallabat and the surface water of Zarqa River. The disposed treated wastewater was used for irrigating forage crops and tree plantations in nurseries, although some violations were observed on some farms on the sides of the river. Further analysis of the irrigation map showed that 77% of irrigation was taking place in Mafraq-Hallabat, indicating that groundwater was the main source for irrigating these drylands. The area to the north of KTD had limited irrigation, as the high rainfall encouraged rainfed agriculture in this mountainous area. In terms of irrigated areas in the basin, figures from this study were less than those reported by a previous study [26], which used unsupervised classification of a single image of Landsat ETM+ and reported a total irrigated area of 27 thousand ha. This could be attributed to the different crops cultivated during these periods and the time difference between this study and previous ones.

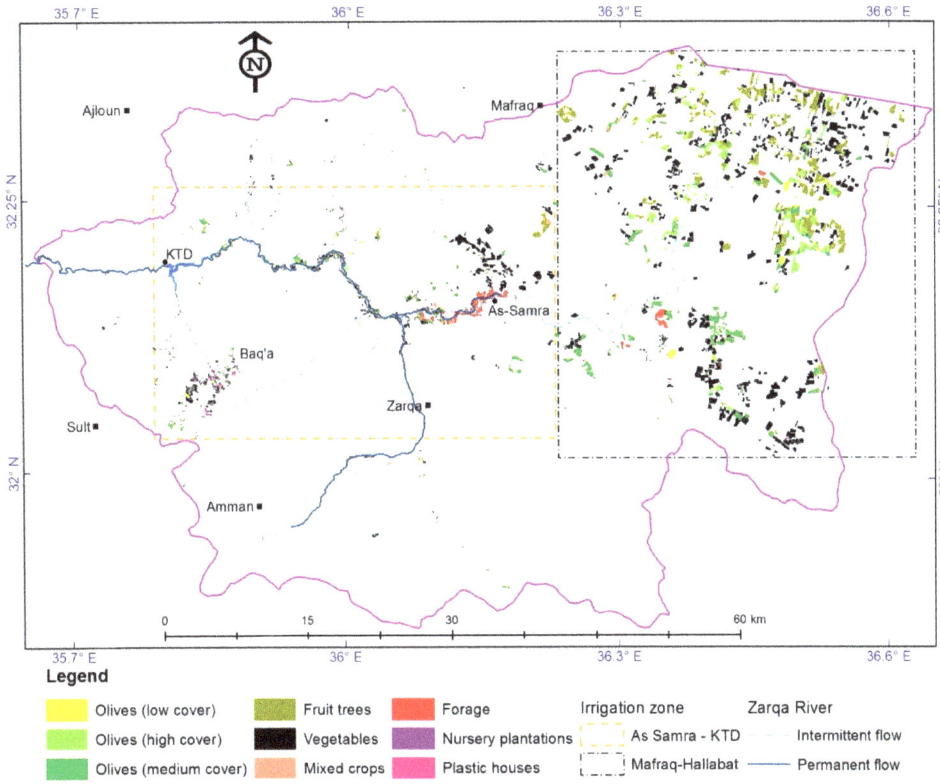

Figure 7. Map of irrigation in the Amman-Zarqa basin as derived from remote sensing data.

4.2.3. Irrigation in Azraq

Analysis of the irrigation map showed that the total irrigated area in Azraq basin was 7.8 thousand ha, mainly distributed in two strips in the basin, Azraq depression and northwest of Azraq (Figure 8). The reason behind this spatial distribution could be attributed to the characteristics of groundwater in these two parts, where the irrigated areas were located in the unconfined aquifer of quaternary basalt outcrop overlaying the aquifer A7/B2. The A7/B2 aquifer, which has two joined and fissured formations (Amman formation (B2) and the older Wadi Es-Sir formation (A7)), is characterized by high conductivity in the form of solution channels and karstic features and extends into the northern parts of Jordan [24,55]. The absence of irrigation in the southern parts of the basin would be attributed to the distribution of B4 (very low transmissivity aquifer) in these areas [24,55]. Analysis of the crop map showed that the main irrigated crops were olives (39%), followed by vegetables (32%) and fruit trees (17%), while other crops were alfalfa and mixed crops (Table 4). In terms of irrigated areas, figures from this study agreed with those obtained in previous work that used visual interpretation of satellite images [24]. On the other hand, an estimate that was based on administrative boundaries of the basin reported that the total irrigated area in Azraq would be in the range of 11.5 thousand ha [50]. These findings clearly indicated the improvement in mapping irrigated areas when multi-temporal remote sensing data were used.

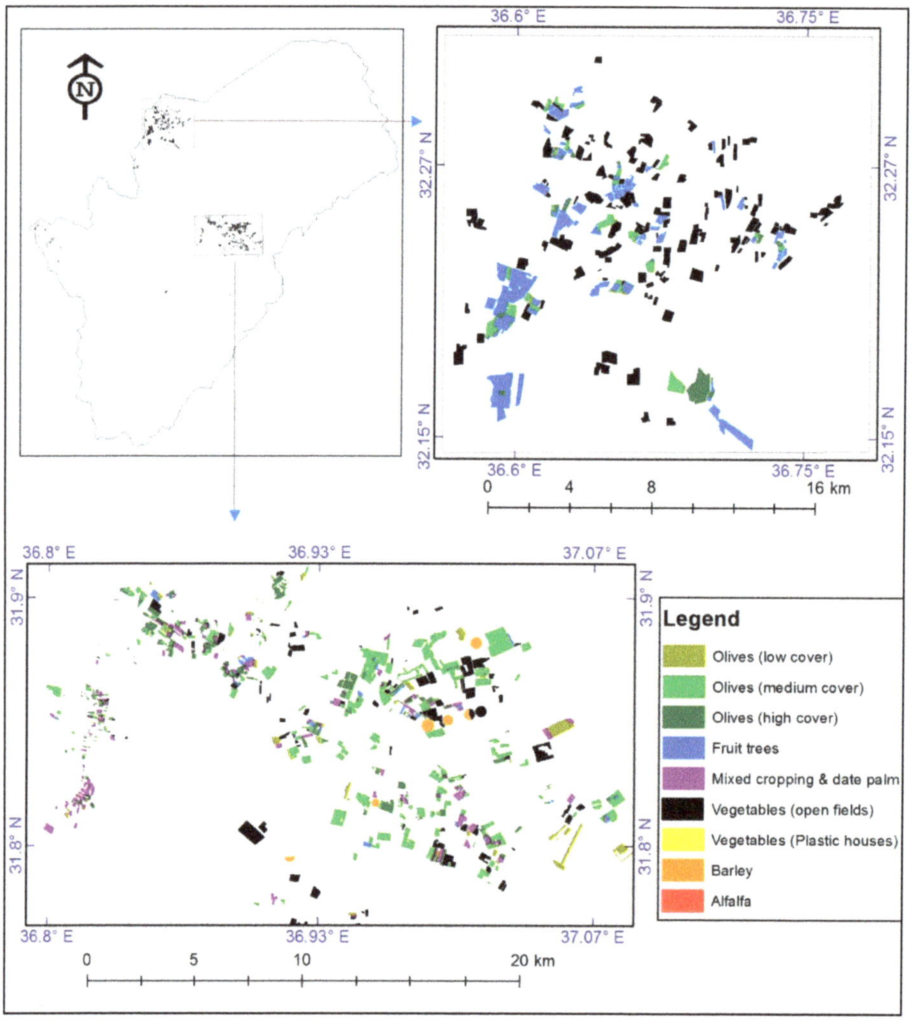

Figure 8. Map of irrigation in Azraq basin as derived from remote sensing data.

4.3. Mapping Accuracy

Although remote sensing data could provide important maps for irrigated lands, their accuracy should be assessed before being used. Comparing field observations (310 points) with maps of irrigated crops showed a good agreement between irrigation maps and ground data, with an overall accuracy of 87% (Table 5). The minimum accuracy was observed for fruit trees, which were mixed, in terms of classification, with olives. Considering the calculations of NCWR from ETc (Table 6), this would mean that NCWR (Table 7) for fruit trees was nearly 10% more than olives. For other irrigated crops, the level of mapping accuracy was relatively high and would be acceptable for the purpose of mapping and for estimating agricultural water consumption in the three basins.

Table 5. Analysis of irrigated areas in the three basins, as mapped from remote sensing data for the period October 2013–September 2014.

		Reference Data							
	Class	**Olives**	**Alfalfa & Forage**	**Fruit Trees**	**Vegetables**	**Barley**	**Mixed**	**Totals**	**Mapping Accuracy (%)**
Classified Image Data	Olives	89		6	2	-	-	97	92
	Alfalfa & Forage	2	32	-	-	-	-	34	94
	Fruit Trees	24	-	44	-	-	-	68	65
	Vegetables	-	1	-	75	-	3	79	95
	Barley	-		-	-	9	1	10	90
	Mixed	-	2	-	-		20	22	91
	Totals	**115**	**35**	**50**	**77**	**9**	**24**	**310**	**86.8**

Table 6. Monthly evapotranspiration for the main irrigated crops in Azraq area.

Month *	**ETo (mm)**	**Crop Evapotranspiration (ETc) in mm**								
		Olives	**Fruit Trees**	**Mixed**	**Alfalfa**	**Vegetables**				
October	104	67	67	80	97	-	-	-	-	77
November	62	39	25	45	55	-	-	-	-	-
December	36	10	–	15	41	-	-	-	-	-
January	45	15	–	10	43	-	-	-	-	-
February	65	22	–	28	75	26	-	-	-	-
March	102	62	49	72	88	72	41	-	-	-
April	158	102	110	122	147	158	110	63	-	-
May	176	102	116	129	157	101	176	123	-	-
June	203	119	183	158	190	41	117	203	81	-
July	242	149	218	188	226	–	48	139	162	97
August	237	130	190	184	221	–	–	47	213	158
September	156	96	109	121	145	–	–	–	130	140
Total	**1586**	**915**	**1067**	**1151**	**1484**	**397**	**493**	**576**	**586**	**472**

* Starting from November 2013 to September 2014, corresponding to one year of water budget at MWI.

Table 7. Summary of irrigated crops and their net annual water consumption.

Class	Yarmouk		Amman-Zarqa		Azraq	
	ETc * (mm)	NCWR (MCM)	ETc * (mm)	NCWR (MCM)	ETc * (mm)	NCWR (MCM)
Olives	700	14.1	726	36.2	887	27.0
Fruit trees	714	6.0	765	31.5	762	9.8
Alfalfa and forage crops	1213	2.3	1206	7.7	1320	7.8
Mixed cropping	-	-	888	3.9	1148	4.9
Vegetables (Open fields)	538	14.5	441	34.1	494	12.2
Vegetables and nursery plantations (Plastic houses)	800	0.9	800	1.7	-	–
Total		**37.8**		**115.1**		**61.7**

* Weighted average for all crop sub-categories.

4.4. Assessment of Groundwater Abstraction Records

One important tool for water management would be the use of outputs from crop mapping for water accounting and auditing. This was possible by deploying crop maps and weather records to calculate evapotranspiration, which is considered as the main component of net crop water requirements (NCWR). Results showed relatively high ETc values for forage, olives and tree crops in the three basins, particularly Azraq (Table 6). For vegetable crops, the NCWR varied according to cultivation season, which was accurately detected by the NDVI profiles. Although not investigated in this study, these profiles could have higher contribution to NCWR by deriving Kc with reasonable accuracy [56].

Variations in the ETc level among the three basins could be attributed to several factors including climatic conditions, the length of growing seasons and management practices that included different tree spacing in the three basins. Considering ETc and the area of each irrigated crop, the NCWR was calculated in the three basins (Table 7). Results showed that olives and fruit trees were the main consumers of water in the three basins. The NCWR for vegetables was variable according to planting and harvesting dates (Table 6). These results emphasized that cropping patterns in the three basins should be revised and changed to sustain groundwater resources. Proposed actions could include the prohibition of irrigating olives and alfalfa with fresh water resources of the three basins and the gradual replacement of both crops with other crops that consume less water. In addition, cultivation of vegetables during the summer season should be discouraged as it would increase water consumption by 30% (as implied from the ETc calculations). Another aspect of water management could be the use of flexible water allocation and pricing according to crop type and cropping season. A previous study in Jordan highlands showed that there was a potential to decrease water consumption and to reallocate it in an optimal way that considered cropping pattern and income from the irrigated area [57].

Outputs from crop mapping and NCWR calculations were used to assess groundwater abstraction records obtained by the MWI. The assessment was made by comparing MWI records with the NCWR, calculated by the FAO56 method after modifying Kc and the length of the growing season. This water auditing task was achieved by considering NCWR, irrigation efficiency and water resources used in irrigation to calculate GCWR. Results are summarized in Table 8. For the three basins, the GCWR were higher than the safe yield of groundwater. In Jordan, the term of groundwater safe yield is based on water balance models calibrated with data from groundwater monitoring wells [58].

Table 8. Groundwater abstraction records compared with estimates from remote sensing.

Basin	Safe Yield (MCM)	Groundwater Abstraction for Irrigation		Agricultural Abstraction/Safe Yield (%)	Abstraction/Safe Yield * (%)
		MWI Records	Remote Sensing		
Yarmouk	40	36.4	48	120	144
Amman-Zarqa	88	63.9	104	118	224
Azraq	24	37.6	67	279	367

* Including non-agricultural water uses (Table 1) and calculated as abstraction/safe yield.

In terms of agricultural water use, irrigation consumed considerable amounts of water that exceeded the safe yields in the three basins. Considering all uses of water, the over abstraction of groundwater was in the range of 144% in Yarmouk to 367% in Azraq. These results were in line with findings from the study on a groundwater aquifer in the north of Jordan [58], which indicated a decline of groundwater levels by 40–60 m during 1986–2014. This decline resulted in changing water flow directions from Zarqa towards the east of Mafraq and to the north in Yarmouk basin, *i.e.*, an opposite direction to its movement 40 years before [58]. Further, the results from Azraq indicated an over-abstraction of groundwater that could result in negative consequences resulting from groundwater depletion. The results for Azraq were also supported by findings from a previous study [59] that indicated a decline and lowering of the water table of the upper aquifer by 20 m during 1983–2003, the period during which irrigation had expanded in this basin.

Findings from crop mapping and GCWR calculations were alarming to MWI as they indicated that expansion in irrigation and adoption of inappropriate cropping patterns could result in future loss of groundwater resources. In addition, the shift between MWI records and remote sensing estimates indicated uncovered violations pertaining to groundwater use. Therefore, the work of crop mapping was extended to study the spatial distribution of irrigated areas in relation to water sources. The disagreement between water sources and irrigated areas was used to highlight areas with possible violations pertaining to water use or even possible water theft. The first map was prepared for the area around Ramtha City in Yarmouk basin (Red boxes in Figure 9).

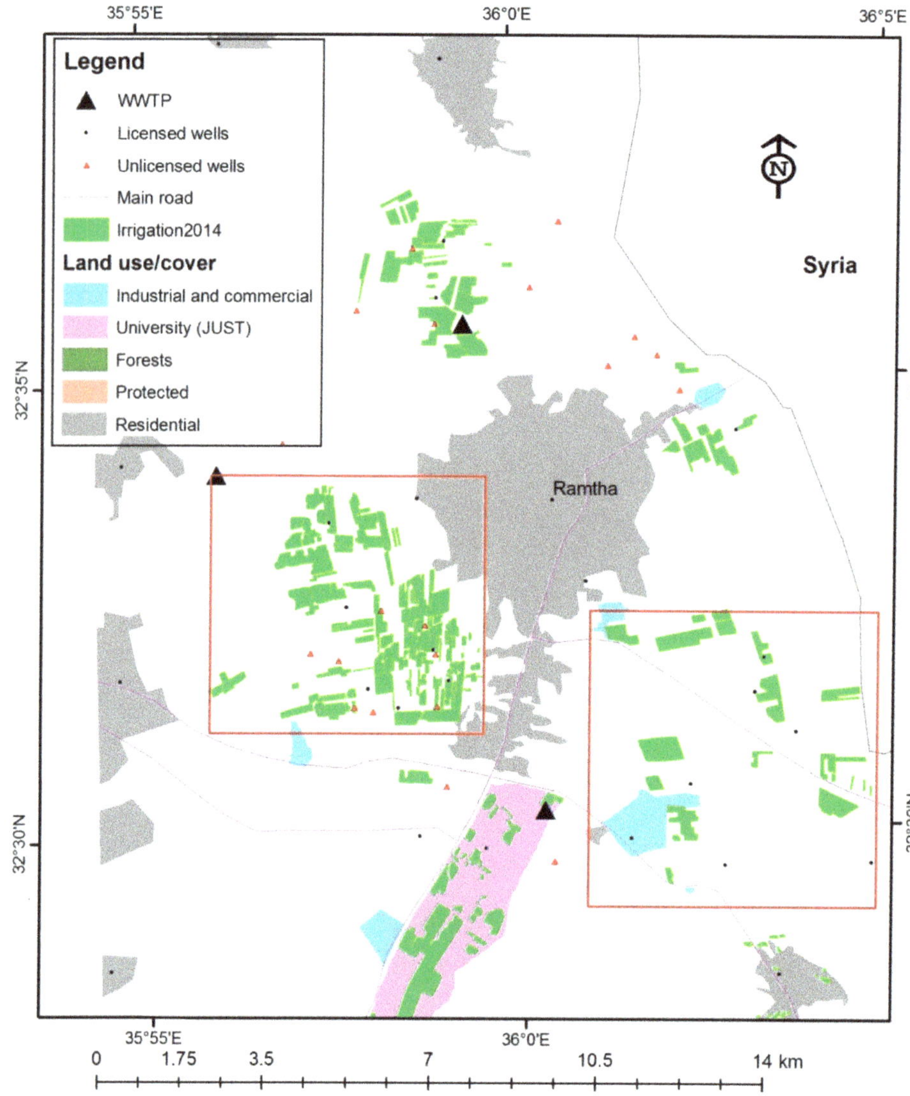

Figure 9. Map of irrigated areas and distribution water sources around Ramtha City.

Results from crop mapping showed that the irrigated area was nearly twice the area estimated by the field crew of MWI, while GCWR values were nearly three times higher than the recorded abstraction. Overlaying maps of irrigated crops and groundwater sources (Figure 9) showed that many irrigated fields were located in areas without groundwater wells or available surface water sources, indicating an illegal practice or violation to water use. In such cases, the concept of violation or illegal irrigation would be related to water pumping and conveyance over long distances rather than to cultivation of certain crops. Therefore, ground visits were carried out in fields that were located far away from water resources to uncover these violations, which were mainly in the form of water conveyance with pipelines for long distances or unlicensed groundwater wells [21]. In some areas, irrigation pipelines were illegally connected to the domestic water network and water was transferred to irrigate fields of vegetables. These findings urged MWI to carry out official campaigns to uncover these violations and to announce findings through mass media [54].

The outputs from this work were reflected in MWI decisions to improve water management and to activate water accounting and auditing. At present, amendments made to the "Water Law" state that satellite images and remote sensing techniques are officially adopted for auditing and estimating amounts of abstracted groundwater through crop mapping and estimation of water requirements for the different irrigated crops [60]. According to Article 4 of the new "Water Law", satellite images are among the accredited means that can be used by MWI to map irrigated areas and crop type and to estimate amounts of groundwater abstraction for irrigated farms. Therefore, it is hoped that the robust methods used in this study would contribute to MWI efforts in improving water management by applying remote sensing methods to uncover violations related to irrigational water use and to audit annual records of water abstraction. Adoption of these methods towards water accounting systems, however, would require capacity building in the use and utilization of remote sensing data and geospatial techniques.

5. Conclusions

Remote sensing data and geospatial techniques act as good sources and tools for providing data needed for managing the scarce water resources of drylands. The study showed that the improved techniques for identifying irrigated areas and crops using remote sensing would include the use of multi-temporal imagery and ancillary data. The contribution of geospatial techniques was mainly in water auditing as they provided crucial information on irrigated areas, cropping patterns and estimates of groundwater abstraction in relation to available water resources and irrigated crops. In terms of water use, the study showed intensive irrigation in the drylands of Jordan, particularly in Amman-Zarqa basin. At present, irrigation is consuming invaluable water resources and resulting in groundwater depletion. It can be concluded that inappropriate cropping patterns may threat scarce water resources of drylands. However, information provided through remote sensing and geospatial techniques may provide solutions that optimize cropping pattern and water use. This fact was recognized by decision makers in Jordan and amendments were made to the "Water Law" so that water accounting and auditing by means of these contemporary techniques were approved. The remaining challenge, however, will be the capacity building needed for adopting and implementing these techniques. Once this target is achieved, then it is hoped that remote sensing and geospatial techniques are taken one step further towards the development of water accounting systems for the different basins in Jordan.

Acknowledgments: This publication and crop mapping work were supported by Ministry of Water and Irrigation (MWI), Jordan. The work was carried out through a project "Regional Coordination on Improved Water Resources Management and Capacity Building Project" funded by the Global Environment Facility (GEF), managed by the World Bank (Project ID P117170), and jointly implemented in Cooperation with NASA. The authors acknowledge the cooperation of MWI and its staff, particularly the Secretary General Eng. Basim Telfah and the Ministry Technical Assistant Eng. Ali Subh for their full support and cooperation. The authors appreciate the efforts and cooperation of Su'ad Nasser, Ali Hayjneh and Issa Al-Obra. The authors would like to thank partners from NASA, particularly Dr. Shahid Habib.

Author Contributions: Jawad Al-Bakri worked on the methodology for crop mapping and NCWR comparisons. All authors were involved in calculations and comparisons made in the manuscript. Jawad Al-Bakri led the manuscript writing with contribution from other authors, who were also involved in field data collection and image processing stage.

Conflicts of Interest: The authors declare no conflict of interest.

References

1. Safriel, U.; Adeel, Z.; Niemeijer, D. Drylands Systems. In *Ecosystems and Human Wellbeing: Current State and Trends*; Hassan, R., Scholes, R., Ash, N., Eds.; Island Press: Washington, DC, USA, 2005; Volume 1, pp. 623–662.
2. IUCN Global Drylands Initiative. Available online: http://www.iucn.org/drylands (accessed on 25 August 2015).
3. Al-Bakri, J.T.; Salahat, M.; Suleiman, A.; Suifan, M.; Hamdan, M.R.; Khresat, S.; Kandakji, T. Impact of climate and land use changes on water and food security in Jordan: Implications for transcending "The Tragedy of the Commons". *Sustainability* **2013**, *5*, 724–748.
4. MWI (Ministry of Water and Irrigation, Jordan). *Water for Life: Jordan's Water Strategy, 2008–2022*; MWI: Amman, Jordan, 2009.
5. Al-Bakri, J.T.; Duqqah, M.; Brewer, T. Application of remote sensing and GIS for modeling and assessment of land use/cover change in Amman/Jordan. *J. Geogr. Inf. Syst.* **2013**, *5*, 509–519.
6. Department of Statistics, Jordan. Available online: http://www.dos.gov.jo (accessed on 20 September 2015).
7. UNHCR (The UN Refugee Agency). A Briefing Note on Syrian Refugees, 2013. Available online: http://www.unhcr.org/ (accessed on 20 September 2015).

8. DoS (Department of Statistics, Jordan). *Iraqis in Jordan: Their Number and Characteristics*; DoS, Fafo and UNFPA: Amman, Jordan, 2007.
9. Schacht, K.; Gönster, S.; Jüschke, E.; Chen, Y.; Tarchitzky, J.; Al-Bakri, J.; Al-Karablieh, E.; Marschner, B. Evaluation of soil sensitivity towards the irrigation with treated wastewater in the Jordan River region. *Water* **2011**, *3*, 1092–1111.
10. Ozdogan, M.; Yang, Y.; Allez, G.; Cervantes, C. Remote sensing of irrigated agriculture: Opportunities and challenges. *Remote Sens.* **2010**, *2*, 2274–2304.
11. Inglada, J.; Arias, M.; Tardy, B.; Hagolle, O.; Valero, S.; Morin, D.; Dedieu, G.; Sepulcre, G.; Bontemps, S.; Defourny, P.; Koetz, B. Assessment of an operational system for crop type map production using high temporal and spatial resolution satellite optical imagery. *Remote Sens.* **2015**, *7*, 12356–12379.
12. Bastiaanssen, W.G.; Karimi, P.; Rebelo, L.-M.; Duan, Z.; Senay, G.; Muthuwatte, L.; Smakhtin, V. Earth observation based assessment of the water production and water consumption of Nile Basin agro-ecosystems. *Remote Sens.* **2014**, *6*, 10306–10334.
13. Gowda, P.H.; Chávez, J.L.; Howell, T.A.; Marek, T.H.; New, L.L. Surface Energy balance based evapotranspiration mapping in the Texas High Plains. *Sensors* **2008**, *8*, 5186–5201.
14. Liou, Y.-A.; Kar, S.K. Evapotranspiration estimation with remote sensing and various surface energy balance algorithms—A Review. *Energies* **2014**, *7*, 2821–2849.
15. Godfrey, J.M.; Chalmers, K. *Water Accounting: International Approaches to Policy and Decision-Making*; Edward Elgar Publishing: Cheltenham, UK, 2012; p. 336.
16. Dost, R.; Bastidas; Obando, E.B.; Bastiaanssen, W. *Water Accounting through Remote Sensing*; A Background Report on Water Accounting Plus (WA+) in the Awash River Basin; Land and Water Division, Food and Agriculture Organization of the United Nations (FAO): Rome, Italy, 2013; p. 56.
17. Karimi, P.; Bastiaanssen, W.; Molden, D. Water Accounting Plus (WA+)—A water accounting procedure for complex river basins based on satellite measurements. *Hydrol. Earth Syst. Sci.* **2013**, *17*, 2459–2472.
18. Karimi, P.; Bastiaanssen, W.; Molden, D.; Cheema, M. Basin-wide water accounting using remote sensing data: The case of transboundary Indus Basin. *Hydrol. Earth Syst. Sci.* **2013**, *17*, 2473–2486.
19. Karimi, P.; Bastiaanssen, W.G.M.; Sood, A.; Hoogeveen, J.; Peiser, L.; Dost, R. Spatial evapotranspiration, rainfall and land use data in water accounting—Part 2: Reliability of water accounting results for policy decisions in the Awash basin. *Hydrol. Earth Syst. Sci. Discuss.* **2015**, *19*, 1125–1167.
20. Al-Bakri, J.T. *Crop Mapping and Validation of ALEXI-ET in Azraq and Mafraq Areas*; A Report for Regional Coordination on Improved Water Resources Management and Capacity Building; Ministry of Water and Irrigation: Amman, Jordan, 2015; pp. 19–31.
21. Al-Bakri, J.T. *Crop Mapping for Azraq and Ramtha Areas*; A Report for the Advanced on-job Training on the Use of Remote Sensing in Crop Mapping and Evapotranspiration; Ministry of Water and Irrigation: Amman, Jordan, 2014.
22. Vörösmarty, C.J. Global water assessment and potential contributions from Earth Systems Science. *Aqua. Sci.* **2002**, *64*, 328–351.

23. Bastiaanssen, W.G.M.; Molden, D.J.; Makin, I.W. Remote sensing for irrigated agriculture: Examples from research and possible applications. *Agric. Water Manag.* **2000**, *46*, 137–155.
24. GIZ (Gesellschaft für Internationale Zusammenarbeit). *The Azraq Dilemma: Past, Present and Future Groundwater Management*; German-Jordanian Programme "Management of Water Resources", MWI (Ministry of Water and Irrigation, Jordan): Amman, Jordan, 2012; pp. 6–12.
25. MWI (Ministry of Water and Irrigation, Jordan). *Water Budget for Year 2013*; Ministry of Water and Irrigation: Amman, Jordan, 2014; pp. 14–17.
26. Al-Wer, I. Zarqa River Rehabilitation and Sustainable Management. Master's Thesis, Department of Land and Water Resources Engineering, KTH Stockholm, Sweden, 2009; TRITA-LWR Degree Project. ISSN 1651-064X ISRN LWR-EX-09-32.
27. Abdulla, F.; Eshtawi, T.; Assaf, H. Assessment of the impact of potential climate Change on the water balance of semi-arid watershed. *Water Resour. Manag.* **2009**, *23*, 2051–2068.
28. Al-Bakri, J.T.; Suleiman, A.; Abdulla, F.; Ayad, J. Potential impacts of climate change on the rainfed agriculture of a semi-arid basin in Jordan. *Phys. Chem. Earth* **2011**, *35*, 125–134.
29. Al-Bakri, J.T.; Brown, L.; Dedalof, Z.; Berg, A.; Nickling, W.; Khresat, S.; Salahat, M.; Saoub, H. Modelling desertification risk in the north-west of Jordan using geospatial and remote sensing techniques. *Geomatics Nat. Hazards Risk* **2014**.
30. The Highland Water Forum: Dialogue for Sustainability. Available online: https://highlandwaterforum. wordpress.com/ (accessed on 17 October 2015).
31. Jensen, J.R. *Introductory Digital Image Processing: A Remote Sensing Perspective*, 3rd ed.; Prentice Hall: Upper Saddle River, NJ, USA, 2006; pp. 337–406.
32. El-Magd, I.A.; Tanton, T.W. Improvements in land use mapping for irrigated agriculture from satellite sensor data using a multi-stage maximum likelihood classification. *Int. J. Remote Sens.* **2003**, *24*, 4197–4206.
33. Simonneaux, V.; Duchemin, B.; Helson, D.; Raki, E.; Olioso, A.; Chehbouni, A. The use of high-resolution image time series for crop classification and evapotranspiration estimate over an irrigated area in central Morocco. *Int. J. Remote Sens.* **2008**, *29*, 95–116.
34. Mclver, D.K.; Friedl, M.A. Estimating pixel-scale land cover classification confidence using non-parametric machine learning methods. *IEEE Trans. Geo-Sci. Remote Sens.* **2001**, *39*, 1959–1968.
35. Mclver, D.K.; Friedl, M.A. Using prior probabilities in decision-tree remotely sensed data. *Remote Sens. Environ.* **2002**, *81*, 253–261.
36. Gumma, M.K.; Thenkabail, P.S.; Nelson, A. Mapping irrigated areas using MODIS 250 meter time-series data: A study on Krishna River Basin (India). *Water* **2011**, *3*, 113–131.
37. Gumma, M.K.; Thenkabail, P.S.; Hideto, F.; Nelson, A.; Dheeravath, V.; Busia, D.; Rala, A. Mapping irrigated areas of Ghana using fusion of 30 m and 250 m resolution remote-sensing data. *Remote Sens.* **2011**, *3*, 816–835.
38. Zhu, Z.; Woodcock, C.E. Object-based cloud and cloud shadow detection in Landsat imagery. *Remote Sens. Environ.* **2012**, *118*, 83–94.

39. Smith, M.; Milton, E. The use of Empirical Line Method to Calibrate Remotely Sensed Data to Reflectance. *Int. J. Remote Sens.* **1999**, *20*, 2653–2662.
40. Du, Y.; Teillet, P.; Cihlar, J. Radiometric Normalization of Multitemporal High-resolution Satellite Images with Quality Control of Land Cover Change Detection. *Remote Sens. Environ.* **2002**, *82*, 123–134.
41. Jensen, J.R. *Introductory Digital Image Processing: A Remote Sensing Perspective*, 3rd ed.; Prentice Hall: Upper Saddle River, NJ, USA, 2005; pp. 337–406.
42. Pulighe, G.; Baiocchi, V.; Lupia, F. Horizontal accuracy assessment of very high resolution Google Earth images in the city of Rome, Italy. *Int. J. Digital Earth.* **2015**.
43. Eastman, J.R.; Filk, M. Long sequence time series evaluation using standardized principal components. *Photogramm. Eng. Remote Sens.* **1993**, *59*, 991–996.
44. Congalton, R.G. A review of assessing the accuracy of classifications of remotely sensed data. *Remote Sens. Environ.* **1991**, *37*, 35–46.
45. Bastiaanssen, W.; Menenti, M.; Feddes, R.; Holtslag, A. A remote sensing surface energy balance algorithm for land (SEBAL) 1. Formulation. *J. Hydrol.* **1998**, *212*, 198–212.
46. Bastiaanssen, W.; Pelgrum, H.; Wang, J.; Ma, Y.; Moreno, J.; Roerink, G.; van der Wal, T. A remote sensing surface energy balance algorithm for land (SEBAL) 2. Validation. *J. Hydrol.* **1998**, *212*, 213–229.
47. Allen, R.G.; Pereira, L.A.; Raes, D.; Smith, M. *Crop Evapotranspiration*; FAO Irrigation and Drainage Paper 56; Food and Agriculture Organization of the United Nations (FAO): Rome, Italy, 1998; p. 293.
48. Allen, R.G.; Pereira, L.A. Estimating crop coefficients from fraction of ground cover and height. *Irrig. Sci.* **2009**, *28*, 17–34.
49. FAO (Food and Agriculture Organization of UN), Land and Water Division: ETo Calculator. Available online: http://www.fao.org/nr/water/eto.html (accessed on 22 October 2015).
50. Abu-Awwad, A.; Blair, S. Economic Efficiency of Water Use by Irrigated Crops in Al'Azraq Area. *Jordan J. Agric. Sci.* **2013**, *9*, 525–543.
51. Mazahrih, N.T. Evapotranspiration Measurements and Modeling for Bermuda Grass, Alfalfa, Cucumber, and Tomato Grown under Protected Cultivation in the Central Jordan Valley. Ph.D. Thesis, The University of Jordan, Amman, Jordan, 2011.
52. Suwwan, M.; Judah, O. Influence of Plastic Mulching on Growth, yield and Soil Moisture Conservation in Plastic House Tomatoes. *Dirasat* **1985**, *12*, 21–32.
53. Al-Bakri, J.T.; Taylor, J.C.; Brewer, T.R. Monitoring land use change in the Badia transition zone in Jordan using aerial photography and satellite imagery. *Geogr. J.* **2001**, *167*, 248–262.
54. JTN (The Jordan Times). Satellite Remote Sensing Uncovers violations on Underground Water Resources. Available online: http://www.jordantimes.com/news/local/satellite-remote-sensing-uncovers-violations- underground-water-resources#sthash.f4mMEUaF.dpufhttp://www.jordantimes.com/news/local/satellite-remote-sensing-uncovers-violations-underground-water-resources (accessed on 25 October 2015).

55. Hamdan, I. *Hydrogeological and Hydrological Investigation for Azraq Basin—North East Jordan*; Deutsche Gesellschaft für Internationale Zusammenarbeit (GIZ): Amman, Jordan, 2010.
56. Campos, I.; Neale, C.M.U.; Calera, A.; Balbontín, C.; González-Piqueras, J. Assessing satellite-based basal crop coefficients for irrigated grapes (*Vitis vinifera L.*). *Agricul. Water Manag.* **2010**, *98*, 45–54.
57. Salman, A.Z.; Al-Karablieh, E. Measuring the willingness of farmers to pay for groundwater in the highland areas of Jordan. *Agricul. Water Manag.* **2004**, *68*, 61–76.
58. Margane, A.; Al-Qadi, M.; Al-Kurdi, O. *Updating the Groundwater Contour Map of the A7/B2 Aquifer in North Jordan*; Technical Cooperation Project 'Syrian Refugee Response', Technical Report No. 1; Bundesanstalt für Geowissenschaften und Rohstoffe (BGR) and Ministry of Water and Irrigation, Jordan (MWI): Amman, Jordan; p. 129.
59. El-Naqa, A.; Al-Momani, M.; Kilani, S.; Hammouri, N. Groundwater Deterioration of Shallow Groundwater Aquifers Due to Overexploitation in Northeast Jordan. *Clean* **2007**, *35*, 156–166.
60. Law Amending the Law of the Water Authority (In Arabic). Law number 22, 2014. Jordan Prime Ministry, Official Gazette, Amman, Jordan, 2014; pp. 3385–3390.

Partitioning of Cotton Field Evapotranspiration under Mulched Drip Irrigation Based on a Dual Crop Coefficient Model

Fuqiang Tian, Pengju Yang, Hongchang Hu and Chao Dai

Abstract: Estimation of field crop evapotranspiration (ET_c) and its partitioning into evaporation and transpiration, are of great importance in hydrological modeling and agricultural water management. In this study, we used a dual crop coefficient model SIMDualKc to estimate the actual crop evapotranspiration ($ET_{c\ act}$) and the basal crop coefficients over a cotton field in Northwestern China. A two-year field experiment was implemented in the cotton field under mulched drip irrigation. The simulated $ET_{c\ act}$ is consistent with observed $ET_{c\ act}$ as derived based on the eddy covariance system in the field. Basal crop coefficients of cotton for the initial, mid-season, and end-season are 0.20, 0.90, and 0.50, respectively. The transpiration components of $ET_{c\ act}$ are 96% (77%) and 94% (74%) in 2012 and 2013 with (without) plastic mulch, respectively. The impact of plastic mulch cover on soil evaporation is significant during drip irrigation ranging from crop development stage to mid-season stage. The extent of the impact depends on the variation of soil moisture, available energy of the soil surface, and the growth of the cotton leaves. Our results show that the SIMDualKc is capable of providing accurate estimation of $ET_{c\ act}$ for cotton field under mulched drip irrigation, and could be used as a valuable tool to establish irrigation schedule for cotton fields in arid regions as Northwestern China.

Reprinted from *Water*. Cite as: Tian, F.; Yang, P.; Hu, H.; Dai, C. Partitioning of Cotton Field Evapotranspiration under Mulched Drip Irrigation Based on a Dual Crop Coefficient Model. *Water* **2016**, *8*, 72.

1. Introduction

Crop evapotranspiration (ET_c) consumes a large amount of irrigation water, especially in arid areas; thus, accurate estimation of evapotranspiration is the basis of hydrological modeling and agricultural water management [1]. Transpiration (T) through plant stomata is a desirable component because this process is usually associated with plant productivity; evaporation (E) over bare soil is usually considered as water loss but sometimes also provides a benefit for maintaining a micro-climate around the crop under conditions of high irrigation levels [2]; therefore, ET_c partitioning is essential for agriculture water resource management as well as hydrological modeling [3,4]. Common ways to partition ET_c include experimental methods or simulation methods. For the experimental methods, lysimeters [5,6], soil water budget [7], sap flow [8,9], and stable isotope [10,11] methods are usually adopted to evaluate *E* and *T* individually in previous studies.

However, these experimental methods usually require costly equipment and can only be carried out at point scale, which provides poor spatial representation [10]. Therefore, a number of models have been developed to predict *E* and *T* conveniently and to simulate evapotranspiration for agricultural management [12–14]. In these models, the FAO-56 crop coefficient reference evapotranspiration methods [12], including single and dual crop coefficient methods, are commonly used to calculate ET_c [6,15–17]. The FAO-56 dual crop coefficient method separately calculates transpiration and evaporation, which has been widely used in agriculture science [5,18–23]. The SIMDualKc model is based on the FAO-56 dual crop coefficient approach and combined with the hydrological extension for complete water balance, which can be used to support general irrigation scheduling needs [24]. The modelemploys a graphic- and menu-driven user interface that can serve as a convenient and effective tool to calculate actual crop ET_c [24]. The model has been used worldwide for different crops, such as wheat, maize, barley, soybean, cotton, vineyard, and other kinds of ecosystems [22,23,25–30]; the model also exhibits good adaptation and efficiency under various irrigation methods [22–24,26,31] and ground cover [32–34]. Hence, the model was utilized in our study to implement the use of the dual crop coefficient of cotton under mulched drip irrigation in Northwest China. The model should be properly calibrated and validated before use when management options have not been initially tested, and the observed soil moisture and ET_c are common variables used for this purpose [22,30,35].

The eddy covariance (EC) system, which can directly measure ET_c, is widely used in the field and is recognized as the standard method to measure ET_c [1,36–38]. This system can accurately and continuously estimate crop coefficients in real time [11]. Studies on ET_c and crop coefficient by EC, have been conducted on cropland, as well as other ecosystems [39,40]. In our study, the EC system was used to measure ET_c in 2012 and 2013, and the measurements were used to calibrate the SIMDualKc model.

Cotton is an important and widely cultivated fiber crop in the United States, India, Pakistan, Uzbekistan, and China [41]. In 2012, the cultivated area for cotton production in Xinjiang Uygur Autonomous Region occupied more than 50% of China [42]. Xinjiang is in an arid area, thus, irrigation is critical for cotton growth [43–45]. Increased water consumption for irrigation leads to groundwater overexploitation and surface water overuse, which accelerates the deterioration of ecological environment especially in the downstream of inland rivers; excessive irrigation in arid areas can also induce secondary salinization which is harmful to the growth of crops [46–48]. The limited water resources and salinization constrain agriculture development in the region; thus, drip irrigation and plastic mulch are developed to conserve water [3]. Mulched drip irrigation is a useful and economic way to improve the soil micro-climate condition and increase the water efficiency [15,27,49,50]. Agricultural water consumption accounts for more than 90% of all water withdrawal from the Kaidu-Kongqi River Basin (a source basin of Tarim River in Xinjiang), and most cotton fields in the Kaidu-Kongqi River Basin are cultivated under mulched drip irrigation [38]. Mulched drip irrigation is a potential water-saving method also in other districts in China [20], as well as in Central Asia, where cotton is also heavily grown; these regions also exhibit a similar dry climate [41,51,52].

A few researchers have investigated dual crop coefficients under drip irrigation with mulching [15,27,49,53]. Nevertheless, research on cotton dual crop coefficient under mulched drip irrigation has been rarely reported. This study aimed to use an EC system and a SIMDualKc model to partition cotton field evapotranspiration under mulched drip irrigation and to evaluate the effect of plastic mulch on ET_c. The SIMDualKc model was calibrated on the basis of the observed evapotranspiration data; recommended parameters, including crop coefficients proposed by the FAO [12], were also evaluated to determine applicable values in our study area.

The objectives of this study are as follows: (a) to validate the SIMDualKc model by using the observed data obtained by eddy covariance system; (b) to analyze the temporal variations of dual crop coefficients and the partition of cotton field ET_c under mulched drip irrigation in different growth periods based on the simulated results; and (c) to estimate the effect of plastic mulch on ET_c.

2. Materials and Methods

2.1. Experimental Site

The study area is located in the Tsinghua University-Korla Oasis Eco-hydrology Experimental Research Station, which is 22 km away from the town of Xiborni in Korla City, Xinjiang Uygur Autonomous Region (Figure 1). This area lies on the alluvial plain of Kaiqu-Kongqi River, at the southern foot of the Tian Shan Mountains. The average elevation is 897–902 m. The study area has a continental desert climate with a warm temperate zone, scarce precipitation and intense potential evapotranspiration. The annual mean precipitation is approximately 60 mm, and annual mean potential evaporation is approximately 2800 mm. Average annual temperature is 11.5 °C and sunshine duration is 3036 h. The average relative humidity, net radiation, and wind speed during the cotton growth period were 40%, 110 $w \cdot m^{-2}$, and 1.90 $m \cdot s^{-1}$, respectively. We carried out a two-year (2012–2013) experiment in this field. The reference evapotranspiration (ET_o) is calculated based on the Peman-Monteith equation (as suggested by the FAO-56 [12]), using the observations from an automatic weather station within the experiment field. The variations of ET_0 are shown in Figure 2.

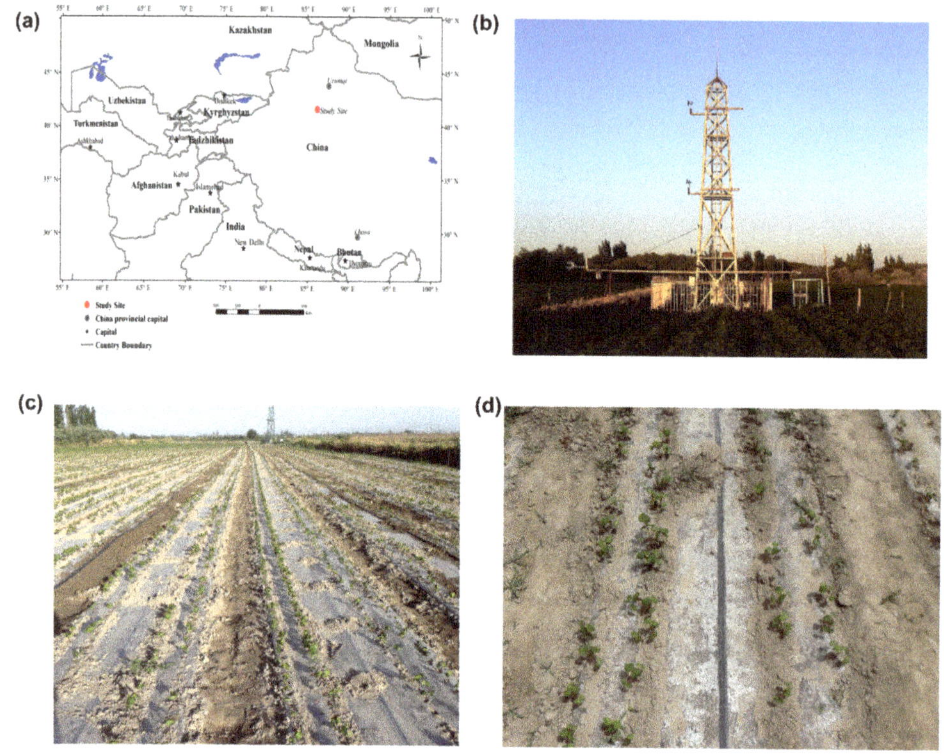

Figure 1. The location and eddy covariance system of the study area: (**a**) the location of the study areas; (**b**) the observing tower in the field; (**c**,**d**) the cotton under mulched drip irrigation and one-film, one-pipe, four-row mode.

The groundwater table level was measured using an automatic water depth sensor (model HOBO U20 Titanium Water Level Data Logger, Onset Computer Corporation, Inc., Pocasset, MA, USA) installed in a groundwater well near the observation tower. The groundwater level varied from 1.0 m to 4.0 m throughout the entire growth period, reaching a high value at the beginning of cotton growth after flood irrigation. The flood irrigation was usually about two weeks before seeding, and from 25–29 March 2012 and from 22–26 March 2013 [54]. The total amount of the spring irrigation is about 375 mm for both the two years. The texture of the soil is loam, which is made up of 30% sand, 5% silt, and 65% loam. The soil bulk density of the experiment field is from 1.40 g·cm^{-3} to 1.64 g·cm^{-3} in the 1.5 m soil profile. The depth of frozen soil is approximately 60 cm. The saturated water content of soil is nearly 0.42 [50].

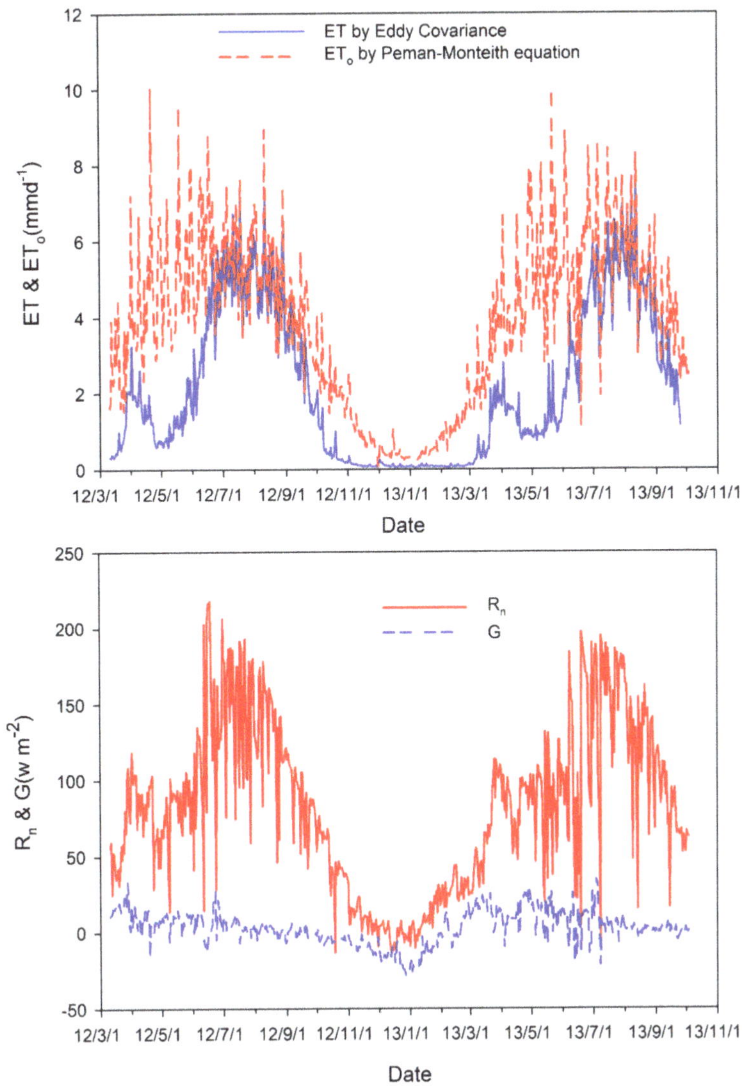

Figure 2. Evapotranspiration and energy flux measured by eddy covariance.

2.2. Cotton Planting

Cotton (*Gossypium hirsutum* L.) is planted under mulched drip irrigation in the entire growth period and the experimental field covers an area of 3.48 ha (see Figure 1). Mulched drip irrigation involves plastic mulch covering the drip tape and the surface soil. The cotton planting and drip irrigation tape employs one-film, one-pipe, four-row mode [38,44]. The drip irrigation tape is located beneath the middle of the mulch. Two cotton rows are symmetrically distributed on both sides of the tape. The mulch width is 110 cm, and the inter-mulch zone width is 40 cm. The widths of the cotton row spaces are 20, 44, and 20 cm (Figure 1). The irrigation schedules in the two-year experimental period are summarized in Table 1. Irrigation was performed roughly once a week starting in mid-June and ending in late August for both years. The irrigation lasts for about 12 h for each irrigation event and the flow is about 15 $m^3 \cdot h^{-1}$ with the meter measurement. The irrigation tape is made of plastic and the external diameter of the irrigation tape is 16mm, the wall thickness is about 0.2 mm, the space between the drip holes is about 0.3 m, the work pressure is 0.10 MPa, and the flow rate of each tape is 3.2 $L \cdot h^{-1}$.

Table 1. Irrigation schedules adopted for experiments in 2012 and 2013.

Growth Stage		Squaring Stage				Flowering Stage				Bolls Stage			
2012	Irrigation date	6–10 & 6–14 *	6–21	6–28	7–6	7–15	7–26	8–4	8–8	8–12	8–17	8–22	8–27
	Volume (mm)	65.2	34.4	35.3	36.8	33.3	44.1	40.0	59.3	46.7	42.2	50.8	52.2
2013	Irrigation date	6–13	6–20	6–28	7–3	7–9	7–16	7–27	8–2	8–8	8–13	8–18	8–22
	Volume (mm)	48.5	32.3	30.7	39.2	76.0	46.5	39.0	53.1	63.2	51.8	52.1	58.8

Note: * The total irrigation on 10 and 14 June 2012 was about 65.2 mm.

Cotton was sown in mid-April and harvested in early October. The growth stages of cotton are shown in Table 2. The seeds were sown with a spacing of 0.1 m between rows, and the planting densities were approximately 89,800 and 95,700 plants ha^{-1} in 2012 and 2013, respectively. The planting density in 2012 was approximately 6% less than that in 2013 because of sandstorm and freezing damage that caused a low emergence rate of cotton. The crop height, root depth, and leaf area index (LAI) were measured at an interval of two weeks, and the main crop physiological parameters of the growth stage are presented in Table 3. All of the leaves were stripped from each plant, and the leaf area was then obtained by directly scanning all of the leaves using a leaf area meter (model Yaxin-1241, Beijing Yaxinliyi Science and Technology Co., Ltd., Beijing, China). The LAI was calculated by dividing the leaf area by the area that each plant occupied [38].

Table 2. Cotton growth stages of 2012 and 2013.

Cotton Growth Stages	Phenological Growth Stages	2012	2013
Planting/initiation	Emergence & Squaring stage	23 April–23 May	22 April–5 June
Rapid growth	Squaring & Flower stage	24 May–6 July	6 June–14 July
Midseason	Flower & Boll stage	7 July–2 September	15 July–28 August
Maturity	Boll stage	3 September–7 October	29 August–4 October

Table 3. Main crop parameters of the growth stage.

Year	Physiological Parameter	Planting	Start Crop Development	Start Mid-Season	Start Late-Season	Harvest
2012	Root depth (m)	0	0.4	0.70	0.70	0.70
	Crop height (m)	0	0.20	0.76	0.76	0.76
	Fraction of ground cover	0	0.2	0.50	0.95	0.85
2013	Root depth (m)	0	0.4	0.62	0.62	0.65
	Crop height (m)	0	0.4	0.62	0.62	0.67
	Fraction of ground cover	0	0.20	0.50	0.95	0.85

2.3. EC System

The EC system was installed in a 10 m-high stationary tower (see Figure 1 for more details). The main components of EC system are as follows: a fast response open-path infrared gas (H_2O and CO_2) analyzer (model EC150, Campbell Scientific Inc., Logan, UT, USA), a fast response 3D sonic anemometer (model CSAT3, Campbell Scientific Inc.), air temperature/humidity sensor (model HMP155A, Vaisala Inc., Woburn, MA, USA), a micro logger (model CR3000, Campbell Scientific Inc.) and net radiometer (model LITE2, Kipp and Zonen, Delft, The Netherlands). The abovementioned equipment was installed at a height of 2.25 m. The soil heat flux plates (model HFP01SC, Hukseflux, The Netherlands) were imbedded 0.05 m below the ground surface under the film-mulched zone and inter-film zone to obtain the soil heat flux (G). Data processing and energy closure have been discussed in the literature [38], and the data over two years from the EC system were used in this study. The latent heat flux was calculated by multiplying vertical velocity fluctuations by a scalar concentration fluctuation [55]:

$$\lambda ET = \lambda \rho_a \overline{w'q'} \tag{1}$$

where $\lambda\ ET$ is the latent heat flux ($W \cdot m^{-2}$), λ is the latent heat of vaporization ($J \cdot kg^{-1}$), ρ_a is the air density ($kg \cdot m^{-3}$), and $\overline{w'q'}$ is the covariance between fluctuations of vertical wind speed w' ($m \cdot s^{-1}$) and air humidity q' ($kg \cdot kg^{-1}$).

The azimuth angle of the CSAT3 sensor is 50° from true north. The maximum height of the crop is approximately 70 cm, which ensures that the EC systems maintain an appropriate footprint. The data were measured at a frequency of 10 Hz and the fluxes were computed in half-hour. Energy closure was used to evaluate the quality of the measurement data. The net radiation (R_n), G, sensible heat (H) and latent heat (LE) were all obtained through the EC systems. The slope of the energy balance equation ($LE + H = R_n - G$) for this site in 2012 and 2013 was 0.72 ($r^2 = 0.90$, $n = 21,886$) [38]. The ratio was similar to the values obtained in previous studies (0.70–0.90) [7]. Thus, the data can be regarded as reliable, considering the influence of the plastic mulch to the energy transport between the soil and atmosphere [56]. The calculated ET and variation of radiation and soil heat flux are shown in Figure 2.

2.4. SIMDualKc Model

The actual ET was simulated by using the SIMDualKc model, which is an irrigation scheduling simulation model. The actual crop ET is estimated by the dual crop coefficient equation as follows [12,24]:

$$ET_{c\ act} = (K_S K_{cb} + K_e) ET_o \quad (2)$$

where K_{cb} is the basal crop coefficient, K_s is the water stress reduction coefficient, K_e is the soil evaporation coefficient, $ET_{c\ \mathrm{act}}$ is the actual crop evapotranspiration and ET_o is the reference evapotranspiration (mm· day^{-1}), which is calculated using the FAO Penman-Monteith equation [12,18].

The model considers the different influence of irrigation and precipitation on the variation of soil evaporation. The K_e is divided into two parts, $K_e = K_{ei} + K_{ep}$, where K_{ei} is the soil evaporation coefficient induced by irrigation and precipitation that raise soil moisture, and K_{ep} is the soil evaporation coefficient induced by the precipitation only [57]. The effects of crop height, crop density, and canopy architecture on K_{cb} were calculated through the density coefficient (K_d) [24,32].

Crop management is also considered in the model, such as using mulch to decrease the evaporation; that is, the model considers the fraction of soil covered by the plastic sheet and estimated the influence compared to the fraction of ground cover [24].

The input data of the model consist of the following [24]:

(a) Soil data: the total available water (TAW, mm· m^{-1}). It can be calculated with the field soil content and wilting point soil moisture or use the suggested the values; amount and depth of the soil layers; effective depth of the evaporation layer (Z_e, m); readily and total evaporable water (REW and TEW, mm); and textural classes of that layer when the values are calculated by the model.

(b) Meteorological daily data: minimum and maximum air temperature, T_{max} and T_{min} (°C); reference evapotranspiration (ET_o, mm); minimum relative humility (RH_{min}, %); precipitation (P, mm); and wind speed at 2 m height (u_2, ms^{-1}).

(c) Crop data: data for the initial, crop development, mid-season, late-season and harvest or end-of-season growth stages; initial and end-of-season data of frozen soil; basal crop coefficient (K_{cb}) for the initial, mid-season and harvest growth stages; soil water depletion fraction without stress (p), and the fraction of ground cover (f_c), for all growth stages; root depths (Z_r, m) and crop height (h, m).

(d) Irrigation data: irrigation system; irrigation data; fraction of soil surface wetted by irrigation (f_w) and the depth of each irrigation.

(e) Other data: data used for the capillary rise and deep percolation equations; mulch data, including related management; active ground cover characteristics; and runoff data.

In this study, the effects of crop density, height, and canopy architecture on K_{cb} are evaluated using a K_d and the effect of using mulches is also included for assessing the water saving effect of plastic mulch covering [12]. The model contains a module which can consider the influence of the plastic mulch according to the actual situation. The model provides two alternative methods *i.e.*, a simplified procedure described in Doorenbos and Pruitt [58] and the parametric equation proposed by Liu *et al.* [1] to evaluate the deep percolation and capillary rise. Considering the available data (soil water parameters, LAI and water table depths) (Figure 3), the parametric equation proposed by Liu *et al.* [34] was chosen to calculate the deep percolation and capillary rise.

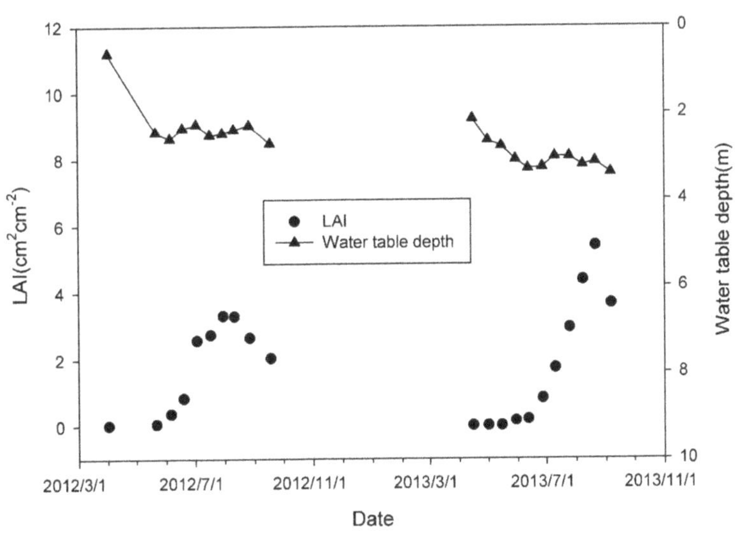

Figure 3. LAI and water table depth during the growing season for two years.

2.5. Model Calibration and Validation

The SIMDualKc model initially simulated the ET_{cct} through the table values of the crop (K_{cb} and p), the soil (Z_e, TEW and REW), and the percolation equation (a_p and b_p) suggested by the FAO-56 [12] and other previous studies [34]. The model was further calibrated to minimize the differences between the simulated and observed $ET_{c\ act}$. Model calibration utilized a trial and error procedure and a gradual change in the parameters from crop to soil, is described with more details in Rosa *et al.* [35]. The initial and calibrated values for the soil and crop parameters of the growth season are presented in Table 4. In our study, the data observed during the 2012 growth season was used to calibrate the model, and the observed data of 2013 was used for model validation. The real irrigation amount and respective data were used in simulation for calibration and validation.

Table 4. The initial and calibrated values of the crop and soil parameters of the growth stage.

Parameter	Initial Values [12,34]	Calibrated
	Crop coefficients	
$K_{cb\ \mathrm{ini}}$	0.15	0.20
$K_{cb\ \mathrm{mid}}$	1.15	0.90
$K_{cb\ \mathrm{end}}$	0.50	0.50
	Depletion fraction	
p_{ini}	0.65	0.70
p_{mid}	0.65	0.60
p_{end}	0.65	0.60
	Soil evaporation	
REW (mm)	8	8
TEW (mm)	20	33
Z_e (cm)	10	15
	Deep percolation	
a_p	408	390
b_p	−0.0173	−0.0173
	Capillary rise	
a_1	320.8	320.8
a_2	303.2	303.2
a_3	−0.15	−0.15
a_4	7.55	7.55
b_1	−0.16	−0.16
b_2	−0.54	−0.54
b_3	2.1	2.1
b_4	−2.03	−2.03

Several goodness-of-fit indicators were applied in previous studies to evaluate the model predictions [35,59]. The simulated and observed ET_{cct} were compared through the figure, and regression was also calculated throughout the growth season. The linear regression between simulate and observed $ET_{c\ act}$ was firstly been obtained. The determination coefficient r^2 of the regression is:

$$r^2 = \left[\frac{\sum_{i=1}^{n} (O_i - \overline{O})(S_i - \overline{S})}{\sqrt{\sum_{i=1}^{n} (O_i - \overline{O})^2} \sqrt{\sum_{i=1}^{n} (S_i - \overline{S})^2}} \right]^2 \quad (3)$$

where O_i and S_i ($i = 1, 2, \ldots, n$) represent the observed and simulated ET, and $\overline{O}$ and $\overline{S}$ are the corresponding mean values, when the value is close to 1.0, the predicted values are statistically close to the observed ones.

The indicators to evaluate the estimation errors are calculated as follows:

(a) The Nash–Sutcliffe model efficiency coefficient NSE, which is the ratio of the mean square error to the variance in the observed data [27]:

$$NSE = 1 - \frac{\sum_{i=1}^{n} (O_i - S_i)^2}{\sum_{i=1}^{n} (O_i - \overline{O})^2} \tag{4}$$

(b) The root mean square error, which characterizes the variance of the errors:

$$RMSE = \left[\frac{\sum_{i=1}^{n} (S_i - O_i)^2}{n} \right]^{0.5} \tag{5}$$

(c) The ratio RSR of the RMSE to the standard deviation of observed data (sd) that standardizes RMSE using the sd of observations:

$$RSR = \frac{\left[\sum_{i=1}^{n} (S_i - O_i)^2 \right]^{0.5}}{\left[\sum_{i=1}^{n} (\overline{S}_i - O_i)^2 \right]^{0.5}} \tag{6}$$

(d) The average absolute error, which expresses the magnitude of estimation errors in alternative to $RMSE$:

$$AAE = \frac{1}{n} \sum_{i=1}^{n} |S_i - O_i| \tag{7}$$

(e) The average relative error, which indicates the size of errors in relative terms and is expressed as a percentage:

$$ARE = \frac{100}{n} \sum_{i=1}^{n} \left| \frac{S_i - O_i}{O_i} \right| \tag{8}$$

(f) The percent bias, which indicates measures the average tendency of the simulated data to be larger or smaller than their corresponding observations:

$$PBIAS = 100 \frac{\sum_{i=1}^{n} |S_i - O_i|}{\sum_{i=1}^{n} O_i} \tag{9}$$

3. Results and Discussion

3.1. Model Calibration and Validation

The initial soil related parameters were set according to local conditions. In common farming practice, the spring flush was employed two weeks before cultivation to leach soil salt, which significantly increases soil moisture. Afterward, however, intensive potential evaporation causes a remarkable depletion of surface soil water [17,50]. Accordingly, the initial depletion of the evaporable layer was set at 80% of TEW, and the initial depletion of the root zone of TAW was estimated at 20% in 2012 and 2013.

The capillary rising parameters (a_1 = 320.8, a_2 = 303.2, a_3 = −0.15, a_4 = 7.55 and b_1 = −0.16, b_2 = −0.54, b_3 = 2.1, and b_4 = −2.03) and deep percolation equation (a_p = 408, b_p = −0.0173) are adopted from Liu *et al.* [34]. LAI and water table depth are shown in Figure 3. The spring flush was applied in March, and the initial water table depth was approximately 1–2 m in April, decreased to approximately 3–4 m in June and varied in the drip irrigation period from June to August during the two experimental years.

The daily variation of the simulated (calibration and validation) and observed $ET_{c\ \mathrm{act}}$ (mm· day^{-1}) are shown in Figure 4. The results show that the simulation results fit the observed data well and no significant biases were detected. The regression coefficient is 0.95 and 1.08 for 2012 and 2013, respectively. The calibrated parameters of the model are presented in Table 5. The parameters p are close to the suggested values by Allen *et al.* [12]. $K_{cb\ \mathrm{ini}}$, $K_{cb\ \mathrm{mid}}$ is 33% more, 25% less than the suggested values by Allen *et al.* [12], while the $K_{cb\ \mathrm{end}}$ is nearly the same. The minimal variation in the parameters of deep percolation equation in the calibration showed the feasibility of the adopted equation.

The indicators of goodness-of-fit relative to the model tests are presented in Table 5. The observed $ET_{c\ \mathrm{act}}$ was 3.3 mm· day^{-1} and 3.2 mm· day^{-1} in 2012 and 2013, respectively, and the simulated $ET_{c\ \mathrm{act}}$ was 3.4 mm· day^{-1} and 3.2 mm· day^{-1}, which were very close to the observed $ET_{c\ \mathrm{act}}$. NSE, r^2, $RMSE$, AAE, ARE, RSR, and $PBIAS$ in 2012 were 0.89, 0.87, 0.68 mm· day^{-1}, 0.50 mm· day^{-1}, 21.8%, 0.38, and 15.0%, respectively, and 0.84, 0.87, 0.72 mm· day^{-1}, 0.50 mm· day^{-1}, 18.1%, 0.40 and 15.4% in 2013. The goodness-of-fit of the model in this area is close to the study for the maize and wheat in Northern China [22] and the study for peach orchard in Portugal [28].

Table 5. Indicators of goodness of fit relative to the model tests of the crop evapotranspiration for the cotton under mulched drip irrigation.

Year	b *	r^2	RMSE (mm· day^{-1})	NSE	RSR	ARE (%)	PBIAS (%)	AAE (mm· day^{-1})
2012 (calibration)	0.95	0.87	0.68	0.89	0.38	21.8	15.0	0.50
2013 (validation)	1.03	0.87	0.72	0.84	0.40	18.1	15.4	0.50

Note: * The b is the slope of regression line.

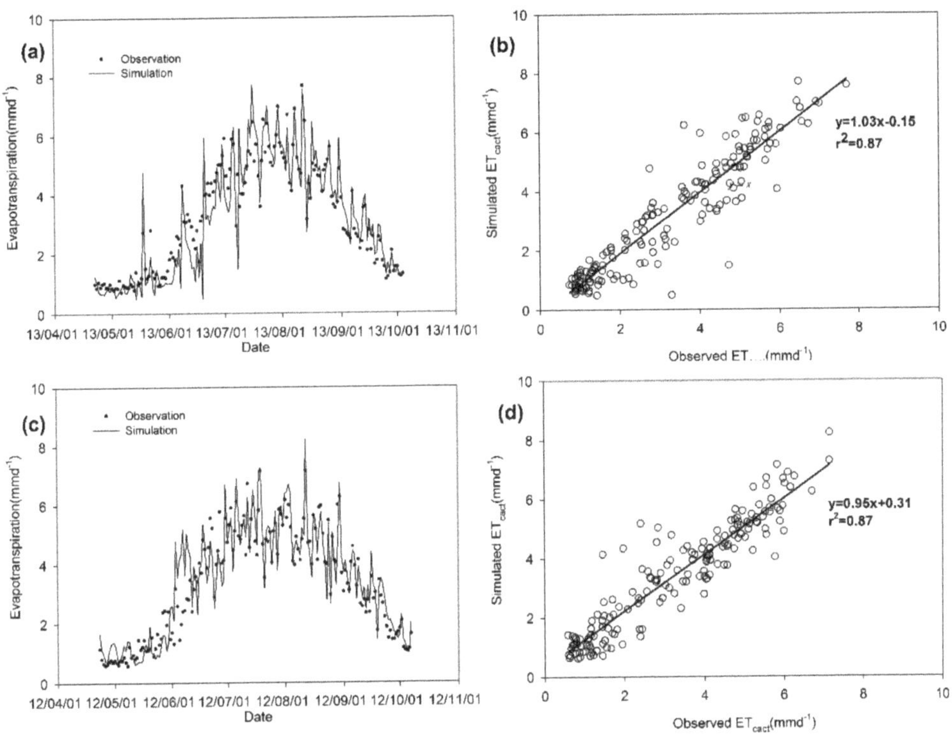

Figure 4. The comparison between observed (EC) and simulated $ET_{c\ \mathrm{act}}$: (**a**,**b**) the results of 2012 (calibration); and (**c**,**d**) the results of 2013 (validation).

3.2. Crop Coefficients

The seasonal variations of K_{cb}, K_{cbct}, and K_e are presented in Figure 5. The calibrated basal crop coefficient $K_{cb\ \mathrm{ini}}$ = 0.20 was higher than the value of 0.15 (Table 4) proposed by Allen *et al.* [12]. In fact, $K_{cb\ \mathrm{ini}}$ is sensitive to irrigation management [16]. In our study area flood irrigation is usually implemented for approximately two weeks before sowing which rapidly increases the soil moisture and, thus, evaporation. Moreover, the soil was ploughed in the sowing period which increased the moisture of the soil surface. The calibrated mid-season basal crop coefficient $K_{cb\,\mathrm{ini}} = 0.90$ was lower by 20% than the proposed value of 1.10–1.15. The difference can be attributed to the influence of plastic mulch which may decrease the suggested K_{cb} by 10%–30% [12]. The end basal crop coefficient $K_{cb\,\mathrm{end}} = 0.50$ which is within the range of proposed values of 0.40–0.50. The influence of plastic mulch on $K_{cb\,end}$ is insignificant compared to $K_{cb\,\mathrm{mid}}$. In the end growth stage, the ground was fully covered with the cotton, and when the irrigation ended, the influence of plastic mulch was negligible. The K_{cb} increased with the growth stage from start to end, whereas the $K_{cb\,\mathrm{act}}$ maintained the same values as the calibrated K_{cb} except in the initial stage when the irrigation started. The soil moisture was lower in this periods which may cause the soil water stress of the cotton. The K_e values of the cotton were high in the initial stage when the soil moisture was high during sowing (Figure 5).

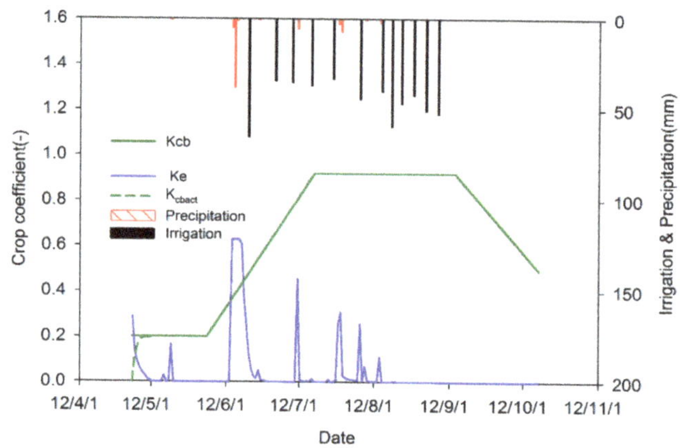

(a)

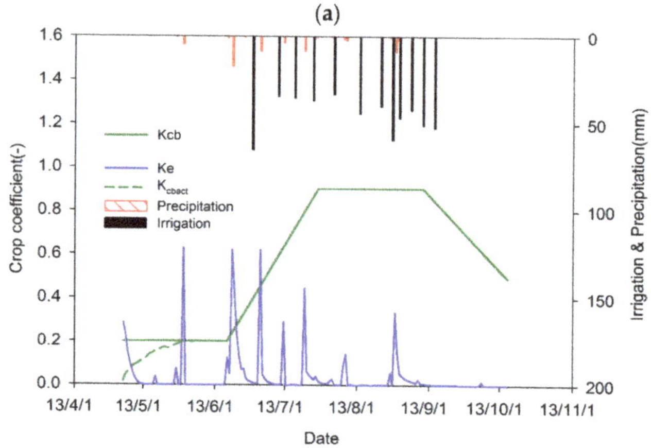

(b)

Figure 5. Seasonal variation of K_{cb}, $K_{cb\,\mathrm{act}}$, K_e, irrigation, and precipitation for cotton: (**a**) in calibration period (2012); and (**b**) validation period (2013).

The parameter K_e was divided into two components in the model, namely, K_{ei} which was related to the exposed fraction of soil wetted by both irrigation and precipitation, and K_{ep}, which was related to the exposed fraction of soil wetted by precipitation only [24]. Considering drip irrigation, the soil wetted conditions with irrigation and precipitation are different. The variations of K_e with the precipitation and irrigation are shown in Figure 5. The large variation range of K_e before the irrigation is mainly because of the precipitation. K_e was sensitive to the irrigation in the initial period which is due to the rapid increase in the surface soil moisture when the irrigation started. The influence of irrigation on K_e decreased in the late middle stage as it was affected by the growth of the cotton leaves. The inter-film zone was covered by the leaves in late July, decreasing the soil evaporation. After this stage, the K_e value stayed at low levels (close to zero), which implied that almost no soil evaporation occurred in that stage.

Allen [18] reported that $K_{cb\,\mathrm{mid}}$ equals to 1.0 in Turkey and suggested that the 15% reduction of the tabled values in the FAO-56 report were due to low planting density and non-uniform irrigation. Rosa *et al.* [24] proposed the $K_{cb\,\mathrm{mid}}$ equals to 1.15 with furrow irrigation in Uzbekistan. Howell *et al.* [60] proposed that $K_{cb\,\mathrm{mid}}$ equal to 1.23 with the lysimeters in the Northern Texas High Plains of the USA with sprinkle irrigation. $K_{cb\,\mathrm{mid}}$, in our study, was less than the two cases above by approximately 10% and 25%, respectively. The differences can be mainly attributed to the influence of plastic mulch and location.

3.3. Partitioning of Evapotranspiration

The $ET_{c\,\mathrm{act}}$ (mm$\cdot$ day^{-1}) and its components in different growth stages during the two-year experimental period which simulated by the model are shown in Table 6 and Figure 6. In the initial stage, T comprised 80%–90% of the entire $ET_{c\,\mathrm{act}}$. In the crop development stage, T accounted for 90% of the entire $ET_{c\,\mathrm{act}}$, increasing with crop growth to 100% during the late season. For the full growth season, T under plastic mulch averaged 96% and 94% in 2012 and 2013, respectively. The components of $ET_{c\,\mathrm{act}}$, without mulch, are presented in Table 6 according to simulation. The transpiration components of initial stage, crop development stage, mid-season, and late season in the two-year experimental period (2012–2013) are 77%, 61%, 79%, and 94%, respectively. The ratio of $E/ET_{c\,\mathrm{act}}$ of cotton without mulch was significantly higher than that with mulch, with the highest ratio observed during the crop development stage. The plastic mulch is the main reason for the rapid decrease in soil evaporation. In the initial stage, the soil moisture is low (Figure 7) and the values of E and T were both low. When irrigation started in the crop development stage, T and T both increased, with T increasing more slowly than E because of undeveloped leaves and, therefore, the transpiration component becomes lower in this stage. In the mid-season, the soil was covered by leaves, and the available energy of the soil evaporation decreased, thereby decreasing $E/ET_{c\ \mathrm{act}}$. Irrigation usually ends by the end of the mid-season, and the lower soil moisture further decreases the ratio of $E/ET_{c\ \mathrm{act}}$.

Table 6. $ET_{c\ \mathrm{act}}$ and their components for different growth stages.

Growth Stages	Observation	2012								Observation	2013							
		Plastic Mulch				No Mulch					Plastic Mulch				No Mulch			
	ET *	ET	T	E	T/ET	ET	T	E	T/ET	ET	ET	T	E	T/ET	ET	T	E	T/ET
Initial	0.9	1.0	0.9	0.1	88.3	1.1	0.9	0.2	78.4	1.3	1.1	0.9	0.2	81.8	1.2	0.9	0.3	76.1
Crop development	3.5	3.5	3.2	0.3	90.3	4.9	3.2	1.8	64.3	4.0	3.6	3.1	0.5	87.0	5.3	3.1	2.2	58.4
Mid-season	5.0	5.1	5.0	0.1	98.3	6.2	4.9	1.2	80.0	5.2	5.4	5.3	0.1	97.5	6.8	5.3	1.5	78.2
Late season	2.5	2.7	2.7	0.0	100.0	2.8	2.6	0.2	93.4	2.4	2.7	2.6	0.0	99.6	2.7	2.5	0.2	94.2
Full crop season	3.3	3.4	3.3	0.1	95.9	4.2	3.3	1.0	77.0	3.2	3.2	3.0	0.2	93.6	4.0	3.0	1.0	74.1

Notes: * The units of the value are below: ET: Evapotranspiration (mm· day^{-1}); T: Transpiration (mm· day^{-1}); E: Evaporation (mm· day^{-1}); T/ET: Fraction of transpiration to evapotranspiration (%).

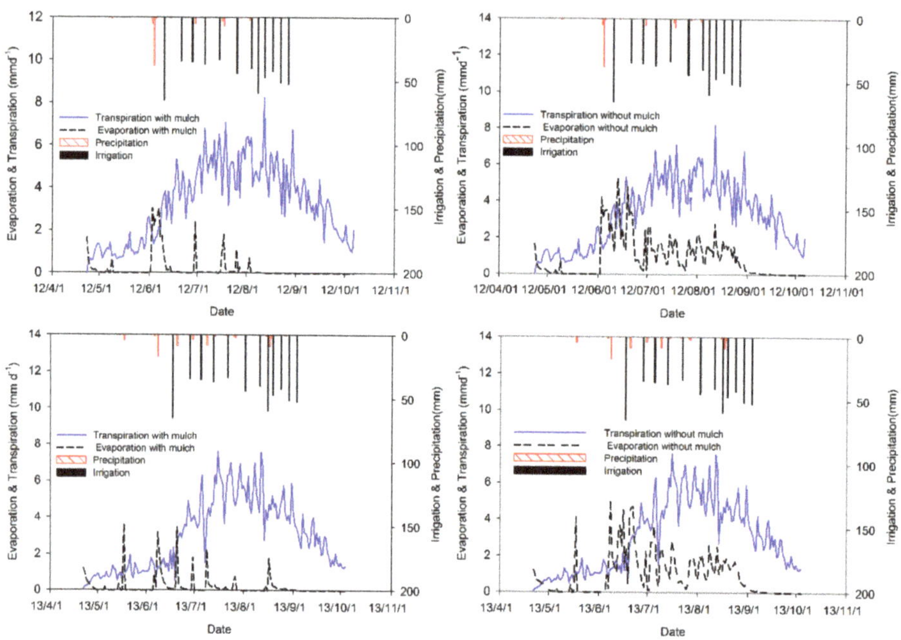

Figure 6. Daily variation of evaporation and transpiration for cotton in 2012–2013 (with and without mulch).

Sap flow gauges were used to measure individual plant transpiration in our experimental station in 2012, and the obtained ratio $T/ET_{c\ \mathrm{act}}$ is approximately 87% in June, 82% in August, and nearly 100% in September [38]. Martins *et al.* [26] indicated that E of $ET_{c\ \mathrm{act}}$ in sprinkler and drip experiments under mulched soil for maize ranged from approximately 91% to 94%, and Rosa *et al.* [35] reported that $T/ET_{c\ \mathrm{act}}$

in furrow-irrigated cotton were 90% and 83% in two different years. These results are similar to those of our study in Xinjiang, and are higher than the other studies in the areas with similar climate, e.g., ~80% in Uzbekistan by Qureshi *et al.* [52] and 56%–68% by Forkutsa *et al.* [61] also in Uzbekistan.

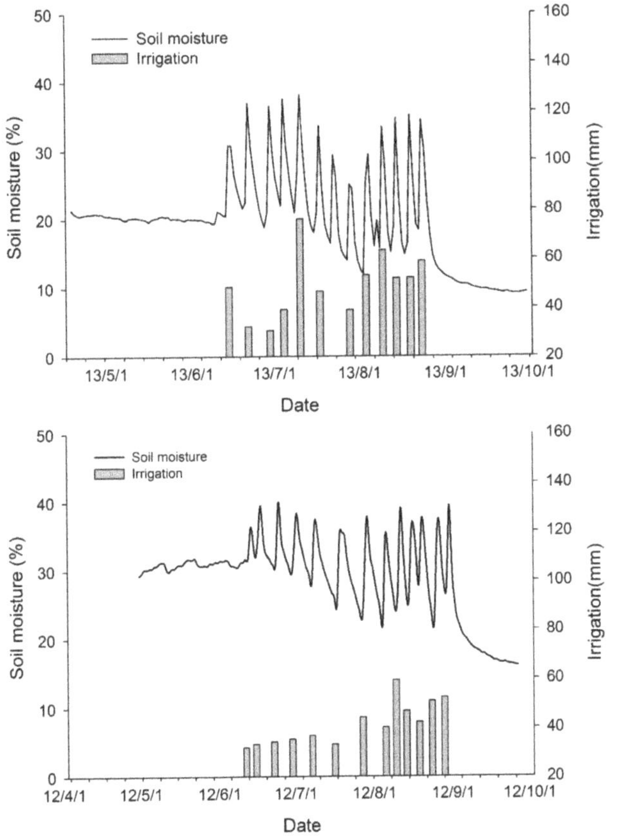

Figure 7. Daily variation of soil moisture under irrigation for cotton in 2012–2013.

3.4. *Influence of Plastic Mulch*

The influence of plastic mulch cover on $ET_{c\ \mathrm{act}}$ was simulated by the model, and the results are shown in Table 6 and Figures 6 and 8. The observation $ET_{c\ \mathrm{act}}$ data, with no mulch, was not obtained in our research. The calibrated and validated parameters with mulch was used to simulate the $ET_{c\ \mathrm{act}}$ with no mulch. The result is used to analyze the influence of mulch on $ET_{c\ \mathrm{act}}$. Plastic mulch has been proven to be able to increase soil temperature and moisture [36], as well as to conserve water [15,26].

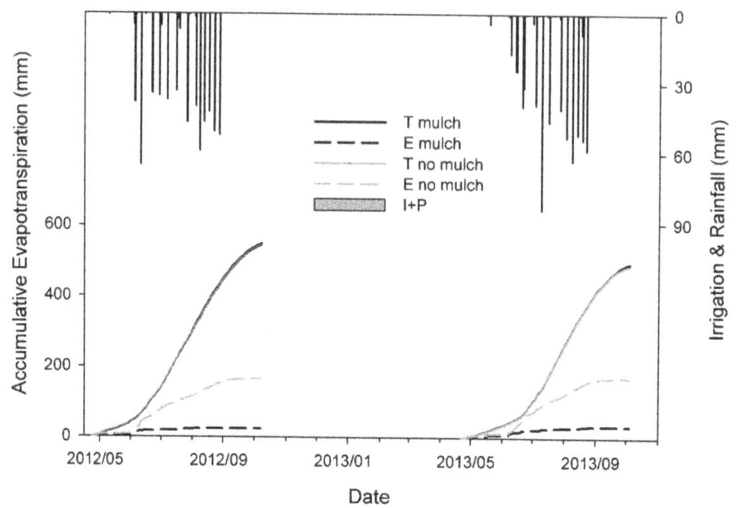

Figure 8. Evaporation and transpiration by the SIMDualKc during growing season for two years.

T values under plastic mulch cover were 549.2 mm and 495.0 mm in 2012 and 2013, respectively, and 545.2 mm and 493.2 mm without mulch cover, respectively. No significant difference was detected. E values under plastic mulch cover were 23.5 mm and 33.9 mm, in 2012 and 2013, respectively, and 163.1 mm and 170.8 mm without mulch cover, respectively. Thus, E increased by approximately 139.6 mm and 137.0 mm without mulch cover. For the entire growing season, the T components of $ET_{c\ \mathrm{act}}$ were approximately 96% under plastic mulch and 77% without mulch in 2012; the T components of $ET_{c\ \mathrm{act}}$ were approximately 94% and 74% in 2013, respectively.

The influence of plastic mulch cover on E is significant during the drip irrigation from the crop development stage to the mid-season stage. These phenomenon can be attributed to the variation of soil moisture and the available energy of the soil surface with the growth of the cotton leaves. Soil moisture is low in the initial stage and E is at a low rate; therefore, no remarkable disparity was detected between mulched and unmulched soil. Under initial irrigation, the soil moisture increases rapidly to promote soil evaporation, which decreases with plastic mulch, compared with unmulched soil. In the mid-season stage, the cotton leaves were enlarged, and the ground was covered by these leaves. The available energy of the soil surface was the key factor affecting soil evaporation instead of soil moisture. The sensitivity of soil evaporation to irrigation with and without mulch decreased, and the evaporation of the mulched soil largely declined to nearly 0 in combination with the impact of the plastic mulch.

The amount of percolation during the drip irrigation was approximately 152.1 mm and 223.6 mm for 2012 and 2013, respectively. The exchange water fluxes between soil and groundwater reservoir at 90 cm were 133.4 and 252.5 mm, respectively, as indicated in the water balance model [38]. These results indicate that simulated percolation is reasonable.

4. Conclusions

ET measurements using the eddy covariance system were conducted in a cotton field for two years, and the dual crop coefficient model SIMDualKc was successfully applied in the study area.

The goodness of fit showed that the predicted and observed values matched quite well. The regression coefficients were 0.95 and 1.08 for the two years, respectively. The $T/ET_{c\,\mathrm{act}}$ ratio increased with mulched drip irrigation; the ratios were equal to 96% and 94% during the two-year growth seasons, respectively. T occupied approximately 100% in mid-season and maturity stages. The SIMDualKc model can be used to assess ground cover, and the influence of plastic mulch is simulated via two setups; namely, with mulch and without mulch. The simulation results showed that the mulched cover reduced E by approximately 139.6 and 137.0 mm in 2012 and 2013, but barely affected T. The study showed that percolation amounts were approximately 152.08 and 223.62 mm in 2012 and 2013, respectively.

According to our results, the SIMDualKc model can be used to support irrigation schedules for cotton under mulched drip irrigation in Northwest China and other areas with similar climate and irrigation methods. The model can also be a useful tool to evaluate the influence of water saving methods and plastic mulches. Moreover, the validated parameters and basal crop coefficient in this study can be helpful and valuable for the further applications in arid land of Central Asia.

Acknowledgments: This research was support by the National Science Foundation of China (NSFC 51190092, 51222901) and the Foundation of the State Key Laboratory of Hydroscience and Engineering of Tsinghua University (2014-KY-01, 2016-KY-03). We gratefully appreciate their support. We acknowledge the staff at Tsinghua Universtiy-Korla Oasis Eco-hydrology Experimental Research Station for their kindly help and assistant.

Author Contributions: Fuqiang Tian performed the analyses, summarized the results, and outlined the manuscript. Pengju Yang conducted the field experiment, prepared and ran the SIMDualKc model, and prepared and revised the manuscript. Hongchang Hu designed the study, calibrated SIMDualKc model simulations and preprocessed the input data, and drafted the introduction section of the manuscript. Chao Dai conducted the field experiment.

Conflicts of Interest: The authors declare no conflict of interest.

References

1. Li, S.; Kang, S.; Li, F.; Zhang, L. Evapotranspiration and crop coefficient of spring maize with plastic mulch using eddy covariance in northwest China. *Agric. Water Manag.* **2008**, *95*, 1214–1222.
2. Burt, C.M.; Mutziger, A.J.; Allen, R.G.; Howell, T.A. Evaporation research: Review and interpretation. *J. Irrig. Drain.* **2005**, *131*, 37–58.
3. Hu, X.; Chen, H.; Wang, J.; Meng, X.; Chen, F. Effects of Soil Water Content on Cotton Root Growth and Distribution under Mulched Drip Irrigation. *Agric. Sci. China* **2009**, *8*, 709–716.
4. Kool, D.; Agam, N.; Lazarovitch, N.; Heitman, J.L.; Sauer, T.J.; Ben-Gal, A. A review of approaches for evapotranspiration partitioning. *Agric. For. Meteorol.* **2014**, *184*, 56–70.
5. Benli, B.; Kodal, S.; Ilbeyi, A.; Ustun, H. Determination of evapotranspiration and basal crop coefficient of alfalfa with a weighing lysimeter. *Agric. Water Manag.* **2006**, *81*, 358–370.
6. López-Urrea, R.; de Santa-Olalla, F.M.; Montoro, A.; López-Fuster, P. Single and dual crop coefficients and water requirements for onion (*Allium cepa* L.) under semiarid conditions. *Agric. Water Manag.* **2009**, *96*, 1031–1036.
7. Wilson, K.B.; Hanson, P.J.; Mulholland, P.J.; Baldocchi, D.D.; Wullschleger, S.D. A comparison of methods for determining forest evapotranspiration and its components: Sap-flow, soil water budget, eddy covariance and catchment water balance. *Agric. For. Meteorol.* **2011**, *106*, 153–168.
8. Poblete-Echeverría, C.; Ortega-Farias, S.; Zuñiga, M. Estimation of Dual Crop Coefficients over a Drip-Irrigated Merlot Vineyard Using Sap Flow Sensors and Eddy Covariance System. *ISHS Acta Hortic. VIII Int. Symp. Sap Flow* **2011**, *951*, 269–275.
9. Zhang, Y.; Shen, Y.; Sun, H.; Gates, J.B. Evapotranspiration and its partitioning in an irrigated winter wheat field: A combined isotopic and micrometeorologic approach. *J. Hydrol.* **2011**, *408*, 203–211.
10. Yepez, E.A.; Williams, D.G.; Scott, R.L.; Lin, G. Partitioning overstory and understory evapotranspiration in a semiarid savanna woodland from the isotopic composition of water vapor. *Agric. For. Meteorol.* **2003**, *119*, 53–68.
11. Zhang, F.; Zhou, G.; Wang, Y.; Yang, F.; Nilsson, C. Evapotranspiration and crop coefficient for a temperate desert steppe ecosystem using eddy covariance in Inner Mongolia, China. *Hydrol. Process.* **2012**, *26*, 379–386.
12. Allen, R.G.; Pereira, L.S.; Raes, D.; Smith, M. *Crop Evapotranspiration-Guidelines for Computing Crop Water Requirements—FAO Irrigation and Drainage Paper 56*; Food and Agriculture Organization of the United Nations: Rome, Italy, 1998.
13. Monteith, J.L.; Unsworth, M.H. *Principles of Environmental Physics*; Academic Press: New York, NY, USA, 2008.
14. Utset, A.; Farre, I.; Martinez-Cob, A.; Cavero, J. Comparing Penman–Monteith and Priestley–Taylor approaches as reference-evapotranspiration inputs for modeling maize water-use under Mediterranean conditions. *Agric. Water Manag.* **2004**, *66*, 205–219.

15. Amayreh, J.; Al-Abed, N. Developing crop coefficients for field-grown tomato (Lycopersicon esculentum Mill.) under drip irrigation with black plastic mulch. *Agric. Water Manag.* **2005**, *73*, 247–254.
16. Bezerra, B.G.; da Silva, B.B.; Bezerra, J.R.C.; Sofiatti, V.; dos Santos, C.A.C. Evapotranspiration and crop coefficient for sprinkler-irrigated cotton crop in Apodi Plateau semiarid lands of Brazil. *Agric. Water Manag.* **2012**, *107*, 86–93.
17. Han, S.; Xu, D.; Wang, S.; Yang, Z. Water requirement with irrigation expansion in Jingtai Irrigation District, Northwest China: The need to consider irrigation-induced local changes in evapotranspiration demand. *J. Irrig. Drain. Eng.* **2014**, *140*.
18. Allen, R.G. Using the FAO-56 dual crop coefficient method over an irrigated region as part of an evapotranspiration intercomparison study. *J. Hydrol.* **2000**, *229*, 27–41.
19. Cancela, J.J.; Fandiño, M.; Rey, B.J.; Martínez, E.M. Automatic irrigation system based on dual crop coefficient, soil and plant water status for Vitis vinifera (cv Godello and cv Mencía). *Agric. Water Manag.* **2015**, *151*, 52–63.
20. Ding, R.; Kang, S.; Zhang, Y.; Hao, X.; Tong, L.; Du, T. Partitioning evapotranspiration into soil evaporation and transpiration using a modified dual crop coefficient model in irrigated maize field with ground-mulching. *Agric. Water Manag.* **2013**, *127*, 85–96.
21. Paço, T.A.; Pôças, I.; Cunha, M.; Silvestre, J.C.; Santos, F.L.; Paredes, P.; Pereira, L.S. Evapotranspiration and crop coefficients for a super intensive olive orchard. An application of SIMDualKc and METRIC models using ground and satellite observations. *J. Hydrol.* **2014**, *519*, 2067–2080.
22. Zhang, B.; Liu, Y.; Xu, D.; Zhao, N.; Lei, B.; Rosa, R.D.; Paredes, P.; Paço, T.A.; Pereira, L.S. The dual crop coefficient approach to estimate and partitioning evapotranspiration of the winter wheat-summer maize crop sequence in North China Plain. *Irrig. Sci.* **2013**, *31*, 1303–1316.
23. Zhao, N.; Liu, Y.; Cai, J.; Paredes, P.; Rosa, R.D.; Pereira, L.S. Dual crop coefficient modelling applied to the winter wheat-summer maize crop sequence in North China Plain: Basal crop coefficients and soil evaporation component. *Agric. Water Manag.* **2013**, *117*, 93–105.
24. Rosa, R.D.; Paredes, P.; Rodrigues, G.C.; Alves, I.; Fernando, R.M.; Pereira, L.S.; Allen, R.G. Implementing the dual crop coefficient approach in interactive software. 1. Background and computational strategy. *Agric. Water Manag.* **2012**, *103*, 8–24.
25. Fandiño, M.; Olmedo, J.L.; Martínez, E.M.; Valladares, J.; Paredes, P.; Rey, B.J.; Mota, M.; Cancela, J.J.; Pereira, L.S. Assessing and modelling water use and the partition of evapotranspiration of irrigated hop (Humulus lupulus), and relations of transpiration with hops yield and alpha-acids. *Ind. Crop. Prod.* **2015**, *77*, 204–217.
26. Martins, J.D.; Rodrigues, G.C.; Paredes, P.; Carlesso, R.; Oliveira, Z.B.; Knies, A.E.; Petry, M.T.; Pereira, L.S. Dual crop coefficients for maize in southern Brazil: Model testing for sprinkler and drip irrigation and mulched soil. *Biosyst. Eng.* **2013**, *115*, 291–310.
27. Nash, J.; Sutcliffe, J.V. River flow forecasting through conceptual models part I—A discussion of principles. *J. Hydrol.* **1970**, *10*, 282–290.

28. Paço, T.A.; Ferreira, M.I.; Rosa, R.D.; Paredes, P.; Rodrigues, G.C.; Conceição, N.; Pacheco, C.A.; Pereira, L.S. The dual crop coefficient approach using a density factor to simulate the evapotranspiration of a peach orchard: SIMDualKc model *versus* eddy covariance measurements. *Irrig. Sci.* **2012**, *30*, 115–126.
29. Pereira, L.S.; Paredes, P.; Rodrigues, G.C.; Neves, M. Modeling malt barley water use and evapotranspiration partitioning in two contrasting rainfall years. Assessing AquaCrop and SIMDualKc models. *Agric. Water Manag.* **2015**, *159*, 239–254.
30. Wu, Y.; Liu, T.; Paredes, P.; Duan, L.; Wang, H.; Wang, T.; Pereira, L.S. Ecohydrology of groundwater-dependent grasslands of the semi-arid Horqin sandy land of inner Mongolia focusing on evapotranspiration partition. *Ecohydrology* **2015**.
31. Wei, Z.; Paredes, P.; Liu, Y.; Chi, W.W.; Pereira, L.S. Modelling transpiration, soil evaporation and yield prediction of soybean in North China Plain. *Agric. Water Manag.* **2015**, *147*, 43–53.
32. Allen, R.G.; Pereira, L.S. Estimating crop coefficients from fraction of ground cover and height. *Irrig. Sci.* **2009**, *28*, 17–34.
33. Fandiño, M.; Cancela, J.J.; Rey, B.J.; Martínez, E.M.; Rosa, R.G.; Pereira, L.S. Using the dual-K c approach to model evapotranspiration of Albarino vineyards (Vitisvinifera L. cv. Albarino) with consideration of active ground cover. *Agric. Water Manag.* **2012**, *112*, 75–87.
34. Liu, Y.; Pereira, L.S.; Fernando, R.M. Fluxes through the bottom boundary of the root zone in silty soils: Parametric approaches to estimate groundwater contribution and percolation. *Agric. Water Manag.* **2006**, *84*, 27–40.
35. Rosa, R.D.; Paredes, P.; Rodrigues, G.C.; Fernando, R.M.; Alves, I.; Pereira, L.S.; Allen, R.G. Implementing the dual crop coefficient approach in interactive software: 2. Model testing. *Agric. Water Manag.* **2012**, *103*, 62–77.
36. Yaghi, T.; Arslan, A.; Naoum, F. Cucumber (*Cucumis sativus*, L.) water use efficiency (WUE) under plastic mulch and drip irrigation. *Agric. Water Manag.* **2013**, *128*, 149–157.
37. Yu, G.-R.; Wen, X.-F.; Sun, X.-M.; Tanner, B.D.; Lee, X.; Chen, J.-Y. Overview of ChinaFLUX and evaluation of its eddy covariance measurement. *Agric. For. Meteorol.* **2006**, *137*, 125–137.
38. Zhang, Z.; Tian, F.; Hu, H.; Yang, P. A comparison of methods for determining field evapotranspiration: Photosynthesis system, sap flow, and eddy covariance. *Hydrol. Earth Syst. Sci.* **2014**, *18*, 1053–1072.
39. Er-Raki, S.; Chehbouni, A.; Boulet, G.; Williams, D.G. Using the dual approach of FAO-56 for partitioning ET into soil and plant components for olive orchards in a semi-arid region. *Agric. Water Manag.* **2010**, *97*, 1769–1778.
40. Facchi, A.; Gharsallah, O.; Corbari, C.; Masseroni, D.; Mancini, M.; Gandolfi, C. Determination of maize crop coefficients in humid climate regime using the eddy covariance technique. *Agric. Water Manag.* **2013**, *130*, 131–141.
41. Baffes, J. *Cotton: Market Setting, Trade Policies, and Issues*; World Bank: Washington, DC, USA, 2004.

42. China National Cotton Information Center (CNCIC). *China Cotton Almanac 2012/2013*; China Financial & Economic Publishing House: Beijing, China, 2014.
43. Han, S.; Hu, H.; Yang, D.; Liu, Q. Irrigation impact on annual water balance of the oases in Tarim Basin, Northwest China. *Hydrol. Process.* **2011**, *25*, 167–174.
44. Hu, H.; Tian, F.; Hu, H. Soil particle size distribution and its relationship with soil water and salt under mulched drip irrigation in Xinjiang of China. *China Technol. Sci.* **2011**, *54*, 1568–1574.
45. Shen, Y.; Li, S.; Chen, Y.; Qi, Y.; Zhang, S. Estimation of regional irrigation water requirement and water supply risk in the arid region of Northwestern China 1989–2010. *Agric. Water Manag.* **2013**, *128*, 55–64.
46. Chen, M.; Kang, Y.; Wan, S.; Liu, S. Drip irrigation with saline water for oleic sunflower (*Helianthus annuus* L.). *Agric. Water Manag.* **2009**, *96*, 1766–1772.
47. Chen, W.; Hou, Z.; Wu, L.; Liang, Y.; Wei, C. Evaluating salinity distribution in soil irrigated with saline water in arid regions of northwest China. *Agric. Water Manag.* **2010**, *97*, 2001–2008.
48. Wang, R.; Kang, Y.; Wan, S.; Hu, W.; Liu, S.; Liu, S. Salt distribution and the growth of cotton under different drip irrigation regimes in a saline area. *Agric. Water Manag.* **2011**, *100*, 58–69.
49. Shukla, S.; Shrestha, N.K.; Jaber, F.H.; Srivastava, S.; Obreza, T.A.; Boman, B.J. Evapotranspiration and crop coefficient for watermelon grown under plastic mulched conditions in sub-tropical Florida. *Agric. Water Manag.* **2014**, *132*, 1–9.
50. Zhang, Z.; Hu, H.; Tian, F.; Hu, H.; Yao, X.; Zhong, R. Soil salt distribution under mulched drip irrigation in an arid area of northwestern China. *J. Arid Environ.* **2014**, *104*, 23–33.
51. Forkutsa, I.; Sommer, R.; Shirokova, Y.I.; Lamers, J.P.A.; Kienzler, K.; Tischbein, B.; Martius, C.; Vlek, P.L.G. Modeling irrigated cotton with shallow groundwater in the Aral Sea Basin of Uzbekistan: II. Soil salinity dynamics. *Irrig. Sci.* **2009**, *27*, 319–330.
52. Qureshi, A.S.; Eshmuratov, D.; Bezborodov, G. Determining optimal groundwater table depth for maximizing cotton production in the Sardarya province of Uzbekistan. *Irrig. Drain.* **2011**, *60*, 241–252.
53. Shrestha, N.K.; Shukla, S. Basal crop coefficients for vine and erect crops with plastic mulch in a sub-tropical region. *Agric. Water Manag.* **2014**, *143*, 29–37.
54. Zhang, Z.; Hu, H.; Tian, F.; Yao, X.; Sivapalan, M. Groundwater Dynamics under Water Saving Irrigation and Implications for Sustainable Water Management in an Oasis: Tarim River Basin of Western China. *Hydrol. Earth Syst. Sci.* **2014**, *18*, 3951–3967.
55. Van Dijk, A.; Moene, A.F.; De Bruin, H. *The Principles of Surface Flux Physics: Theory, Practice and Description of the ECPACK Library*; University of Wageningen: Wageningen, The Netherlands, 2004.
56. Zhou, S.; Wang, J.; Liu, J.; Yang, J.; Xu, Y.; Li, J. Evapotranspiration of a drip-irrigated, film-mulched cotton field in northern Xinjiang, China. *Hydrol. Process.* **2012**, *26*, 1169–1178.

57. Allen, R.G.; Pereira, L.S.; Smith, M.; Raes, D.; Wright, J.L. FAO-56 dual crop coefficient method for estimating evaporation from soil and application extensions. *J. Irrig. Drain.* **2005**, *131*, 2–13.
58. Doorenbos, J.; Pruitt, W.O. *Crop Water Requirements. Irrigation and Drainage Paper No. 24*; Food and Agriculture Organization of the United Nations: Rome, Italy, 1977.
59. Popova, Z.; Pereira, L.S. Modelling for maize irrigation scheduling using long term experimental data from Plovdiv region, Bulgaria. *Agric. Water Manag.* **2011**, *98*, 675–683.
60. Howell, T.A.; Evett, S.R.; Tolk, J.A.; Schneider, A.D. Evapotranspiration of full-, deficit-irrigated, and dryland cotton on the Northern Texas High Plains. *J. Irrig. Drain. Eng.* **2004**, *130*, 277–285.
61. Farahani, H.J.; Izzi, G.; Oweis, T.Y. Parameterization and Evaluation of the AquaCrop Model for Full and Deficit Irrigated Cotton. *Agron. J.* **2009**, *101*, 469–476.

Winter Irrigation Effects in Cotton Fields in Arid Inland Irrigated Areas in the North of the Tarim Basin, China

Pengnian Yang, Shamaila Zia-Khan, Guanghui Wei, Ruisen Zhong and Miguel Aguila

Abstract: Winter irrigation is one of the water and salt management practices widely adopted in arid irrigated areas in the Tarim Basin located in the Xinjiang Uygur Autonomous Region in the People's Republic of China. A winter irrigation study was carried out from November 2013 to March 2014 in Korla City. A cotton field was divided into 18 plots with a size of 3 m × 3 m and five winter irrigation treatments (1200 m^3/ha, 1800 m^3/ha, 2400 m^3/ha, 3000 m^3/ha, and 3600 m^3/ha) and one non-irrigation as a control were designed. The results showed that the higher winter irrigation volumes allowed the significant short-term difference after the irrigation in the fields with the higher soil moisture content. Therefore, the soil moisture in the next sowing season could be maintained at the level which was slightly lower than field capacity and four times that in the non-irrigation treatment. The desalination effect of winter irrigation increased with the increase of water irrigation volume, but its efficiency decreased with the increase of water irrigation volume. The desalination effect was characterized by short-term desalination, long-term salt accumulation, and the time-dependent gradually decreasing trend. During the winter irrigation period, air temperature was the most important external influencing factor of the soil temperature. During the period of the decrease in winter temperatures from December to January, soil temperature in the 5-cm depth showed no significant difference in all the treatments and the control. However, during the period of rising temperatures from January to March, soil temperature in the control increased significantly, faster than that in all treatments. Under the same irrigation volume, the temperature difference between the upper soil layer and the lower soil layer increased during the temperature drop period and decreased during the temperature rise period. In this paper, we proposed the proper winter irrigation volume of 1800–3000 m^3/ha and suggested that the irrigation timing should be delayed to early December or performed in several stages in the fields with the drainage system. Under the current strict water management and fixed water supply quota situation, the methods are of great practical significance.

Reprinted from *Water*. Cite as: Yang, P.; Zia-Khan, S.; Wei, G.; Zhong, R.; Aguila, M. Winter Irrigation Effects in Cotton Fields in Arid Inland Irrigated Areas in the North of the Tarim Basin, China. *Water* **2016**, *8*, 47.

1. Introduction

Cotton is one of the most important economic crops in Xinjiang Uygur Autonomous Region, China. The average annual planting area of cotton in Xinjiang is 1,564,000 ha, accounting for nearly 1/3 of total crop area in Xinjiang. The cotton production in Xinjiang ranks first in China and accounts for 10% of global cotton production [1]. The cotton planting mode commonly used is winter or spring irrigation plus drip irrigation under plastic mulch. Under this cropping pattern, drip irrigation ensures the crop water requirement which can leach the salinity of the root zone. However, due to the lesser water quantity in drip irrigation, salinity mainly accumulates at the edge of the wet zone, thus reducing the rate of emergence in the coming year. To solve this problem, flood irrigation is applied to leach salt from October to November every year. Flood irrigation is now considered as the pre-requisite for a high yield, and it is generally believed that winter irrigation has more beneficial impacts. For example, it leaches the salts from the plant root zone, and stores winter water for spring. Flood irrigation in winter, namely winter irrigation, is often considered as the "improvement irrigation", which can improve cotton sprouting rate and survival rate. Regular winter irrigation volume is generally between 3000 and 4500 m^3/ha and accounts for 75% of cotton water consumption [2]. It leads to a situation of efficient water utilization in the development stage and inefficient water utilization in the non-development stage. Under the current strict water management and fixed water supply quota situation, it is necessary to improve the inefficient water utilization in winter irrigation.

Since the 1950s, winter irrigation has been widely applied in cotton, wheat, and fruit trees in Xinjiang and has become one of the routine measures of water management and soil improvement. Chen *et al.* [3] studied the impacts of different winter irrigation modes of cotton fields in northern Xinjiang on soil moisture content, soil salinity, and temperature distribution. They concluded that both drip irrigation of high discharge rate and flood irrigation could reduce salinity. Moreover, drip irrigation of 3000 m^3/ha was more beneficial to moisture retention and desalination. Sun *et al.* [4] analyzed the effects of winter irrigation on salt transport and proposed that winter irrigation of the larger quota allowed the better salt leaching effect. However, at the same time, the high water volume applied during winter irrigation may increase the soil pH and the concentration of HCO_3^-, thus causing the rise of underground water level and accelerating salt return. Therefore, Sun recommended the winter irrigation volume of 3500 m^3/ha. Feng Cao *et al.* [5] studied the effects of leaching salts and alkali reduction of different irrigation modes in No. 30 Regiment, Xinjiang Agricultural Reclamation No. 2 Division and proposed that winter irrigation allowed the higher sprouting rate and yield than spring irrigation.

The above results were mainly focused on the winter irrigation volume, which was relatively high. Moreover, the transport mechanisms of water and salt under freeze-thaw conditions were explored. Guisheng Fan *et al.* [6] carried out moisture infiltration tests of naturally frozen soil in the fields with different groundwater depths under the freeze-thaw conditions and analyzed the influence of groundwater depth on soil moisture infiltration capability and relatively stable infiltration rate. Similarly, Jihong Jing *et al.* [7] conducted water transport tests in the fields with artificially controlled groundwater at various depths on the plain in the northern foot of the Tianshan Mountains. Jing found that soil water transport towards the frozen layer led to groundwater evaporation and resulted in the decline in the groundwater level.

With the increase in the drip irrigation planting years of cotton under film covering in Xinjiang, the groundwater level declined and the mode and intensity of salt accumulation in fields began to differ from those formed under the ditch irrigation mode. As the area under plastic mulch drip irrigation increases in Xinjiang, the irrigation method of groundwater resulted in the decline in the groundwater level. The traditional flood irrigation requires a large amount of water and will be restricted by limited resources. Therefore, the water-saving irrigation schedule and the management of salt has become a hot issue in the cotton fields. In order to save water resources, leach salts, reduce alkali and improve farmland soil environment without increasing the pressure on the ecological environment, it is necessary to further study winter irrigation effects and applicability in cotton fields. The paper not only aims to find suitable irrigation time and water quantity of winter irrigation but to provide experimental evidence for farmers to change traditional irrigation methods gradually.

2. Materials and Methods

2.1. Study Area

The study area is Korla Irrigation Experiment Station which belongs to the Xinjiang Agricultural University located in Xi'ni'er Town of Korla City (Figure 1). Its geographic coordinates are 41°35′ N–41°37′ N and 86°09′ E–86°12′ E. The study area is characterized by arid conditions, less precipitation, and large temperature differences. Average annual precipitation is 53.3–62.7 mm and average annual evaporation is 2273–2788 mm (obtained with the evaporation pan with a diameter of 20 cm). The evaporation-precipitation ratio reaches 43.6. The altitude is between 895 m and 903 m. The climate in the study area belongs to warm temperate continental desert climate. Soil texture is loamy sand soil with lower soil fertility [8].

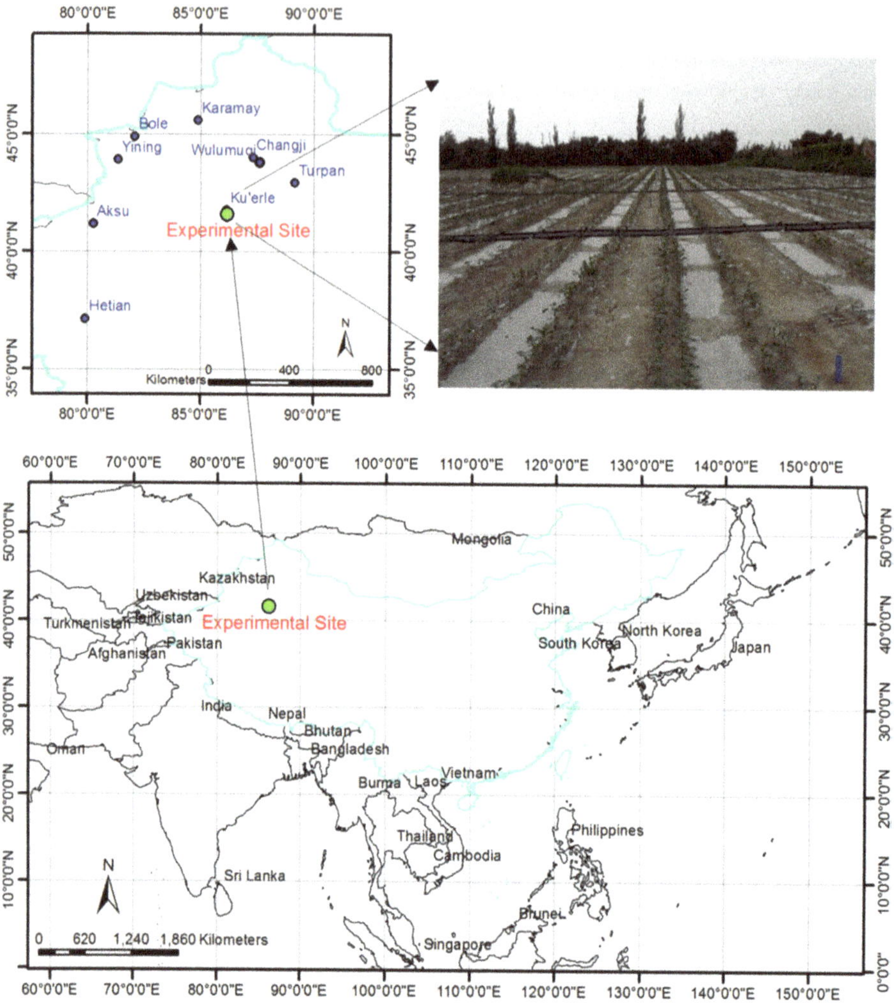

Figure 1. Geographic location of the study area.

2.2. Experimental Design

Winter irrigation tests were performed from 5 November 2013 to 20 March 2014. A total of five irrigation treatments and one control were respectively designed as 1200 m^3/ha, 1800 m^3/ha, 2400 m^3/ha, 3000 m^3/ha, 3600 m^3/ha, and 0 m^3/ha (control). Total dissolved solid and electrical conductivity in the irrigation water was 1185.1 mg/L and 1935 μs/cm, respectively. The pH value was 7.3. The size of the experimental plot was 3 m × 3 m. Irrigation was performed from 10 November to 11 November. Soil samples were obtained before and after winter irrigation to determine soil moisture content and soil electrical conductivity. Sampling depth was

1.0 m. Four soil samples were obtained on 6 November, 21 November, 5 December and 9 March up to a depth of 1.0 m. The total duration was 119 days after irrigation. To avoid spatial heterogeneity, two soil sampling methods were adopted. First, in order to acquire undisturbed soil samples, a small gouge with the small borehole diameter (φ = 3.0 cm) was used to acquire soil samples. Second, more soil samples were obtained. In the first and last soil sampling, 6 samples were obtained. In the second and third sampling, 3 samples were obtained. To determine soil moisture content and salt content, soil samples were obtained for every 10-cm soil depth. The soil profile in the experimental field showed two soil texture classes. According to the particle analysis results, soil from the ground to the depth of 60–80 cm was silt and the soil below the depth of 60–80 cm was the transition silt sand. The field capacity (mass) in the field was between 16.0% and 18.0%.

Soil moisture content was measured gravimetrically with the drying method. The water–soil solution mixture was prepared according to the ratio of 1:5 and then the electrical conductivity values were measured by Leici DDS-307 conductivity meter ((DDS-307, INESA Scientific Instrument Co., Ltd., Shanghai, China)). Soil temperature was measured by installing the Micro Lite U-disc temperature data loggers (Fourier Co. Tel Aviv, Israel) at the depths of 5 cm, 15 cm and 25 cm in two treatments (1200 m^3/ha and 3600 m^3/ha) and control treatment. The sensor's logging interval was 2 h. The resolution of the instrument was 0.06 °C and the accuracy was 0.3 °C. Air temperature, relative humidity, wind speed, and solar radiation were monitored by Davis Vantage Pro 2 automatic weather station ((Davis Instruments, CA, USA,). Groundwater depth was measured twice during the experiment period. Groundwater depths at 4.5 m and 4.6 m were observed on 5 December 2013 and 9 March 2014 and showed no change or little change.

3. Results and Analysis

3.1. Relationship between Total Salt Content and Conductivity Values

According to the analysis results of 103 soil samples collected from the experimental plots in December 2011, the correlation between total salt content and the conductivity of the extraction solution prepared according to the ratio of 1:5 is expressed as: Y = 0.3911x − 0.0227 (R^2 = 0.9), where Y is total salt content in soils (g/100 g) and x is the conductivity of soil solution (mS/cm). It can be seen from the above formula that when the conductivity of soil solution (soil-water ratio = 1:5) is increased by 1 dS/m, total salt content is increased by 0.37%. We established the corresponding relationship between the conductivity value and total dissolved solids and found that the salt content in soils could be indicated by the conductivity value, which could be easily measured. According to the method by B· A· Kovda [9] the chemical types of soil salts can be classified based on anions. The salt type in surface soil belongs to Mg-Na-SO_4, and the salt type in the soils with the depth of 10–30 cm is Mg-Na-Cl-SO_4 or Ca-Na-Cl-SO_4.

3.2. Environmental Effects of Winter Irrigation

3.2.1. Effect of Winter Irrigation on Water Storage

Winter irrigation can be completed within several days. After that, it is followed by a long winter period. Under the influences of gravity and thermal potential, soil moisture firstly continuously moves towards the frozen layer to form the maximum frozen layer depth. In the subsequent thawing period, under the influence of temperature rise, soil moisture in the frozen layer moves upward and downward. The results showed that the soil moisture content profile varied with time. This variation could be considered as the subsequent effect of winter irrigation. The variations of soil moisture contents are shown in Figure 2.

As shown in the profile curves (Figure 2), soil moisture content increased significantly after irrigation but gradually subsided with time. For the former two, post-irrigation determinations of soil moisture content were taken within 1 month after winter irrigation. The determination results were highly affected by irrigation. The last determination was taken on the 118th day after irrigation. Soil moisture experienced long-term freezing and was largely changed in the soil profile. With the obtained results, we calculated the average soil moisture content within 40 cm and 100 cm depth in different treatments (Table 1). Up until 9 March, soil moisture contents in different treatments were basically the same and slightly lower than field moisture capacity, indicating that the subsequent effect of winter irrigation was maintained at the field water capacity level and was nearly four times that of the non-irrigation blank control.

In the non-irrigation treatment, in the whole overwintering period, the soil moisture content remained stable. The background moisture contents in each treatment were different. In order to avoid the interference from different background values among different treatments, we subtracted corresponding average soil moisture content obtained before irrigation from that obtained after irrigation. The calculation results could be considered to be the indications of the water storage effect of winter irrigation. We subtracted the background value obtained in 6 November, respectively, from the soil moisture contents obtained on 21 November, 5 December, and 9 March and obtained the increased soil moisture content in the 1.0-m deep soil after winter irrigation, which could be used to quantify winter irrigation effects, as shown in Figure 3.

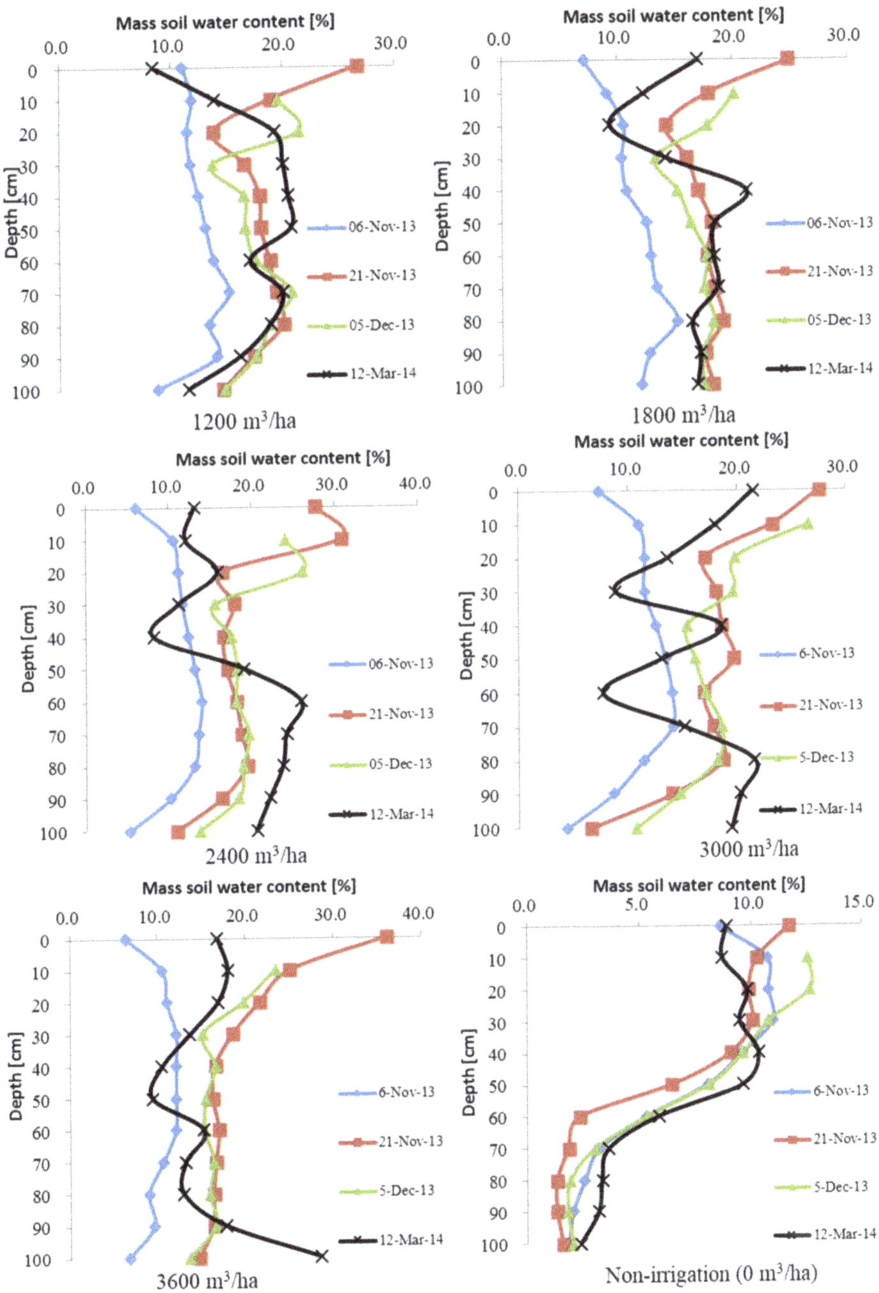

Figure 2. The dynamic soil profile of soil moisture content of all treatments under winter irrigation.

Table 1. The dynamics of average soil moisture content in soil depths of 0–40 cm and 0–1.0 m cm before and after winter irrigation (unit: %).

Dates		6 November 2013 (BI)	21 November 2013 (AI)	5 December 2013 (AI)	9 March 2014 (AI)
Non-irrigation		10.6/3.8	9.8/2.5	11.4/3.7	9.6/4.7
Volume of winter irrigation (m^3/ha)	1200	11.9/13.1	16.9/18.1	17.8/17.7	18.5/17.5
	1800	10.2/13.1	16.3/18.2	16.6/17.5	14.3/17.7
	2400	11.4/11.5	20.4/16.7	20.8/17.8	11.9/22.6
	3000	11.6/11.0	19.2/15.6	20.3/15.9	14.8/16.2
	3600	11.4/10.0	20.4/16.2	18.7/15.6	14.8/16.2
Average of winter treatment		11.3/11.7	18.6/17.0	18.9/16.9	14.8/18.1

Notes: Slash (/) indicates the average quantity of water content of 0–0.4 m and 0–1.0 m depth; BI indicates that the data are obtained before irrigation; AI indicates that the data are obtained after irrigation.

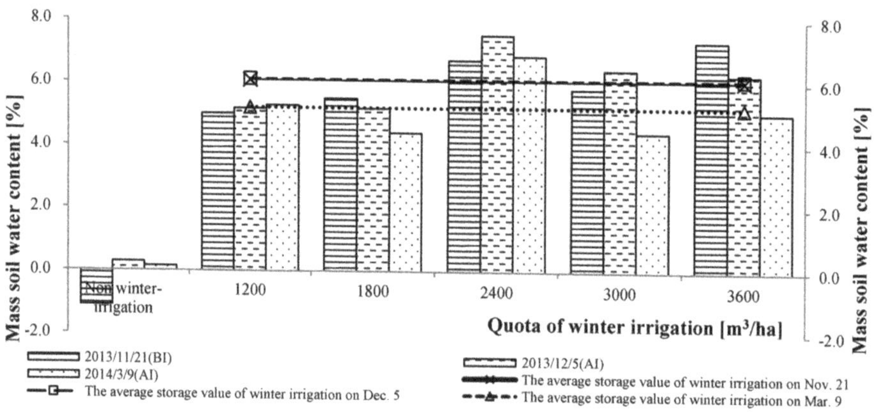

Figure 3. The comparison chart of water storage effects of winter irrigation in 1.0-m deep soil in six treatments.

As shown in Figure 3, in the non-fertility stage, soil moisture shows three distinct characteristics. Firstly, the variation of soil moisture content in the non-irrigation treatment was less than 1%. Therefore, it is considered that soil moisture content is essentially unchanged in the overwintering period. Secondly, the increase of soil moisture content in the winter irrigation treatments was significantly higher than that in the non-irrigation treatment, indicating a water storage effect of winter irrigation. With the increase of winter irrigation quota, the soil moisture content was also increased (21 November). Thirdly, the water storage effect of winter irrigation gradually decreased with time. Even the higher quota had no effect on the higher water storage in the later winter irrigation stage which could be used for groundwater supply. Therefore, water utilization efficiency was reduced the following spring [10].

3.2.2. Desalination Effect of Winter Irrigation

The dynamic variation of salt contents in the soil profiles in different treatments after winter irrigation is shown in Figure 4.

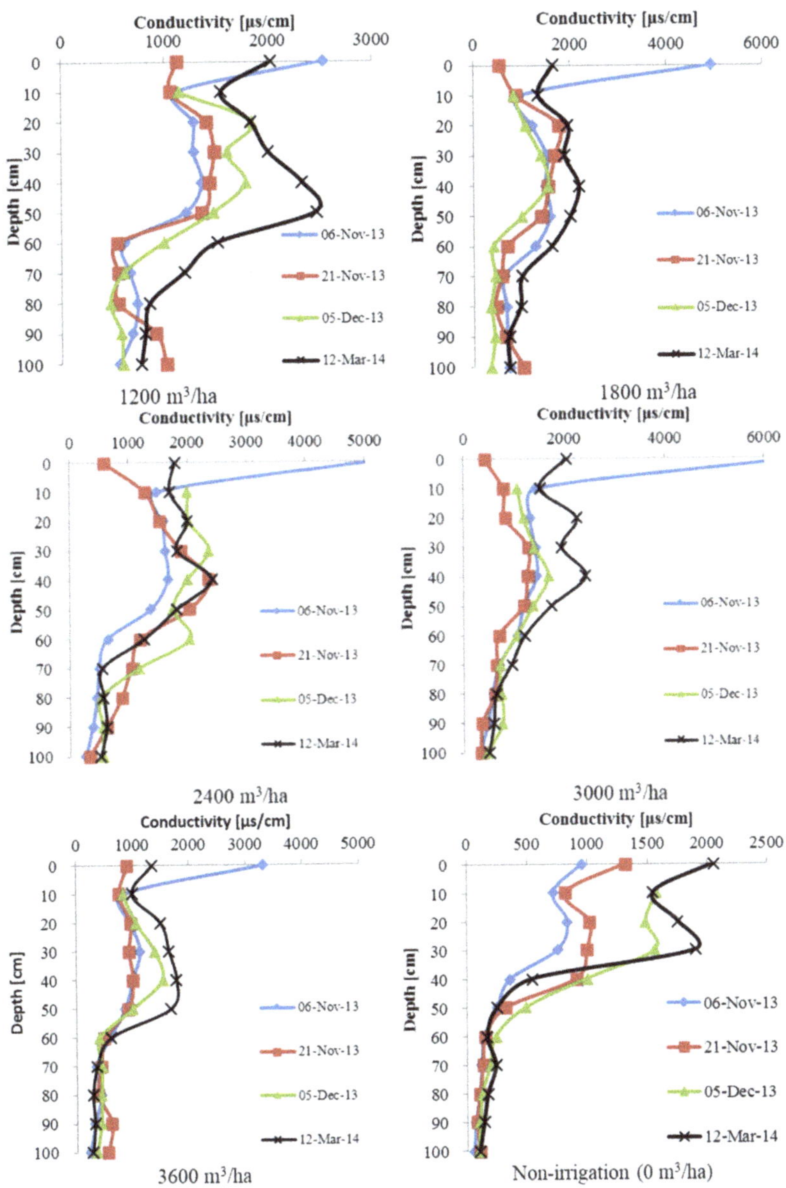

Figure 4. The dynamic variation of salt contents in the soil profiles in six treatments.

As shown in Figure 4, the variation of salt contents in soil profile shows three characteristics. Firstly, the significant surface accumulation phenomenon was observed. The salt content was very high in the 0–2 cm depth and decreased abruptly in the depth below 2 cm. With the increase of soil depth, salt content was decreased. Secondly, salt content in soil profile showed the significant boundary at 60-cm depth. The depth above 60 cm could be named as the drastically changed zone and the depth below 60 cm named as the stable zone of salt content. The salt leaching effect of winter irrigation was decreased with the increase of soil depth (0–60 cm). Thirdly, during the winter irrigation, salt content was increased gradually.

According to the traditional calculation method of salt leaching rate, the leaching rate was calculated by dividing the salt content difference before and after irrigation by the salt content before irrigation. If the leaching rate is a positive value, winter irrigation shows the salt accumulation effect; if the leaching rate is a negative value, winter irrigation shows the desalination effect. The salt content variations in soil profiles of different treatments were shown in Figure 5.

According to the analysis results of the samples obtained in the short term (10 day) after winter irrigation, the desalination depth was increased with the increase of the irrigation volume. When irrigation volume was 1800 m^3/h, the desalination depth reached 10.0 cm. When it was 2400 m^3/h, the desalination depth reached 20.0 cm. When irrigation volume was larger than 3000 m^3/h, the desalination depth was larger than 40.0 cm.

However, after 119 days of winter irrigation, desalination depths in all the treatments were about 10 cm. The soil depth below 10 cm mainly showed the salt accumulation effect. The soil depth between 10 cm and 60 cm was the main salt accumulation depth. The average desalination rates obtained respectively on 21 November 2013 and 12 March 2014 are shown in Figure 6.

Figure 6 shows the following two characteristics. Firstly, according to the short-term trend from winter irrigation to 21 November, the larger winter irrigation volume indicates the more significant desalination effect. When the irrigation volume was lower than 2400 m^3/ha, the desalination effect at the depth of 10–60 cm was not significant even though salt accumulation effect could be observed. However, when irrigation volume was higher than 2400 m^3/ha, the desalination effect at the depth of 10–60 cm depth was observed. Secondly, in the later winter irrigation stage (12 March), salt accumulation was evident in all the treatments, indicating the salt accumulation process during the winter period. The salt accumulation rate in the control treatment was the highest, while it was relatively lower in other treatments except in the 1200 m^3/ha to 3600 m^3/ha treatments.

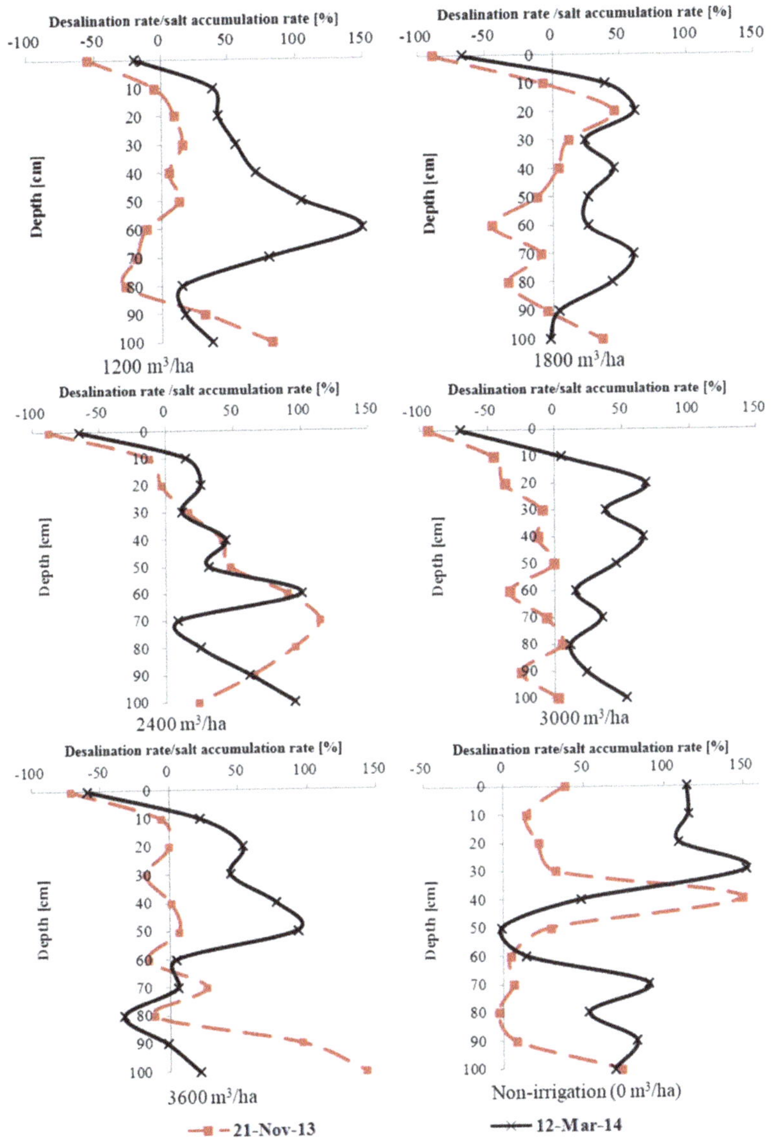

Figure 5. The dynamic variation of salt content in soil profiles in six treatments.

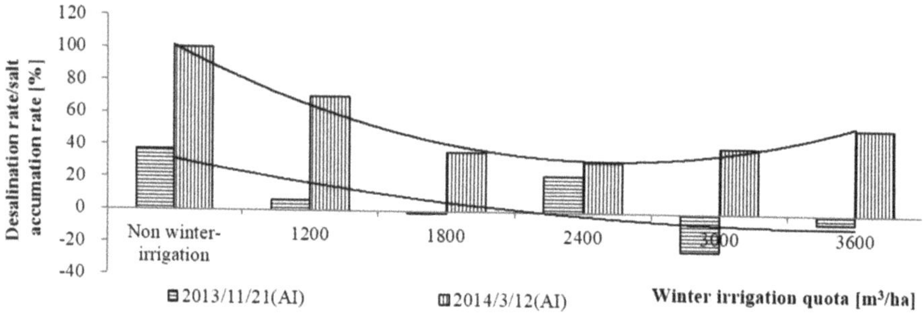

Figure 6. Average desalination rate/salt accumulation rate in the soil profile of 0.1–0.6 m in 6 treatments after winter irrigation.

Desalination effect of winter irrigation can be divided into short-term effect and long-term effect. It was found that the short-term desalination effect was significant and that the long-term desalination effect gradually became weak and salt accumulation was observed. Therefore, even the larger winter irrigation volume would not prevent salt accumulation effect in the next year. On the contrary, the high irrigation volume resulted in high water consumption, raised the groundwater level, and led to the loss of nitrate nitrogen. It is important to note that salt accumulation after winter irrigation has a great influence on the cotton sprouting rate. As shown in Figure 6, treatments of 1800 m^3/ha, 2400 m^3/ha and 3000 m^3/ha have lower salt contents in soil in spring. Therefore, the proper winter irrigation volume should be within this range. In addition, arid zone climatic characteristics also determine salt accumulation trends on the soil surface. Irrigation is an important means to adjust this trend. Owing to the role of winter irrigation, unidirectional salt migration towards the surface is interrupted and then re-distributed at the depth range of 0–60 cm, thus decreasing the salt accumulation on the surface. However, due to the influences of freezing and thawing, the upper soil profile shows a certain degree of salt accumulation effect, but it is still lower than that before irrigation, indicating a subsequent effect of winter irrigation. In this way, the inhibition effect on cotton sprouting caused by continuous salt accumulation in the overwintering stage can be avoided [11].

3.3. *The Dynamic Characteristics of Soil Temperature during the Wintering Period*

Soil temperature was monitored in different treatments at the depth of 5 cm from 10 November 2013 to 11 March 2014. This is shown in Figure 7. However, soil temperature dynamics at different depths in the same irrigation treatment measured from 11 December 2011 to 21 March 2012. Monitoring depths were respectively 5 cm, 15 cm, 25 cm, 35 cm, 45 cm, and 55 cm. The soil temperature dynamics in different depths is shown in Figure 8. The soil temperature recorded every 2 h. Therefore,

12 sets of temperature data were available in a day. The 12 sets of temperature data could be averaged and used as the daily average soil temperature.

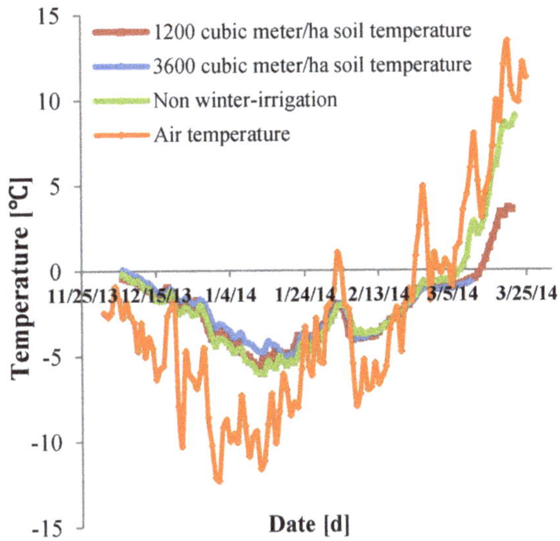

Figure 7. Soil temperature dynamics in the depth of 5 cm in different winter irrigation treatments.

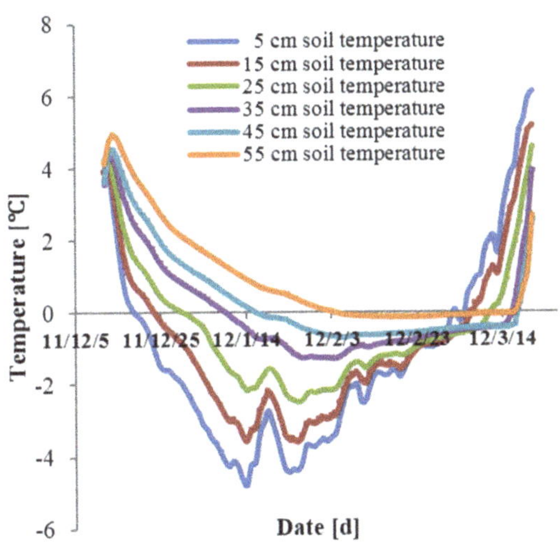

Figure 8. Soil temperature dynamics in different depths in the same winter irrigation treatment.

As shown in Figure 7, three soil temperature curves showed a V-shaped synchronous variation characteristic. The lowest soil temperature at the 5-cm depth recorded on 12 January 2014 was −6 °C. From this time point, the graph could be divided into two parts. The left part of the graph shows a temperature drop. The non-irrigation treatment had the lowest soil temperature and irrigation treatment of 3600 m^3/ha has the highest soil temperature. However, the temperature difference among different treatments was small, indicating that the influence of winter irrigation on soil temperature was not apparent. The right part of the graph showed the increasing trend of temperature. Soil temperature in the non-irrigation treatment was the highest and attained a positive value five days earlier than that in other treatments. It was observed that when the soil temperature in the non-irrigation treatment was above 0 °C, the temperature differences between the non-irrigation treatment and other treatments became more significant. However, the temperature differences among winter irrigation treatments were not significant. Similarly, it was also found that the influence of air temperature on soil temperature in the non-irrigation treatment was more significant than that in other treatments. This might be ascribed to the difference of soil thermal capacity caused by winter irrigation [12].

As shown in Figure 8, in different depths of certain treatment, the dynamic soil temperature drop process showed the characteristics of the lower surface temperature and the higher temperature in the deep. In the temperature drop process, temperature potential was gradually enhanced and the potential of water and soil was also in the increasing distribution from top to bottom. Therefore, the potential allowed the upward movement of soil moisture. During the temperature rise process, the temperature difference was gradually decreased. In early March, soil temperatures in all the soil layers were concentrated in the vicinity of −1 °C and were maintained for several days. Thereafter, temperature in all the soil layers rose quickly. Therefore, the temperature difference was increased gradually and showed the gradient distribution which was opposite to that in the temperature drop process. After this, the frozen soil entered the thawing state [13].

3.4. Appropriate Timing of Winter Irrigation

In southern Xinjiang, cotton winter irrigation was generally performed during the period from late October to late November. The irrigation time was selected according to the following two factors. Firstly, in order to minimize evaporation, irrigation time should be postponed as far as possible. Secondly, in order to ensure the canal engineering security, the selected irrigation deadline should prevent the freezing failure of the main ditches network. The two above factors should be considered in the cotton planting area, which relies on surface water irrigation. However, in the planting areas where wells and supporting pipeline networks were available, well water could flow into the fields through underground pipeline networks. Therefore, the second factor might be omitted and winter irrigation time

MDPI AG
St. Alban-Anlage 66
4052 Basel, Switzerland
Tel. +41 61 683 77 34
Fax +41 61 302 89 18
http://www.mdpi.com

Water Editorial Office
E-mail: water@mdpi.com
http://www.mdpi.com/journal/water

www.ingramcontent.com/pod-product-compliance
Lightning Source LLC
Chambersburg PA
CBHW061619220926
51850CB00005B/550